HERBERT FEIGL

INQUIRIES AND PROVOCATIONS

SELECTED WRITINGS 1929–1974

VIENNA CIRCLE COLLECTION

VOLUME 14

EDITOR: ROBERT S. COHEN

C. G. HEMPEL H. FEIGL

THE LAST TWO EMPIRICISTS
12 MAY, 1973

HERBERT FEIGL

INQUIRIES AND PROVOCATIONS

SELECTED WRITINGS 1929–1974

Edited by

ROBERT S. COHEN

D. REIDEL PUBLISHING COMPANY

DORDRECHT : HOLLAND / BOSTON : U.S.A.

LONDON : ENGLAND

Library of Congress Cataloging in Publication Data

Feigl, Herbert.
Inquiries and provocations.

(Vienna circle collection ; v. 14)
Includes bibliographies and index.
1. Philosophy — Collected works. 2. Logical positivism — Collected works. 3. Science — Philosophy — Collected works. I. Cohen, Robert Sonné. II. Title. III. Series.
B29.F33 191 80–17721
ISBN-13: 978-90-277-1102-1 e-ISBN-13: 978-94-010-9426-9
DOI: 10.1007/978-94-010-9426-9

Translations of 'Probability and Experience' and 'Meaning and Validity of Physical Theories' by Gisela Lincoln and R.S.C.

Published by D. Reidel Publishing Company,
P.O. Box 17, 3300 AA Dordrecht, Holland

Sold and distributed in the U.S.A. and Canada
by Kluwer Boston Inc.,
190 Old Derby Street, Hingham, MA 02043, U.S.A.

In all other countries, sold and distributed
by Kluwer Academic Publishers Group,
P.O. Box 322, 3300 AH Dordrecht, Holland

D. Reidel Publishing Company is a member of the Kluwer Group

TABLE OF CONTENTS

PREFACE

The title is his own. Herbert Feigl, the *provocateur* and the soul (if we may put it so) of modesty, wrote to me some years ago, "I'm more of a catalyst than producer of new and original ideas all my life . . .", but then he completed the self-appraisal: ". . . with just a few exceptions *perhaps*". We need not argue for the creative nature of catalysis, but will simply remark that there are 'new and original ideas' in the twenty-four papers selected for this volume, in the extraordinary *aperçus* of the 25-year-old Feigl in his Vienna dissertation of 1927 on *Zufall und Gesetz*, in the creative critique and articulation in his classical monograph of 1958 on *The 'Mental' and the 'Physical'*; and the reader will want to turn to some of the seventy other titles in our Feigl bibliography appended.

Professor Feigl has been a model philosophical worker: above all else, honest, self-aware, open-minded and open-hearted; keenly, devotedly, and even arduously the student of the sciences, he has been a logician and an empiricist. Early on, he brought the Vienna Circle to America, and much later he helped to bring it back to Central Europe. The story of the logical empiricist movement, and of Herbert Feigl's part in it, has often been told, importantly by Feigl himself in four papers we have included here. First of these, chapter two, the retrospective sketch of 'The Origin and Spirit of Logical Positivism' [1969a] with its astute focus upon the thought and influence of Moritz Schlick. Second, chapter four, a widely-read analytic essay on the intellectual and social history of 'The *Wiener Kreis* in America' [1969d]. Third, chapter sixteen, Feigl's personal account of 'The Mind-Body Problem in the Development of Logical Empiricism' [1950d]. Finally, chapter one, the sweet-tempered but impishly-titled autobiographical essay, 'No Pot of Message' [1974a].

Nearly fifteen years ago, Paul Feyerabend and Grover Maxwell edited a splendid *Festschrift* for Herbert Feigl, *Mind, Matter, and Method* (Minneapolis, University of Minnesota Press, 1966). Feyerabend's biographical sketch for that occasion supplements our first chapter with a lively account of Feigl's life in philosophy, of his posts and his honors, and of his impact on students, colleagues, and the rest of us.

I am grateful to Professor Duane L. Cady of Hamline University in St. Paul,

Minnesota, for kindly permitting us to use his admirable photograph of Herbert Feigl with his dear friend, Carl G. Hempel (taken in 1973 during Professor Hempel's visit to Gustavus Adolphus College when Cady was teaching there). Special thanks also to Carolyn Fawcett, and to Renate Hanauer, and Barbara Nielson for their help, and to Gisela Lincoln for her translations. Most of all, this is Herbert Feigl's own book, and I most warmly express gratitude for his collaboration in the labor of preparing it; as he often shortens such sentiments, 10^6 thanks.

R. S. COHEN

ACKNOWLEDGMENTS

The author, the editor and the publisher are grateful to the following persons and institutions for permission to reprint the papers included in this volume:

'No Pot of Message', from *Mid-Twentieth Century Philosophy: Personal Statements*, Peter A. Bertocci (ed.) (1974), pp. 120–139. © 1974 by Humanities Press, Inc.

'The Origin and Spirit of Logical Positivism', from *The Legacy of Logical Positivism*, Peter Achinstein and Stephen F. Barker (eds.) (1969), pp. 3–24. © 1969 by The Johns Hopkins University Press.

'The Power of Positivistic Thinking', from *Proceedings and Addresses of the American Philosophical Association* **36** (1963), pp. 21–41. Copyright 1963 The American Philosophical Association.

'The *Wiener Kreis* in America', originally published in *Perspectives in American History*, Vol. 2, D. Fleming and B. Bailyn (eds.) (1968), pp. 630–673; reprinted in *The Intellectual Migration 1930–1960*, D. Fleming and B. Bailyn (eds.)(1969). © 1968 by the President and Fellows of Harvard College.

'Scientific Method without Metaphysical Presuppositions', *Philosophical Studies* 5 (1954), pp. 17–32. D. Reidel Publishing Co.

'Probability and Experience'; first published in German under the title: 'Wahrscheinlichkeit und Erfahrung', *Erkenntnis* **1** (1930), pp. 249–259, by Felix Meiner Verlag.

'Meaning and Validity of Physical Theories'; first published in German in *Theorie und Erfahrung in der Physik*, Chapter III (1929), pp. 94–138, by G. Braun.

'Confirmability and Confirmation', *Revue Internationale de Philosophie* 5 (1951), pp. 268–279.

'The Logical Character of the Principle of Induction', *Philosophy of Science* **1** (1934), pp. 20–29. Copyright © 1934 The Williams & Wilkins Co.

'What Hume Might Have Said to Kant', from *The Critical Approach to Science and Philosophy*, Mario Bunge (ed.) (1964), pp. 45–51. Copyright © 1964 by The Free Press of Glencoe, a Division of Macmillan Publishing Co., Inc.

'Operationism and Scientific Method' and 'Rejoinder and Second Thoughts', *Psychological Review* 52 (1945), pp. 250–259 and pp. 284–288, respectively. Copyright 1945 by the American Psychological Association.

'Existential Hypotheses: Realistic Versus Phenomenalistic Interpretations', *Philosophy of Science* 17 (1950), pp. 35–62. Copyright © 1950 The Williams & Wilkins Co.

'Logical Reconstruction, Realism and Pure Semiotic', *Philosophy of Science* 17 (1950), pp. 186–195. Copyright © 1950 The Williams & Wilkins Co.

'*De Principiis Non Disputandum* . . . ? On the Meaning and the Limits of Justification', from *Philosophical Analysis*, Max Black (ed.) (1950), pp. 119–156. © 1950 by Cornell University Press.

'Empiricism at Bay?: Revisions and a New Defense', from *Methodological and Historical Essays in the Natural and Social Sciences* (Boston Studies in the Philosophy of Science, Vol. 14), Robert S. Cohen and Marx W. Wartofsky (eds.) (1974), pp. 1–20. Copyright © 1974 by D. Reidel Publishing Co.

'The Mind-Body Problem in the Development of Logical Empiricism', *Revue Internationale de Philosophie* 4 (1950), pp. 64–83.

'Physicalism, Unity of Science and the Foundations of Psychology', from *The Philosophy of Rudolf Carnap* (The Library of Living Philosophers, Vol. 11), Paul A. Schilpp (ed.) (1963), pp. 227–267. © 1963 by The Library of Living Philosophers, Inc. The Open Court Publishing Company.

'Mind-Body, *Not* a Pseudoproblem', from *Dimensions of Mind*, Sidney Hook (ed.) (1960), pp. 24–36. © 1960 by New York University. New York University Press.

'Some Crucial Issues of Mind-Body Monism', *Synthese* 22 (1971), pp. 295–312 . Copyright © 1971 by D. Reidel Publishing Co.

'Naturalism *and* Humanism', *American Quarterly* 1 (1949), pp. 135–148.

'Validation and Vindication: An Analysis of the Nature and the Limits of Ethical Arguments', from *Readings in Ethical Theory*, Wilfrid Sellars and John Hospers (eds.) (1952), pp. 667–680. © 1952, renewed 1980, by Prentice-Hall, Inc.

'Everybody Talks about the Temperature', from *Festschrift for Jørgen Jørgensen* (1964). © 1964 Munksgaard, Copenhagen.

'Is Science Relevant to Theology?', *Zygon* 1 (1966), pp. 191–199. © 1966 by The University of Chicago. The University of Chicago Press.

'Ethics, Religion, and Scientific Humanism', *The Humanist* 28 (1968), pp. 21–25.

1. NO POT OF MESSAGE

[1974a]

Complying with the flattering invitation to write in a rather personal vein on what I believe, I have taken the liberty of sketching an autobiographical background to the development of my philosophical point of view. I did this in the hope that it might be of some interest, if not of some slight use, to young 'budding' philosophers.

After half a century of studies, searchings and reflections, philosophy still strikes me as a highly questionable enterprise. The Scylla of absurdity and the Charybdis of triviality seem forever to threaten. Or, to switch to a more homely metaphor, philosophy – especially in its more ambitious form of 'systems' – is like a blanket that is too short: if you pull it up to your chin, your feet get exposed (and vice versa). Complete 'coverage', it seems, can be achieved (to use yet another figure of speech) only by means of verbal sedatives. Moreover, as Bertrand Russell has remarked, the solemnity of some philosophers merely conceals humbug. In what follows I hope to discredit and diminish the 'prophet motive' so typical of philosophers throughout the ages.

Having had the advantage of an early and thorough training in the 'exact' sciences (mathematics, physics, chemistry, and astronomy) I could only heartily agree with Kant that (especially) metaphysics is a "groping in the dark." Yet, Kant's own "critical" philosophy seemed to me peculiarly musty and dogmatic. My teen-age idols were the philosophizing scientists Ernst Mach and Wilhelm Ostwald. But even then I objected to their (rather anachronistic) negative attitude toward the atomic theory. In this issue I agreed enthusiastically with the realism of Ludwig Boltzmann and Max Planck. At the age of about sixteen I would wander through the wonderful, extended pine forests of my homeland (i.e., the Sudetenland in northern Bohemia) and visualize electrons, atoms, and molecules in my physical environment. While pure mathematics and geometry were of great interest to me, the fields of study that fascinated me most were theoretical physics, physical chemistry, and astronomy.

As far as I can tell I was one of the first two scholarly-minded persons in the German-Bohemian-Jewish Feigl and Beck families from which I originated. The other one was a second cousin, two years my senior, Arthur Beer, who became a well-known astronomer (astrophysicist and spectroscopist) later in

Cambridge, England. It was he who helped me toward a better understanding, especially of the mathematics of Einstein's general theory of relativity (differential geometry and tensor algebra). I had happened, at the age of sixteen or seventeen, upon an article by an able (I think his name was Emil Abel) Viennese physical chemist in an Austrian chemistry journal. This was, I believe, in 1918 when Einstein's name and fame were still completely new to me. I found this article on the special theory of relativity extremely challenging. My first reaction was the suspicion that the theory must be in error. Promptly I set about to refute it. But in trying to do that I learned a great deal of physics and mathematics and, of course, found out after a few months of diligent work just who was wrong! My attitude then changed completely and Einstein became my number one intellectual hero. Indeed, in 1920 (when I was 17 years old) I made the 60 mile trip from my hometown to Prague where I heard Einstein's lecture on relativity on the first evening and a thrilling discussion of Einstein with some of his critics on the following evening. Einstein's Olympian calm and superiority, coupled with his humility and great sense of humor remain unforgettable to me. It was in 1923 that I finally had the courage to visit the great man in Berlin. He advised me on further studies in theoretical physics; he was also quite outspoken about his philosophical predilections – and aversions. It was not until 1954 (one year before his death) that I was privileged to spend another afternoon in conversation with Einstein in his Princeton home.

After I had read (as did two or three of my classmates in high school) parts of the works of Kant, Schopenhauer and Nietzsche, I happened (in 1919) upon Moritz Schlick's little book on *Space and Time in Contemporary Physics*. There I found a reference to the first edition (1918) of Schlick's great masterpiece, his *Allgemeine Erkenntnislehre* (General Theory of Knowledge.)* Schlick soon became my favorite philosopher. I was also impressed with the book by Alois Riehl on contemporary philosophy. Riehl, whose work I still think was more important than that of the other Neo-Kantians, never received the recognition he deserved. In addition to many other important contributions and clarifications, I was immensely pleased with the sort of critical realism and its supporting arguments I found in both Riehl and Schlick. Impressed as I was with Mach's insistence upon the immediately given (his 'elements') as the ultimate basis of all factual knowledge, I had of course realized that my physicalistic orientation urgently required a coherent and tenable solution of the mind-body problem. The place of mind in nature, with the corollary puzzles regarding the freedom of the will, exercised my curiosity and my imagination. Partly prepared by Schopenhauer's metaphysical approach, the mind-body monism explicated and defended by both Riehl and Schlick struck me like a thunderbolt. (It was many years later that I learned of Bertrand Russell's similar but independent solution, fully expounded in his book *Human Knowledge* of 1948.) Those very early influences and experiences left a permanent mark on the development of my philosophical thought.

Perhaps I should also mention that at the age of about eleven to thirteen I had read ardently most of the science-fiction novels of Jules Verne and Kurd Lasswitz. I 'identified' with Captain Nemo of *Twenty Thousand Leagues Under the Sea* and of *Mysterious Island*. I was thrilled with Lasswitz's *On Two Planets* and with the intellectually, technologically and morally superior Martians described in that story. Very naturally then, physics and chemistry (and to some extent the French language) became my favorite subjects in the *Realschule* (secondary school, age 11–14) of my hometown Reichenberg (in what was then Austria-Hungary).

My two grandfathers had been simple, but moderately prosperous businessmen. My father, having escaped from school at age thirteen, began what was to become a meteoric career, first as a weaver (working 12 hours a day, six days a week in a small town of Saxony). Soon his great technical talents were recognized. He was a skillful and highly ingenious textile designer. He also invented important improvements for the weaving looms of that time. My father was the first in Europe to use rayon patterns in the weaving of woolen fabrics. (He also was among the first few automobilists in Austria.) In the course of the few years before the first world war, he became a leader in the Austrian textile industry, and after 1920 was for a few years the General Director of 'Textilana.' This was perhaps the greatest textile manufacturing (spinning, weaving, and dyeing) corporation in continental Europe. Only the English industry could successfully compete with Textilana. My father's interests lay exclusively on the technical and organizational side. (He left the financial aspects to his associates.) For a few years he was a 'man of property and affluence.' But the (delayed) effects of the business depression did catch up with him and his enterprises, too. Thus he retired from his position and moved to Vienna in 1927. That was the year in which I received my doctorate in philosophy from the University of Vienna. Also in 1927 I began teaching – first astronomy – and later, various courses in philosophy, especially in the philosophy of science at the *Volkshochschule Wien*, i.e., the People's Institute of Vienna (for adult education). Many of its instructors were also faculty members of the University of Vienna.

During his active years my father had hoped that I would become an industrial chemist and thus join the scientific staff of Textilana. But already during the four years of my studies at the excellent vocational high school of Reichenberg (major field: chemistry) my interests shifted strongly toward pure and theoretical (rather than applied and technological) science. The high seriousness of a superb teacher and devoted scientist, Paul Artmann, who had returned to our school from the war in 1916, very definitely influenced my studies, interests and aspirations.

My father, a very practical and extremely energetic man, was at first quite disappointed. But as he watched and recognized my success in school, he reconciled himself with having a budding scholar for a son.

As I see it now, psychologically (if not psychoanalytically), it was necessary for me to find a kingdom of my own in order to obtain the approval of my father. This extremely powerful and successful man expected from his children similarly great achievements. But because in contrast to my father I lacked practical and manual skills, and was of a more contemplative nature ('dreamy' my father used to say), it had to be the realm of ideas in which I could hope for any measure of success.

My father was socially and politically highly alert. He was on excellent terms with his workers; he provided them with adequate and pleasant housing, and though his attitude was somewhat paternalistic, he never suffered the adversities of a strike. To be sure, he had no patience with the Communists, but his mind was open in regard to a humane socialism. My father was clearly a typical 'self-made man.' He would, no doubt, also have been extremely successful in America had he chosen to migrate there. He was an outspoken atheist – this made my own emancipation from Judaism (at age eight) quite easy.

My mother was an excellent spouse and counterpoise for my father. She had a lifelong interest in the arts, literature, and music. Her religion – if it can be so called – was that of a Spinoza-Goethe-type pantheism. Since she played the piano with great understanding (but without technical virtuosity), it was from her that I received what I still consider my life's most precious gift: the love of music and some knack for improvisation on the piano. Even before I reached age ten I asked my mother to play one or the other of the many Beethoven sonatas for me – again and again. And when at the age of ten I heard a performance of *Die Meistersinger* I became for at least fifteen years an ardent admirer of Richard Wagner's music. Having succumbed to that magical spell, it was not before I reached the age of thirty that I began fully to appreciate Bach, Mozart, and Schubert. During my student days in Vienna (1922–27), however, my greatest musical experiences were the symphonies of Anton Bruckner and Gustav Mahler. I have remained faithful to these two Austrian composers, as well as of course, to the classical and a few contemporary ones. Thus I must avow that I consider truly great music *the* supreme achievement of the human spirit. In keeping with the well-known aphorism of Ludwig Wittgenstein ("whereof one cannot speak thereof one must be silent"), I am inclined to think that music expresses (even more than poetry) what is inexpressible in cognitive and especially in scientific language.

My social and political interests were most dramatically and rudely aroused by my experiences (1921–22) at the University of Munich. Twelve years before Hitler's rise to power, the anti-Semitism in Germany was already quite noticeable. My cousin Arthur Beer and I encountered a curiously vicious circle when we tried to become students at Munich's University. In order to enroll at the University we needed a permit of residence from the police (we were, as Czechoslovaks, after the revolution of 1918, 'foreigners' in Germany). But in order to obtain the permit of residence we would have to be admitted as regular students by the University. Both my cousin (by then already known for some contributions to astronomy) and I were politically quite innocent as well as rather ignorant. But the rejection by the University stung us into action. After a long, adventurous, and often distressing fight against the bigoted authorities at the University and the Police we finally won. An elaborate letter of complaint, endorsed by a liberal lawyer, went up through several levels of the Bavarian bureaucracy. After two months of our protestations the Minister of the Interior of Bavaria himself signed our permits of residence. By early March of 1922, however, i.e., at the end of our first semester at Munich, the new regulations excluded even non-Bavarian German students. There was no possible recourse against this. Hence we moved to Vienna where mathematics, physics and philosophy promised to equal, and were soon to excel, the qualities of the University of Munich.

I had learned that Moritz Schlick (whom I had thus far known only as the author of two superb books) was going to accept the famous chair (earlier held by Mach and Boltzmann) in the Philosophy of the Inductive Sciences at the University of Vienna. Thus I was fortunate to become a student of this truly great philosopher, and later even more fortunate in being favored with the friendship of this superbly kind and generous man.

Another event proved of the greatest influence upon my work and philosophical development. Browsing in a bookstore in Munich in early 1922, I came upon Edgar Zilsel's *Das Anwendungsproblem*, a book dealing with issues of probability and induction. It had been published in 1916, but perhaps because of some flaws in its logico-mathematical aspects had not attained the recognition it so richly deserved for its brilliant and challenging originality. Five years before the publication of John Maynard Keynes' *Treatise on Probability* (and two years before C. D. Broad's first essay on the subject) Zilsel formulated in an extremely stimulating manner the problems that are even today unsettled and subject to intensive disputes. After writing (in 1922) a prizewinning (but unpublished) long essay on the philosophical significance of the theory of relativity, I turned my attention increasingly to the topics of determinism, indeterminism, induction, and probability. I still remember the fever of excitement that gripped me in the spring and summer of 1925 when I began making copious notes of the ideas that whirled in my mind. 'Chance and law; an epistemological analysis of the roles of probability and induction in the natural sciences' was the title of my doctoral dissertation (finished early in 1927). Much of this early work of mine has, of course, been superseded by the rapid subsequent developments. Perhaps the most glaring defect was that I was unable at the time to discuss the indeterminism of quantum mechanics. The theories of Heisenberg, Jordan, Born and Schrödinger were formulated precisely during the period in which I wrote my thesis. I could not possibly have penetrated and interpreted those truly revolutionary developments. All I can say for myself in this connection is that I dealt with the issues of determinism vs. indeterminism as judiciously as my lights then permitted. The early quantum theory of Planck, Einstein, and Bohr had already suggested to me that theoretical physics may have to countenance a concept of absolute chance. Still, my (emotional?) preference was the one to which Planck and Einstein tenaciously held throughout their lives, i.e., that of a 'rock-bottom' determinism. In such a conception of the world there is room only for *relative* chance; and the exploration and explication of relative chance formed one of the more original parts of my dissertation. Schlick (who had hardly helped with advice or supervision) was extremely pleased with my work, and proposed publication in the *Schriften zur wissenschaftlichen Weltauffassung* (edited by P. Frank and M. Schlick, Springer, Vienna). Yet, in view of the new turn in physics I decided to revise my dissertation thoroughly. But I had underestimated the task. The interpretation of quantum mechanics has remained an intricate and highly controversial matter to this day.

A small part of the ideas of my thesis did get published in the first volume of *Erkenntnis*. The title of that article was 'Wahrscheinlichkeit und Erfahrung.' Having favored the frequency interpretation of probability, I was nevertheless (I believe) the first to point out its logical difficulties. I had realized that such notions as (i.e., R. von Mises') limit of relative frequency is not definable in terms of convergence criteria such as are used in the pure mathematics of infinite sequences or series. The dilemma in the case of probability is to say either nothing empirical (i.e., the limit of frequency lies in the closed interval between 0 and 1) or something that is (almost) bound to be false (by stating a definite form of convergence). I suggested a pragmatic solution for this problem as well as for the 'grand' problem (Hume's problem) of induction. I was greatly pleased by the approval of my ideas by R. Carnap, and (with some qualifications) by H. Reichenbach.

My dissertation was written during the first two years of the activities of the Vienna Circle. My friend and mentor, Friedrich Waismann (ca. six years my senior) and I had – I believe in 1923 or 1924 – suggested to Schlick the formation of an evening discussion group. Philosophically interested mathematicians and scientists, together with mathematically and scientifically trained philosophers, were to discuss issues in the logical and epistemological foundations of mathematics, physics, and occasionally also of biology, psychology and the social sciences. The subsequent development is by now an important and well-known chapter in the history of recent philosophy. Along with other ('advanced') students I enjoyed the great privilege of being a member of the Vienna Circle from the time of its inception in 1924 up to the time of my emigration in 1930. Besides Schlick, who was its rather self-effacing leader, the Circle comprised such mature and outstanding scholars as the mathematicians Hans Hahn and Kurt Reidemeister, the sociologist and economist Otto Neurath, the philosopher Victor Kraft – and after 1926 its most contributive member, Rudolf Carnap. Among the younger participants were F. Waismann, and a little later Karl Menger, Kurt Gödel, and Gustav Bergmann. There were also stimulating visitors such as Philipp Frank, C. G. Hempel, Eino Kaila, Alfred Tarski, *et al.* During my last year (1929–30) in Vienna, a young American student, Albert E. Blumberg, joined the Circle. He worked on his doctoral thesis under Schlick, and I helped him with the German language as well as philosophically. It was Blumberg who got me interested in American philosophy and encouraged me to visit (and possibly to settle in) the United States. Even before that I had the pleasure and good fortune to get acquainted with Dickinson Miller who was a visiting member of the Circle in 1926. A year later I received from him a charming and generous invitation to join him and his friend Professor Emeritus C. A. Strong in Fiesole (in the hills above Florence, Italy). Both gentlemen spoke German quite fluently so they did not have to suffer my (then) rather poor English. We mostly discussed matters of epistemology, and it was my task to report on developments in the Vienna Circle. I accepted another such invitation to Fiesole during the following year (1928). I cherish a copy of C. A. Strong's *Theory of Knowledge* which he personally inscribed with a friendly dedication to me.

My studies in philosophy under the guidance of Moritz Schlick, Robert Reininger and Heinrich Gomperz; of theoretical physics under Hans Thirring; mathematics under Hans Hahn; psychology under Karl Bühler – together with the tremendously exciting experiences in the Vienna Circle and the continuing private discussions, especially with Waismann, Carnap, Gödel and Natkin, helped in shaping my philosophical outlook. By 1929 we had become convinced of our philosophical mission in the world. The first Unity of Science meeting took place in Prague in the summer of that year. This was also the first public joint effort of the Vienna Circle with the Berlin Society for Scientific Philosophy, led by H. Reichenbach and R. von Mises. I also participated in the even more fascinating second meeting in Königsberg in 1930. Present at that meeting were also Werner Heisenberg, John von Neumann and Arend Heyting.

Ludwig Wittgenstein, whose *Tractatus Logico-Philosophicus* was the subject of many extended and lively discussions in the Circle, never joined our group during the two years (1927–29) of his architectural activities in Vienna. Schlick and Waismann, however, managed to persuade this brilliant but highly eccentric man to meet with them and often also with me, either at Schlick's house, or in cafés. Occasionally Wittgenstein also came to tea at my fiancée's home. I remember that on one occasion he read some poetry by Tagore to us. Wittgenstein was at that time very reluctant to discuss philosophical matters. When asked about one or the other cryptic passage in the *Tractatus* he said he couldn't reconstruct any more just what he meant when writing it. But on a few rare occasions we witnessed his most impressive and highly intuitive approach to various philosophical problems. As a person of great seriousness and ruthless intellectual honesty and self-criticism he remains, next to Einstein, certainly the most fascinating thinker I ever encountered. Schlick and Waismann quickly became ardent disciples of Wittgenstein. To my chagrin Schlick ascribed to Wittgenstein philosophical ideas that he (Schlick) had already expounded much more lucidly in his 1918 book on epistemology. I was also disappointed with Schlick's compromise with positivism (phenomenalistic version) – and the abandonment of his critical realism as 'metaphysically suspect.'

As all the world knows, Wittgenstein's return (in 1929) to Cambridge and the development of his later point of view – or method of philosophizing –

exerted a tremendous influence in English, American, Australian, and most recently, even in German philosophy. In complete agreement with Bertrand Russell and Karl Popper I find little of genuine importance in the work of the later Wittgenstein. In my opinion it must have been the almost hypnotic power of the man that created a coterie of disciples around him and that rendered his 'ordinary language' approach so fashionable for a good many years. A truly brilliant mind and independent, honest searcher for clarity, Wittgenstein still suffered from (I think) a typically positivistic phobia of metaphysical issues, especially of the quandaries of transcendence. Thus he tried to restore the world view of commonsense by describing the uses of words in common speech. This whole approach seemed doomed to sterility to me, mainly because it (largely) ignored the importance of concept formation and theory construction in the sciences. Given my own education and predilections I still consider knowledge at its best, most reliable, and most fascinating in the more advanced sciences. Moreover, the ordinary language analysis can resolve philosophical problems only if they arise out of rather obvious equivocations, ambiguities or syntactic-semantic confusions. The more profound issues of traditional philosophy, such as the nature of truth, the validity and scope of knowledge, the mind-body problem, the problem of universals, etc., etc., can only be repressed, concealed or glossed over, but not solved by the rather lazy and cavalier 'Oxbridge' type of procedures.

As I have indicated above, my 'formative' years were those devoted to the study of the sciences, and later the scientifically oriented philosophical work in the Vienna Circle. Schlick and Carnap, and to a much lesser extent the early Wittgenstein, but also the many discussions with Neurath, Kraft, Gomperz, Zilsel, and later with Reichenbach, decisively influenced my philosophical outlook. It was not before 1929 that I came to know and soon became a friend of Karl Popper's. To him and to Reichenbach I owe a great intellectual debt. Although I had been in the 'loyal opposition' in regard to the positivism of the Vienna Circle, I had a hard time maintaining against them the sort of critical realism that I had originally learned to adopt from Schlick's own early work (also from Planck, Boltzmann, and from Oswald Külpe's *Realisierung*, and from Alois Riehl's *Der philosophische Kritizismus*, etc.). The first impact and shock came to me from Carnap's *Der logische Aufbau der Welt* which I studied in typescript (it was larger and even more technical than the book published in 1928). Under the influence of Carnap and the early Wittgenstein, Schlick and Waismann were converted to a sort of phenomenalistic positivism during the middle twenties. Their brilliant and powerful arguments overwhelmed me temporarily. But encouraged and buttressed by the support of

Popper, Reichenbach, and Zilsel, I regained confidence in my earlier realism and developed it in my first book on *Theorie und Erfahrung in der Physik* (1929), and later in several articles written during my academic career in the United States. That first book received much cherished comments, not only from my friends and colleagues in the Vienna Circle, but also in letters from Albert Einstein and Wolfgang Pauli. I had been able to deal briefly (though I think now only rather inadequately) with the philosophical aspects of quantum mechanics. In addition I expounded ideas of the level structure of scientific explanation, and the hypothetico-deductive method in theory construction generally. In this regard I anticipated in my very informal way some of the views developed more fully by C. G. Hempel and Karl Popper in later years. But I was not aware of William Whewell's work at the time.

Encouraged by Albert Blumberg (and also by those two magnificent American philosophers, Strong and Miller) I applied for an International Rockefeller Research Fellowship early in 1930. It was my great good luck to receive such a fellowship. I had submitted very favorable recommendations from Einstein, Schlick, Carnap, Thirring, and several other Viennese physicists and philosophers. So, immediately after the Königsberg meeting in September of 1930 I made my first voyage to the 'New World' where I stayed for about eight months at Harvard University. My main first contact there was the physicist-philosopher, P. W. Bridgman, the originator of the 'operational' approach in the logic of science. I also became acquainted with A. N. Whitehead, Henry Sheffer, C. I. Lewis, R. B. Perry, Susan Langer; and among the younger scholars or graduate students, Paul Weiss and W. V. O. Quine. On the advice of my Harvard friends I began looking for a teaching position in the United States (by spring of 1931). I enjoyed a reunion with Albert Blumberg during the Christmas season of 1930 in New York. It was there that we wrote our well-known article, 'Logical Positivism: A New Movement in European Philosophy.' We thus provided the philosophical outlook of the Vienna Circle with its internationally accepted label. Our article (published in the spring of 1931 in the *Journal of Philosophy*) started 'the ball rolling', and for at least two decades lively controversies ensued regarding Viennese Positivism. Although I quite openly expressed my critical opposition, especially to the phenomenalistic reductionism, I did behave myself rather aggressively as an antimetaphysician throughout the nineteen-thirties. My role in the United States was thus similar to that of Alfred J. Ayer in England. I still regret that early manner of mine, especially when I remember how cordially and generously I was received in the many academic institutions that invited me for lectures or colloquia (often entire series of such). Since I have told the story of my experiences and impressions in the United States in my memoir in the book *The Intellectual Migration* I shall mention here only how grateful I feel to this country, and especially to the deans and presidents of the State University of Iowa and the University of Minnesota, where I found a peaceful haven well before hell broke loose in Europe. – I now turn to

WHAT I BELIEVE

Space permits only a very brief and condensed survey of whatever philosophical conclusions I have reached. While I feel quite strongly about some of them, the terse, dogmatic tone in which I shall present them is merely due to the brevity demanded on this occasion. The attitude of open-mindedness is the only fruitful one in our age of science.

1. *The Task of Philosophy*

In producing a lucid and coherent account and analysis of human knowledge and valuation, the aim is to make explicit the presuppositions or basic assumptions, to clarify the pivotal concepts and methods of cognition and valuation. The most successful way of doing this still seems to me the procedure of logical reconstruction. The question 'What do we mean by the words and symbols we use?' should precede the questions of justification. The tools of modern logic should be employed wherever they are genuinely helpful. But formalization and axiomatization is needed only for the more exact and exacting purposes. 'Half-way stations' toward ideal language formulations are often quite sufficient for the solution of problems in the general theory of knowledge, moral philosophy – and even in the philosophy of science. While it would be most desirable and commendable if professional philosophers could contribute (on the side of conceptual revisions and innovations) to the progress of the sciences, it seems that only a very few philosophers nowadays have the scientific training to collaborate helpfully with the specialized scientists. The tasks of clarification and understanding are already extremely demanding and ambitious for philosophers in our time.

2. *Language, Meaning and the Demarcation of Science from Metaphysics*

Critical reflection in these regards has shown that philosophy traditionally has suffered from the burdens of conceptual confusion and of the (related) 'disease' of demonstrationism. As already strikingly exemplified in Plato, the necessity of mathematical truths, the conclusiveness and exactitude of mathematical reasoning and proof have been perennially shaping the aspirations of philosophers. Thus we find that exaggerated conception of knowledge – as contrasted with mere opinion – and the conviction of the rationalists that genuine knowledge must either be self-evident to the 'natural light' of human reason, or strictly deducible from self-evident premises. But ever since David Hume's sceptical arguments at least the empiricists have considered all factual knowledge claims as at best highly confirmed (by observational data), and as always in principle open to revision, if not refutation. Thus responsible scientists hold their theories as 'true only until further notice', such notice being given when discordant evidence turns up. Pure logic and pure mathematics, however, rest on postulated premises – which need not be self-evident. I realize I will be considered quite an inveterate and 'reactionary' logical empiricist. But I remain unconvinced by the clever arguments of Quine that are intended to show that there is no sharp line of distinction between the purely formal truths (e.g., of arithmetic) and the factual truths (e.g., of physics). I retain the distinction between the purely formal and the factual type of meaning. But I do not object if the formal type of meaning is regarded as the 'null case', or extreme lower limit of factual significance. The tautologies, i.e., the logical truths, or statements whose form – once *definienda* have been replaced by their *definientia* – boils down to logical truths do not require observations for their validation; certainly not observations of the sort that are indispensable in the factual sciences (natural or social).

If purely formal and factual meanings are the two types of cognitive significance, then cognitive significance itself must – for the sake of clarity – be distinguished from pictorial, emotional and motivative significance of words or sentences. Only in regard to cognitively significant statements is it appropriate to ask whether they are 'true or false.' And while in common communication cognitive and non-cognitive significance are often combined or fused, it is of the utmost philosophical importance to avoid confusing emotive with cognitive significance. Many of the 'guaranteed 100% unsolvable' puzzles of traditional metaphysics rest on exactly this type of confusion. For example, the metaphysical ideas of space, time, substance, causality, vital force (entelechy), soul, etc., are peculiar combinations of absolutely untestable notions with empirically testable concepts. Once the confusion is disentangled, the scientifically meaningful and fruitful core can be separated out.

The much debated and often revised testability criterion of factual meaningfulness seems to me useful and, even, indispensable. Unless some of the concepts appearing in our statements are connected, no matter how indirectly, with some data of immediate experience, those statements would at best have formal significance but they would be devoid of factual meaning. I think the enormous amount of debate and quibbling that concerned the meaning criterion has been largely a waste of time and energy. The logical empiricists (especially Carnap) had an extremely important point there.

According to the liberalized criterion much of traditional metaphysics and even of theology is perfectly meaningful. But, of course, empiricists should never forget to ask as to whether there are any good reasons for accepting the tenets of those speculative and/or intuitive (or purely dogmatic or kerygmatic) doctrines. In my opinion nearly all reasons traditionally given are no good reasons at all. I think that the beliefs in a personal God, or in the metaphysical Absolutes (space, time, substance, etc.), are motivated by wishful or pictorial-emotional thinking. And since I – against the current vogue of obscurantism – maintain that there is a fundamental difference between the subjective motives and the supporting objective reasons (i.e., justifying grounds) of beliefs, I reject as false (or as extremely unlikely) all those 'transcendent' (but meaningful) assertions.

There are, however, fairly clear cases in which the meaning criterion has to be invoked. If a metaphysician or theologian makes his beliefs absolutely 'proof against disproof', and thereby (usually unwittingly) renders his problems absolutely unsolvable, then it is proper to ask as to whether his syntax, semantics, or pragmatics is in good order, or as to whether he is the victim of a misuse of language – be it through simple equivocations; violations of formation rules; or lack of connections, ultimately by ostensive steps, with the data of direct experience. I think that despite its metaphysical-theological framework, Leibniz's arguments in the famous correspondence between him and Clarke regarding Newton's notions of absolute space and time are a classical anticipation of the pragmatist-positivist critique of the 'perniciously' transcendent metaphysics. "A difference must make a difference in order to be a cognitive difference" – this is the way I keep paraphrasing what is common to the views of C. S. Peirce, William James and the Logical Empiricists. The difference between an assertion and its denial must in principle be open to at least partial and indirect observational test, otherwise there is no factual meaning present in the assertion.

3. *The Scope and Validity of Knowledge*

The liberalized meaning criterion allows for hypothetico-deductive tests (i.e., confirmations or disconfirmations), for inductive interpolations and extrapolations, for analogical conceptions and inferences. Hence I think that a critically realist position is justifiable. The epistemological puzzles about the 'external world', 'other minds', the past and future of the universe, the reliability of memory, the occurrence of unconscious mental processes, the existence of unobservable entities (subatomic particles; nuclear forces, etc.) can be resolved affirmatively. In these matters I feel I am closer to the outlook of Reichenbach and Popper (their differences are irrelevant in this connection) than to Carnap's position. Nevertheless, Carnap's distinction of 'external' and 'internal' questions, i.e., questions regarding the categorial frame and questions within the frame, is well conceived.

My own distinction between two types of justification ('vindication' and 'validation') is an exact counterpart of Carnap's distinction. Since in the justification of knowledge-claims (and for that matter also of moral or aesthetic evaluations) we must obviously avoid both the begging of questions and infinite regresses, validation must terminate with premises (basic assumptions, presuppositions) which in the given context are not susceptible to further validation. But these 'ultimate' presuppositions of validation are neither arbitrary conventions, *nor* unquestionable *a priori* truths. They are susceptible to a different type of justification, or 'vindication' as I call it. The vindicative argument can show that the adoption of the validating premises will achieve the ends we pursue in the given context. Or, as in the clearly weaker case of the justification of inductive inference or of the hypothetico-deductive procedure, it can at least show that if any method or procedure yields the desired results (correct explanations, predictions, retrodictions etc.), the usual application of the scientific method will do the same. This idea can be made more plausible by considering observation and experimentation as a sampling procedure, i.e., as the attempt to obtain representative samples of the regularities (deterministic or statistical) of the – possibly limitless – universe. There can, of course, be no guarantee that a given sample is representative, and without special assumptions of background knowledge, there cannot even be any estimates of the probability of the 'fairness' of the sample. Nevertheless it is clearly *rational* to assume – until further notice! – that the given sample *is* representative. For to assume that it is not representative would open the flood gates to limitless possibilities. This would amount to a complete capitulation to scepticism or cognitive pessimism. David Hume has shown once and for all that neither rational nor empirical grounds (nor any combination of the two) can justify the ultimate presupposition of inductive generalization. Hence he maintained a purely psychological view of habits of expectation (later labeled by Santayana as the doctrine of 'animal faith'; even B. Russell at one stage of his philosophical development spoke of 'physiological induction'). Karl Popper, deeply impressed with Hume's arguments, abandoned all efforts toward a justification of induction; he even denied the importance, if not the very occurrence, of induction in the growth of knowledge. But, as perhaps the first to criticize this view of Popper's, I asked the crucial question why we should put our trust in (or 'place our bets on') laws, hypotheses and theories which, despite severe tests, have thus far not been refuted. To this question Popper has never given a satisfactory answer. He used to (and perhaps still does) call such theories that have successfully survived searching criticism 'well corroborated.' But since Popper himself refuses to identify corroboration with confirmation, he does not provide any reason whatever for the generally accepted practice of using a well-corroborated theory as a guide for further research or, in its practical application, for our expectations and actions. Nevertheless there is an important lesson to be learned from Popper's views: Practically all our knowledge-claims are based on background theories. Only within the frame of assumptions about space and time; about the place of experience in the world of nature; about the mechanisms of perception and observations; etc., etc., can we examine (test) all knowledge-claims, be they fully specific descriptions, or very general hypotheses. Science is (contrary to classical positivist dogma) not a compendious and economical summary of experience, but an attempt at understanding (explaining) the facts of nature by means of laws, hypotheses and theories.

In the light of this conception of scientific knowledge (which I wholeheartedly accept) my view of the justification of induction can be more fully and adequately formulated. The experimental samplings of 'nature' are to be considered as confirming or disconfirming pieces of evidence for *theories*; and it is the extrapolation of the (e.g., spatio-temporal) scope of those theories which is genuinely inductive. Thus while no guarantees can be given for even the 'probability' let alone certainty of theories, the only reasonable ('rational' in one of the several senses of this slippery term) procedure is to stick, until further notice, to those theories which have 'proved their mettle' under severe scrutiny, and for which impressive positive evidence is accumulating.

4. *Mind and its Place in Nature*

As I mentioned before, beside the issues of induction and of scientific explanation, the mind-body problems have held my attention ever since my adolescence. I am now inclined to proceed according to the 'divide and conquer' maxim. It seems helpful to distinguish at least three major issues in the cluster of mind-matter problems: (1) sentience, (2) sapience, (3) selfhood. While it is not advisable to disregard the interrelations of these three features of mind, I think the sentience aspect has engendered more profound bafflements than the other two. In any case I have been primarily concerned with this first problem. I have tried to steer clear of the reductive fallacies of materialism and radical behaviorism; but I have equally shunned the seductive fallacies of the mental-substance (let alone of an indestructible soul) doctrines. It is obvious that in regard to this central problem of modern philosophy we still have to combat the tendencies of simple-minded (or tough-minded) as well as muddle-headed (or tender-minded) thinking. The occurrence of immediate experience, as well as the physical structure of the universe and of the human organism and its processes have to be assigned their proper places in a scientifically as well as philosophically tenable account of our world. Fortunately the battle against 'crass' materialism is now as obsolete and unnecessary as is the battle against (subjective, or objective) idealism. What is wanted is clearly a view in which conscious experience is understood as the subjective aspect of certain processes in the central nervous system, perhaps primarily in the cerebral cortex. Of course this formulation is metaphorical and hence needs to be replaced by the straightforward result of a logical analysis. Just this, however, is a difficult and highly controversial task. First of all nothing significant can be achieved without close attention to the best established results of the sciences, in this case of psycho-neurophysiology. If we dismiss on scientific grounds animistic, vitalistic and dualistic-interaction views, we seem to be left with a choice between doctrines of emergent evolution and psychophysiological parallelism or of a materialistically slanted epiphenomenalism. Mind as a product of emergent evolution is still a favorite among some philosophically untutored biologists. Although I think this view could be made (in fact has been made) philosophically more sophisticated and respectable, I am inclined to repudiate it in favor of a more plausible theory. Starting with parallelism, i.e., with a view of a one-one, or many-one correspondence of the physical to the mental, a more incisive reflection upon the meaning of the concepts on each side of the relation leads to a gratifyingly simple, though on first sight, unconvincing solution of the puzzle of the relation between sentience (immediate data and qualities of consciousness) and brain processes. This is the identity theory, currently hotly debated. In an early attempt I conceived of this theory along the lines of a simple translation of mental into physical terms. But since the correspondence (one-one; or many-one) is clearly a matter of empirical investigation rather than of logical synonymy or equivalence, a more adequate solution seems to be suggested by a twofold approach or twofold knowledge theory. We have knowledge of the phenomenally given (the 'raw feels' of sensation, desire, emotion, mood, intention, etc.) by direct acquaintance; whereas our knowledge of physical objects (including our own brain) and the processes in which they are involved (or, if you will, of which they consist) is indirect; it is what B. Russell called "knowledge by description." Such knowledge is 'structural' in the sense that it concerns the logico-mathematical network of relationships only; and that the qualitative content (as in the case of our knowledge of other persons' mental states) has to be conceived and inferred by analogy to our own immediate experience. Within the physical (scientific) conception of the world the qualities of direct experience are indeed 'homeless', unless they are 'attached' or 'introjected' by analogical reasoning. I keep an open mind (and continue to work) on the best formulation of the identity theory. In a long essay of 1958 I argued for a theory of co-designation; i.e., the view that mentalistic (phenomenal) terms designate the same events or processes that are also designated by the terms of the physicalistic (brain-physiological) language. Currently I prefer to put it as a matter of two 'perspectives' (of course this is still metaphorical) namely the egocentric and the intersubjective account of the world. The files are not closed on this intricate problem. Both the approaches of science and those of logical analysis are still far from finished. There is the problem of the complete reducibility of biology (via biophysics, biochemistry, and molecular biology) to basic physics. Much as I am impressed with the progress in this direction, there are on the other side some scruples about the adequacy of the neo-Darwinistic and genetic explanations of evolution. Furthermore, as an empiricist I even have to (at least) go through the motions of an open mind in regard to parapsychological phenomena (ESP). While these strange 'facts' (if they are not products of deception, illusion, delusion, etc.) need not necessarily militate against the basic physicalistic program of science, they might yet force upon us incisive revisions of our total theoretical outlook on mind and nature.

I confess that my outlook is still somewhat 'Victorian' in that I favor the view that mental processes are part of nature. But I admit, nay insist, that the world conception of modern physical science, impressive as it is, can hardly be final. As to whether there is a 'rockbottom' of nature, and if so, as to whether it conforms to strictly deterministic laws (as conceived in classical physics from Newton to and including Einstein) or as to whether there is an ineluctable indeterminacy (à la Heisenberg, Bohr, Born, von Neumann, et al.) must – in principle – remain undecidable. Still, the experimental evidence of quantum physics makes it rather plausible that the physical theories of the future will have to countenance an element of absolute chance. Perhaps (to use Einstein's famous phrase, but contradicting him) God does play dice with nature!

There is, however, no solace to be derived from the micro-indeterminism for the traditional problem of free will. Absolute chance cannot be the basis of free choice or action. As Hume, Mill, Sidgwick, Schlick, and Hobart already very clearly realized, the sort of moral responsibility that presupposes free choice, is not only compatible with determinism. It would be outright impossible if it were not for a fairly large measure of lawful determination of human volition and action. Only if we are the doers of our deeds, i.e., only if our deliberations, intentions, decisions, etc. are causal consequences of our characters and personalities (and are not forced upon us by external or internal compulsion), can we be justly considered praiseworthy or blameworthy for whatever we will or do.

A Few Remarks on Ethics, Religion, and the Problems of Human Affairs

Although I feel strongly about all these matters of 'Practical Philosophy', I have a little to add to what many other recent thinkers (for example, M. Schlick and B. Russell) have said much better than I could possibly say. I repudiate the 'naturalistic fallacy', i.e., the attempt to derive moral value judgments from facts about human-nature-in-the-social-context alone. All the strenuous and tortuous sophistry that has been applied to the contrary has left me unconvinced. Without basic commitments even the most thorough and complete knowledge of matters of fact does not and can not yield moral imperatives. The Hippocratic oath that young medicos have to swear is a good illustration of the point at issue. Without firm resolve to help rather than to harm his patients the doctor could use his medical, psychological (and even commonsense socio-psychological) knowledge quite 'diabolically' – i.e., to the detriment of those whom he treats. Quite generally, what we *ought* (or ought not) to do cannot be inferred from what *is* the case (or from a knowledge of what happens under what conditions) alone. To be sure, factual information concerning the likely consequences of various courses of action is indispensable, and always of the greatest importance. But without commitments to basic aims or values (and/or to basic rules regarding 'right' and 'wrong') we could never justify our morally relevant decisions or actions. 'Saints' as well as 'Scoundrels' can make use of factual information. The difference between them is not intelligence, but the presence or absence of good will.

For anyone with even a minimum of training in logical analysis it is a ludicrous and pathetic spectacle presented by countless philosophers who invariably beg the question by either presupposing tacitly some moral valuation, or by (what really amounts to the same) using persuasive definitions of human nature. Thus, if it is argued that the 'real nature' of being human involves the capacity for being fair, just, helpful, kind, and self-perfecting, then this simply conceals the exhortatory character of this persuasive definition. It seems to me intellectually more honest to state quite explicitly the nature of our moral-social commitments.

I, too, believe (with Aristotle) that man is a 'rational animal'. But I feel that on the whole mankind has made deplorably little use of the capacity for thinking and acting rationally. As Strindberg put it so poignantly in *Dream Play*: "It is a pity for mankind." We are now well on the way to ruining the planet we inhabit, if not to exterminating ourselves by overcrowding, or by environmental pollution, or by a nuclear holocaust. Quite generally our social behavior is barbarous, in many respects closer to that of ferocious beasts than to the ideal that philosophers and the founders of the great religions (Buddhism, Judaism, Christianity, etc.) delineated. Of course, in our age of science, orthodox or dogmatic religions are no longer intellectually acceptable. If I am to label my own outlook it is that of a scientifically oriented humanism. This is best summarized in Bertrand Russell's sentence: "The good life is one inspired by love and guided by knowledge." As to whether mankind can grow up toward genuine humane-ness, as to whether we can stop being crude and cruel blunderers in the art of living together (individually, racially, nationally, and internationally) seems questionable now. As a humanist I do not wish to deprive persons of their religious faith if it is an indispensable support for them and if it does help them to love their fellow human beings. But I entertain only a cautious hope that a truly enlightened mankind will be able to achieve a permanent peace and a harmonious way of life.

In regard to the question concerning the 'meaning of life', I can do no better than to quote Wittgenstein (*Tractatus* 6.521): "The solution of the problem of life is something one becomes aware of when the problem vanishes." A life that is conducted in accordance with the ideals of justice, kindness, brotherhood, freedom, love and self-perfection is the only one (for normal human beings) that promises genuine satisfaction, perhaps even a measure of joy and happiness. There is abundant advice (by wise psychologists, economists, sociologists, etc.) as to how to improve *la condition humaine*. While I have given a good deal of thought to these matters, I do not know of a cure-all for the many evils that beset us.

A Concluding Remark

I may not be a 'philosopher' in the commendatory sense of that word. It would be arrogant and largely unjustifiable for me to designate myself as a wise man. Even after many years of honest searchings, I can only claim to being a lover of wisdom – which is, after all, the original, etymologically genuine meaning of 'philosopher.' I have also sufficient insight into the sources and factors of my intellectual development to realize that my views are largely the outcome of my deep involvement with the scientific outlook. Other philosophers owe their point of view to very different sources of inspiration. Aware of all this I must admit the present essay is indeed no more than a *confessio fidei*. Hence I am far from confident that I could offer a 'message' that would be acceptable to any but kindred spirits.

NOTE

* [Publications mentioned in this article are listed in the bibliography to this volume – Ed.]

2. THE ORIGIN AND SPIRIT OF LOGICAL POSITIVISM

[1969a]

Logical positivism began to form a fairly definite outlook in philosophy about forty years ago. As is well known, it was primarily the influence of Ludwig Wittgenstein and Rudolf Carnap that initiated the early phase of this – then new and radical – departure from the traditional ways of philosophizing. To be sure, some aspects of logical positivism are derived historically from Hume and Comte; but, in contrast, especially to Mill's positivism, a new conception of logic (having its origins in Leibniz, Frege, and Russell) was united with the empiricism of Hume, Mach, and the early Einstein.

The Vienna Circle consisted mainly of scientifically trained philosophers and philosophically interested mathematicians and scientists. Most of the members were tough-minded thinkers, *Weltzugewandt* (as Hans Hahn put it), 'this-worldly' rather than 'other-worldly.' They were radically opposed to metaphysical speculation, especially of the *a priori* and transcendent types. Since the development of the Vienna Circle is by now a familiar chapter in the history of recent philosophy, I propose, after dealing with some of the antimetaphysical doctrines of logical positivism, to concentrate on some of the aspects that are not as well known. I shall refer particularly to the work of Moritz Schlick, the founder and leader of the Vienna Circle. Schlick's early work anticipated a good deal of what in more precise formulations was later developed by Carnap, Reichenbach, and others. In his *Allgemeine Erkenntnislehre* (1918) there were also anticipations of some of the central tenets of Wittgenstein's *Tractatus Logico-Philosophicus.** I think it was Schlick's extremely unassuming character, his great modesty and kindliness, and his deep personal devotion to Wittgenstein that made him forget or suppress the great extent to which his views, independently developed and quite differently stated, already contained very important arguments and conclusions regarding the nature of logical and analytic validity; the semantic explication of the concept of truth; the difference between pure experience (*Erleben*), acquaintance (*Kennen*), and genuine knowledge (*Erkennen*), etc. Indeed, so deeply impressed was Schlick with Wittgenstein's genius that he attributed to him profound philosophical insights which he had formulated much more lucidly long before he succumbed to Wittgenstein's almost hypnotic spell.

In the first flush of enthusiasm of the late 1920's, the Vienna Circle proclaimed its outlook as a philosophy to end all philosophies, as a decisive turn toward a new form of enlightenment. The pamphlet *Wissenschaftliche Weltauffassung: Der Wiener Kreis*, published in 1929, was our declaration of independence from traditional school philosophy. This slender brochure, composed collaboratively by Carnap, Hahn, and Neurath (aided by Waismann and myself), was presented to Schlick upon his return from his visiting appointment at Stanford University. As I remember only too well, Schlick, while appreciative of this token of friendship and admiration, was deeply disturbed by the idea of having originated another 'school of thought.' He was a philosophical individualist; although he promoted group colloquy and believed in the fruitfulness of mutual criticism and searching discussion, he was profoundly convinced that everyone should think for himself. The idea of a united front of philosophical attack was abhorrent to Schlick – notwithstanding the fact that he himself promulgated the Viennese point of view in many of his lectures, in Europe and also in the United States.

Logical positivism became noted, as well as notorious, through its critique and complete rejection of metaphysics. In the spirit of Hume and Comte, but equipped with more fully developed logical tools, the Vienna Circle declared any question (and any answers) of a transempirical sort to be factually meaningless. The original formulations of the criterion of meaningfulness were somewhat brash and careless. The motivation, however, was fairly obvious. Just as Hume considered significant only statements about the relations of ideas (i.e., logic and mathematics) or matters of fact (the empirical sciences), so the logical positivists were adamant in excluding as *nonsensical* any question that, in the light of logical analysis, revealed itself to be absolutely unanswerable. Always granting, and even emphasizing, that pure logic and pure mathematics have standards of meaningfulness and validity of their own, the meaning criterion was designed to separate factual questions and propositions from metaphysical pseudoproblems and pseudosolutions of such problems.

It was only some years later that the Viennese positivists realized their kinship of outlook with that of the American pragmatists, especially C. S. Peirce, and with the operationalist approach of P. W. Bridgman. The pragmatists declared a proposition meaningless if there was no difference that *made* a difference between asserting it and denying it. The difference that Peirce referred to was a difference with respect to observable consequence. Bridgman considered a concept to be genuinely meaningful only if it could be defined by specifiable, observational, mensurational, or experimental procedures. Under the influence of Wittgenstein's and Carnap's early work, Schlick maintained that the factual meaning of empirical statements consists in 'the method of their verification.' Taken literally, this was a regrettably inadequate formulation. No wonder that sharp criticisms from many sides were raised against this slogan. But a closer reading of Schlick's and Carnap's formulations of the late twenties and early thirties clearly shows that they were concerned with testability rather than verifiability. As I recall it, the early discussions with Hans Reichenbach, and a little later with Karl Popper,[1] engendered the more tolerant formulation of the meaning criterion in terms of testability (later, confirmability or disconfirmability). Popper quite rightly pointed out that there can be no conclusive verification of scientific hypotheses or theories. Inasmuch as these are expressed in universally quantified propositions, they are at best falsifiable, that is, refutable. (Popper himself used falsifiability, not as a criterion of meaningfulness – he rejected any and all attempts to formulate meaning criteria – but as a criterion of demarcation enabling us to distinguish between empirical and non-empirical assertions.) In any case, Carnap, as early as 1928, spoke of '*Prüfbarkeit*' or '*Nachprüfbarkeit*' in the sense of *testability*, which he later elaborated in a new and original way in his essay 'Testability and Meaning' (1936).

The essential antimetaphysical attitude of the Vienna Circle is perhaps best understood in terms of a distinction between two major functions of language: the cognitive (or informative) function; and the non-cognitive (or emotive) function. In his monograph *Pseudoproblems in Philosophy* (1928) Carnap differentiated sharply between the cognitive content or knowledge claim conveyed by a linguistic utterance, and the accompanying imagery or emotions. This was the origin of the much-discussed distinction between the cognitive meaning and the emotive (i.e., expressive and/or evocative) significance of words or sentences. In transcendent-metaphysical utterances, the emotive significance was seen to masquerade for genuinely cognitive meaning. In this manner the theologians' talk of an inscrutable deity, the vitalists' talk of entelechies, and the metaphysicians' talk of the absolute (absolute substance, truth, reality, absolute space and time, or the 'Absolute' of the Hegelians) were diagnosed as pieces of verbal magic or as verbal sedatives, devoid not only of any explanatory power but of all cognitive significance. The vital forces or entelechies still defended by the notable biologist-philosopher Hans Driesch were explicitly conceived of as being 'not in space' but as 'acting into space', unknowable before they produce their marvelous effects but somehow inferable *ex post facto*. Hume's analysis of the concept of causality was endorsed in that the regularity in the sequence of events was recognized as the only testable cognitive content of the principle of causality. Any introjection of 'intrinsic necessity' was branded as a confusion either of logical entailment (which holds only between propositions) or of the psychological feeling of compulsion (which is all too anthropomorphic) with the objectively confirmable order of natural events. To put it in the manner of the pragmatists: There is no difference that makes a difference (in the testable consequences) between the statement '*A* is always followed by *B*' and the statement '*A* is always *and necessarily* followed by *B*.' There is no testable surplus meaning in the second statement, over and above the first.

In a similar spirit the Vienna Circle accepted the relational view of space and time, and the kinematic relativity of motion. This nowadays obvious conceptual clarification was achieved by Leibniz in the seventeenth century (in his famous correspondence with Samuel Clarke in which he criticized the absolutism of Newton's conceptions). Once (in 1920) I heard a disciple of Franz Brentano's – Oskar Kraus at the University of Prague – debate Einstein with great excitement. He maintained that the following was a synthetic *a priori* truth: "If two bodies move relatively to each other, then at least one of them moves with respect to absolute space." This illustrates beautifully the intrusion of the pictorial appeal of the Platonic 'receptacle' notion of space or a confusion of a purely definitional truth (regarding three coordinate systems) with genuinely factual and empirically testable statements regarding the motion of bodies. These are obvious matters, of course, and are quite elementary and unquestionable presuppositions of Einstein's special and general theories of relativity.

More impressive and influential in the days of the Vienna Circle was the critique of the ether concept in Einstein's special theory. The ether theory, in its final stage of defense, appeared to us to be a typical instance of a theory made proof against disproof. The hypotheses of Lorentz and Fitzgerald regarding contraction and 'local time' made the ether hypothesis immune to any conceivable test. There was no difference that could conceivably make a difference in the outcome of any mechanical, optical, or electromagnetic experiment by which one might determine the velocity of a body with respect to the ether. The problem that confronted the physicists at the time was not to be solved by further *ad hoc* hypotheses but, as Einstein's genius perceived it, by a critical revision of the concept of simultaneity and, in consequence of it, a revision of the concepts of distance and duration. Einstein solved the problem by transforming the puzzle of the constancy of the speed of light into a postulate of the theory. By choosing appropriate definitions of congruence for space and time, he made the ether hypothesis not only superfluous but demonstrably devoid of empirical content or significance in the theory of Lorentz and Fitzgerald.

In his critique of the pseudoproblems of philosophy, Carnap applied the same criterion of meaningfulness to the traditional problems of realism versus idealism. The so-called problems of 'the existence of the external world' and of the 'reality of other minds' were shown to stem from confusions of the perfectly legitimate (because testable) ascriptions of empirical reality with some intuitive or emotive (and untestable) notions of reality. Carnap, throughout all his pronouncements (including his most recent ones), has rejected the metaphysical doctrines – of both realism *and* idealism – as being cognitively meaningless; but he has come to endorse the conceptual framework of an *empirically* realistic physicalism as a preferable reconstruction of common sense and especially of science. This seems to be essentially the same position as that taken by Wittgenstein in the *Tractatus*: (5.62) "what the solipsist *means* is quite correct; only it cannot be said, but it makes itself manifest"; and (5.64) "it can be seen that solipsism, when its implications are followed out strictly, coincides with pure realism. The self of solipsism shrinks to a point without extension, and there remains the reality co-ordinate with it."

Carnap's and Ernest Nagel's recent pronouncements on realistic versus phenomenalistic or instrumentalistic interpretations of physical theories appear to me to be essentially similar to the antimetaphysical (or ametaphysical) position of the Vienna Circle. Both Carnap and Nagel maintain that there is no factually or logically decidable issue here, and that the difference between these interpretations is merely one of preference for one *language* or another. I shall return to this controversial matter when I discuss Schlick's early critical realism and his subsequent positivism.

In regard also to other problems of philosophy, especially the philosophy of science, the attitude of the Vienna Circle was clearly empiricist. Those of us concerned with induction and probability strongly favored the frequency interpretation. We rejected all notions of probability based on *a priori* intuition, such as the classical one with its 'principle of insufficient reason', and J. M. Keynes's version with its unanalyzable relation between the supporting evidence and the hypothesis that is probable in relation to it. It was the theory of Richard von Mises, especially as elaborated later by Hans Reichenbach, that we found most attractive. I suppose our preference for the frequency interpretation was motivated by the same reasons that C. S. Peirce advanced for his views on probability. The relative frequency of events within a given reference class seemed obviously susceptible to empirical determination. But soon enough (in my own case, as early as 1924) it dawned on us that the talk of 'limits of frequencies' involved a conception radically different from the limits of infinite sequences or series in pure mathematics. There, one could always prove the existence of a limit in the light of criteria of convergence. But in the empirically given statistical ratios this is impossible.

I think I was the first of the Viennese to see quite clearly the dilemma of the frequency interpretations: To speak of a limit of statistical frequency is either to say that that limit is somewhere in the closed interval between zero and one (and that is a tautology, and hence devoid of factual information) or, if one states a limit in terms of a definite relation of n (the ordinal number of elements in a sequence) and ϵ (as used in genuine convergence criteria), to say something that is almost bound to be false. In the language of symbolic logic, the statement of genuine convergence is triply quantified. Thus, for heads or tails, for examples, the limit of the frequency ½ would be formulated by:

$$(\epsilon)\,(\exists N)\,(n)\;[(n > N) \supset (|f_u - \tfrac{1}{2}| < \epsilon)]$$

It is evident that because of the universal quantifiers such statements cannot be conclusively verified and that because of the existential quantifier they can never be conclusively refuted – as long as no specific function of n relating it to ϵ is stipulated. A little later, F. Waismann, influenced mainly by the scanty pronouncements on probability in Wittgenstein's *Tractatus*, attempted a purely logical definition of the probability concept in terms of the ranges (to be determined by appropriate measure functions) of the hypothesis and its relevant evidence. But it was only in the early forties that Carnap, starting from this suggestive idea, began to develop his system of inductive logic involving a definition of degree of confirmation, that is, the degree of probability bestowed upon a hypothesis by the supporting evidence. Because Carnap considered statements of degree of confirmation to be analytically or purely logically true (in virtue of his definition of probability), no synthetic *a priori* presuppositions seemed required. But just this point has remained controversial, and the dispute between the defenders of the subjective, the logical, and the frequency interpretations still continues. There is only one point of general agreement of all these schools of thought with the early position of the Vienna Circle: the calculus of probabilities as such is a branch of pure mathematics. It can do no more than calculate, that is, derive, deductively, the probability of complex events on the basis of given (or assumed) probabilities of elementary events. What is called, with objectionable ambiguity, the 'law of large numbers' is either the theorem of Bernoulli (and that is a purely mathematical truth based on combinatorial analysis and the addition and multiplication theorems of the probability calculus) or an empirical statement about the statistical frequencies 'in the long run', and as such is a logically problematic, but practically useful, conclusion of inductive inference.

The more general philosophical import of the foregoing discussion is that the so-called problem of induction (i.e., Hume's problem of the justification of non-demonstrative inferences) cannot be solved with the help of either of the logical or the statistical concept of probability. Some logical positivists have therefore declared it to be a pseudoproblem; others (especially Reichenbach, myself, and, more recently, Wesley Salmon) have attempted a pragmatic solution of the old puzzle. In my own view this rests on a distinction between the justification of knowledge-claims within the framework of certain guiding principles and the practical justification of these very principles. I have tried to settle this by distinguishing between 'validation' and 'vindication.' This is to say that the credibility of empirical hypotheses (e.g., predictions) can be validated in the light of the principles of inductive logic. But these very principles (while, of course, not open to validation) can be vindicated: their adoption can be justified pragmatically as means that are necessary (but surely never provably sufficient) for the success of scientific research, that is, for the discovery of reliable regularities. As I recall my conversations with Schlick (in 1935, one year before his untimely death), he favored this sort of approach. But I am no longer sure that it really solves the basic problem of induction.

On many occasions the logical positivists of Vienna (before and after their 'diaspora') discussed the program and goals of their 'reform of philosophy', especially in regard to the philosophy of science. Influenced and inspired by the ideas and achievements of Helmholtz, Mach, Poincaré, the early Bertrand Russell, and Hilbert, it was Moritz Schlick who blazed the trail for a new and synoptic approach. In his *General Theory of Knowledge*[2] he proposed as a genuine task of philosophy the clarification of the basic logical and methodological concepts and principles of the sciences. While admitting that the scientific specialist may well produce important results without being aware of, or particularly interested in, those fundamental concepts and principles, Schlick stressed again and again that a full understanding of science is possible only through reflection upon its foundations. To make these basic principles fully explicit, to specify the meaning of the pivotal concepts of science, is, according to Schlick, a proper and worthwhile job for the philosopher. Of course, this is not a new idea; at least from Aristotle down to Kant and the scientist-philosophers of the nineteenth century, it has been an acknowledged aim of philosophy. But renewed emphasis was urgently needed, especially in the German countries that had not yet recovered from the intellectual debaucheries of the post-Kantian romantic metaphysicians and the naive agnosticism (the 'ignorabimus' of Dubois-Reymond) and the equally naive dogmatism of Ernst Haeckel (*The Riddle of the Universe!*). Much of the iconoclastic and, indeed, negativistic spirit of the Viennese positivists may be understood as a reaction against the highflown, pretentious verbiage of metaphysical speculation. Along the lines of Bertrand Russell, Hans Hahn, the great mathematician, in a pamphlet *Überflüssige Wesenheiten* (Superfluous Entities), advocated the use of 'Occam's razor' toward the elimination of all metaphysical, ethical, and political 'absolutes.'

It was Schlick's merit to have set an example of sober, lucid, and scientifically well-informed philosophical analysis. While his approach was mostly informal, he showed his appreciation of the tools of modern logic in his courses at the University of Vienna. I think it is fair to say that he conceived of the philosophy of science as a task of logical reconstruction. He distinguished quite clearly between a psychological (or sociopsychological) account of scientific discovery and a logical analysis of scientific concepts and principles. In his *General Epistemology* (1918) he dealt with the nature of scientific explanation as a prime example of cognition. He maintained that all genuine knowledge consists of a reduction of one kind of entities to another, or, of what is tantamount, the derivation of more specific propositions from more general (lawlike) ones. Thus he repeatedly used such illustrative cases as the identification of light with electromagnetic waves or the identification of the chemical bond with electromagnetic forces. Even for the level of the knowledge-claims of everyday life, he made it clear that genuine cognition amounts to the subsumption of a particular item under a class, as in recognizing a perceived thing as an elm tree.

Schlick drew a distinction (to which he ascribed fundamental philosophical importance) between *Erleben* (or *Kennen*), on the one hand, and *Erkennen*, on the other. This can be rendered in English by the distinction between *acquaintance* (or knowledge by acquaintance) and *knowledge proper* (knowledge by description). Bertrand Russell, who also recognized this indispensable distinction, used the English terms just mentioned. Thus, the mere having, that is, the undergoing (e.g., enjoying or suffering), of an immediate experience, and the ability to recognize and affix consistently the same predicates to the same qualities of immediate experience, may be called, respectively, *acquaintance* and *knowledge by acquaintance*. Both Schlick and Russell considered these to be the ultimate basis for the testing of all scientific (intersubjective, public) knowledge-claims, although they are initially subjective and private.

There are several other features of Schlick's epistemology (developed from about 1908 to 1918) which are very similar to Russell's, particularly the view contained in Russell's final systematically philosophical book, *Human*

Knowledge (1948). Both consider genuine knowledge to be essentially *propositional*, that is, intersubjectively communicable by means of language. Propositions are taken to be the meanings of sentences, no matter in what language, notation, or symbolism the sentences are formulated. Thus we find in Schlick's early work important contributions to the philosophy of linguistic communication. In one of these contributions he came very close to anticipating the by now well-known *semantic* definition of *truth*. This definition, developed more systematically and precisely by Tarski (1936) and Carnap (1942), was at least adumbrated in Schlick's early articles and was fully formulated in his *General Epistemology*. According to Schlick's analysis, the truth of factual statements consists in a one-to-one (or at least a many-to-one) correspondence of the words (names, predicates) of a sentence to the objects and properties or relations denoted by these words. Falsity, no matter how it may arise, ultimately consists in the equivocal use of words through which a one-to-many correspondence results between them and what they denote. Thus, if I call a given tree an oak when it actually is an elm, I am using the word 'oak' ambiguously in that it would then be assigned to both kinds of tree. Schlick rejected any sort of picture view of representative language. All that seemed to matter for him was the abstract relation of correspondence (in German, *Zuordnung*). (Perhaps this is also one possible interpretation of the cryptic pronouncements on language and truth in Wittgenstein's *Tractatus*.)

Schlick realized, of course, that a direct checking of the correspondence of statements to matters of fact is possible only within the narrow scope of the perceptually given. He, like the later Russell, was therefore confronted with the problem of the meaning of truth in regard to scientific statements, most of which transcend by far the limits of immediate perception. It is interesting to note that both Schlick (from 1910 to 1925) and Russell (by 1948, at any rate) were critical realists and thus had to come to grips with the problems of transcendence. And, while they differed sharply in their views on probability and induction,[3] they argued essentially inductively for the existence of entities beyond the scope of the narrow domain of immediate experience. Both Schlick and Russell thus liberalized the radical empiricism of Hume,[4] namely, by asserting the existence of a world of *knowable* things-in-themselves – be they such objects of common life as sticks or stones, rivers or mountains, or be they the fields and particles of modern physics. It was only under the impact of Carnap's and Wittgenstein's ideas and criticisms that Schlick withdrew to what he conceived of as a neutral, non-metaphysical position. Nevertheless, he retained an *empirically* realistic view in that all specific assertions of existence – as of the atoms and electrons in physics, of the genes in biology (and perhaps also of the unconscious motives, etc. assumed in psychoanalytic theory) – were at least susceptible to indirect tests.

In his profoundly searching essay *Erleben, Erkennen, und Metaphysik* (1926) (as well as in his London lectures on *Form and Content* [1932; published in 1938]) Schlick clarified the distinction between empirically testable assertions of reality and empirically untestable ones. The latter, according to Schlick, are at the very core of traditional metaphysics in that they use an intuitive and 'ineffable' notion of reality. This notion could not possibly furnish a surplus of meaning to the cognitive, scientific concept of existence, but belonged to the expressive-evocative function of language. Such expression and evocation, Schlick averred, is not part of scientific knowledge-claims, but belongs to life itself and is best utilized in poetry, music, and the arts in general. It was the mistake of the traditional metaphysician to confuse (or conflate?) the cognitively meaningful and legitimate concept of existence with the intuitive, 'ineffable' notion of reality.

As a consequence of this outlook, both Schlick and Russell (and there are also some hints of this in the early Carnap) maintained that knowledge proper can concern only the *structural* features of the world, and must necessarily remain silent as regards its purely qualitative *contents*. Schlick was aware of the difficulties and the dangers of this formulation. He certainly wished to avoid any type of metaphysics or mysticism regarding the 'ineffable.' Perhaps he did not quite succeed. Nevertheless, this kind of view, especially of physical knowledge, seems defensible. Schlick's view was in agreement not only with the epistemology developed much later (and independently) by Russell but also with the views of the brilliant but all too harshly criticized Eddington;[5] it was a view adumbrated even earlier in the writings of Poincaré. There are also traces of this kind of view in C. I. Lewis' early book *Mind and the World Order*.

How, then, did Schlick specifically interpret the logical edifice and the empirical foundations of scientific theories? Here, I think, he was chronologically the first to recognize clearly how fruitful and illuminating a reconstruction can be if it proceeds in terms of implicit definitions and correspondence rules. Schlick maintained that the *structural* knowledge of the world provided by physical theories, for example, can be understood best by the kind of axiomatization that Hilbert had produced for geometry. The axioms (or postulates) link together a number of at first undefined and uninterpreted concepts (the so-called primitives). To this extent an axiom system – if it fulfills the necessary requirement of consistency (and to the extent that it has the desirable features of completeness and independence) – is no more than a purely logicomathematical structure (at least in the sense of Hilbert's strictly formalistic philosophy of mathematics). It is easily seen that the postulates so conceived do not convey any information whatsoever about the real world. The postulates cannot even be conceived of as definitions in the ordinary sense.[6] For they merely provide, as it were, the rules (and an initial position) in a game played with uninterpreted symbols. Viewed as definitions, they are circular in that the postulates stipulate a system of symbols related only to one another – and to nothing outside of that network. As C. I. Lewis put it so nicely: "A circle is the less vicious the bigger it is." The *bon mot* no doubt referred to the remarkable 'fertility' of certain types of axiom systems, such as that of Peano for arithmetic, and that of Hilbert for geometry: An infinity of theorems (many of them non-trivial) can be derived deductively from sets of axioms rich enough in their initial intrinsic network connections.[7] In Schlick's picturesque account the entire deductive system, that is, the postulates (implicit definitions), together with explicit definitions of equally uninterpreted, purely formal concepts (as well as all derivable theorems), form a 'free-floating' structure. If this structure is to be given empirical significance, it must be anchored by "coordinative definitions" (Reichenbach's term) or "correspondence rules" (Carnap's term) to some data on the level of observation.[8] These rules of interpretation are viewed by Carnap as semantic designation rules. They are attached to some – but by no means necessarily all – of the erstwhile only implicitly defined concepts of the pure-postulate system (or to purely formal concepts that are explicitly defined in terms of the 'primitives'). Carnap and Hempel have therefore spoken of a 'partial interpretation' of the pure-postulate system. But they maintain that the concepts of theoretical physics can be understood adequately in terms of their empirical significance in view of this sort of logical reconstruction. In picturesque language it is the 'upward seepage of the empirical juice' which provides a meaning for the otherwise altogether unvisualizable (non-intuitive) concepts of theoretical physics.

It must be kept in mind that all this is a *logical* reconstruction. It was never intended to be an account of the origin and development of scientific theories. (This is a task for psychologists, sociologists, and historians of science.) The value of the seemingly very artificial *logical* reconstruction consists in the distinction that it allows us to make between logicomathematical and empirical questions that may be asked regarding scientific theories. It is clearly one thing to ask about the consistency of the postulates or the conclusiveness (validity) of the derivation of theorems; it is quite another matter to ask about the empirical significance of theoretical concepts or about the evidential support of a theory.

As I understand the intentions of Schlick, Carnap, and Reichenbach, they were especially interested in providing an adequate account not only of the logical structure of theories but also of their empirical confirmation (or disconfirmation). Schlick, Reichenbach, and Carnap, though highly impressed with Poincaré's genius, repudiated his conventionalism. This doctrine held that the postulates of the physical sciences (including physical geometry) are 'disguised definitions', and that, therefore, such principles as the law of the conservation of energy are basically tautological and hence empirically neither confirmable nor disconfirmable. Schlick repeatedly referred to the clarification of this first law of thermodynamics given by Max Planck, according to which it is not at all a mere convention but the formulation of a most pervasive regularity of nature. Continuing along similar lines, Reichenbach opposed the views of Poincaré and Duhem, according to which experimental tests always concern theories as entireties (i.e., the postulate systems in their totality). Reichenbach, of course, admitted that in testing some postulates others are unquestioningly presupposed. But he also pointed out that it is possible to formulate the postulates of physical theory in such a manner that there is at least some degree of independent testability for each of them. He showed that scientific procedures usually consist in the successive securing (by experimental confirmation) of some postulates which then can be used as presuppositions for the testing of more problematic assumptions. Thus, for example, the geometrical optics of telescopes (or microscopes) can safely be presupposed when it comes to the testing of astrophysical (or biological) hypotheses. Quite generally, what seemed correct to the logical positivists in the views of Poincaré and Duhem, is no more than the obvious logical truth that, given a theory in the form of a conjunction of postulates, and an empirically refuted consequence of that conjunction, any (or even all) members of the conjunction may be false (at least one must be false). But it is precisely the virtue of ingenious experiments to pinpoint the 'culprit' (or 'culprits'), as it were, that is, to identify, and finally to eliminate, those postulates which are responsible for the empirically refuted conclusion.

In the light of the foregoing account of the nature of scientific theories, we may say that the meaning of theoretical concepts consists in the rules according to which they (or rather the symbols representing the concepts) are used. These rules are of at least two radically different types. We have, first, the implicit definitions and, second, the rules of correspondence. (Usually there are also explicit definitions.) If anything is to be retained of the original empiricist criterion of factual meaningfulness, then it is at least the requirement that the purely formal concepts of the postulate system (the 'pure calculus', in Carnap's terminology) be connected by correspondence rules with the concepts of observable things or features.[9] Just what is to be taken as observable is a matter of pragmatic decision in a given reconstruction. Carnap long ago preferred the observables of common life (roughly the objects and their properties and relations as directly perceived) to the sense-data basis of the earlier positivists (including his own position in *Der logische Aufbau der Welt* [1928]).

It was in view of this outlook that Carnap (in the early thirties) formulated his two famous theses of the 'unity of science.' Retrospectively, I am inclined to think that the *first* of these theses is relatively obvious, if not trivial, for it amounts to no more than the assertion of a certain unity of the *language* of the factual sciences. This unity of the natural and social sciences is to be understood in terms of the same confirmation basis for all scientific statements; this basis is formulated in what Carnap called the (intersubjective) physicalistic thing language, that is, the ordinary observation language, or the language of data. Only a non-metrical account of space and time and a qualitative (again topological) description of properties of observable objects are required for this purpose. By an ingenious logical device (the technique of reduction sentences) Carnap attempted to show that concepts of dispositions (capacities or abilities) could be introduced on the basis of terms of observation language. These latter terms were assumed to be understood directly, and hence not to be in need of definition. (Carnap, at least since 1956, has changed his views on this matter, and now prefers an explication of most concepts of scientific theories by means of postulates and correspondence rules.)

The points just discussed in connection with the first thesis of the unity of science pertain to the logical reconstruction of scientific knowledge-claims, and to the explication of the meaning of scientific concepts. The choice of the intersubjective observation language as a basis for the reconstruction implies a (qualified) adoption of the behavioristic outlook in regard to psychology and the social sciences. I say 'qualified' because the data of introspection can well be included to the extent that they themselves are formulated in the language of behavior. The linguistic utterances, that is, the verbal behavior, of human beings quite clearly constitute an important part of the data of psychology, just as the discriminatory responses of animals and humans furnish a basis for the ascription of mental states. According to the early stage of logical positivism, the mentalistic language (as it is used in connection with introspection) was construed as being strictly translatable into the language of behavior. But later analyses showed this view to be grossly oversimplified. (It has the same defects and shortcomings as the phenomenalistic reduction of physical-object statements to statements of immediate experience.) The later view, now adopted by many logical empiricists, is that peripheral behavior is to be taken as a probabilistic indicator of certain central states – the latter to be described (to the extent that the required knowledge is available) in the language of neurophysiology or in the (mentalistic) language of introspection. (The mentalistic language can by metaphorical extension also be made to cover the depth-psychological statements about unconscious states or processes as assumed, e.g., by psychoanalytic theories.)

This leads me, finally, to a brief discussion of the *second* thesis of the unity of science (or of physicalism). This is a much more exciting, but also more problematic, pronouncement. Schlick, Carnap, and Reichenbach, who espoused this thesis, were fully aware of its conjectural and hence precarious character. Essentially it endorses a certain program for the current and future development of science toward a unitary or monistic set of explanatory premises. It is encouraged by the partial but impressive successes in the reductions (in the sense of explanation) of chemistry to physics; of biology to physics and chemistry; and of psychology to neurophysiology. As a distant goal of this program of unitary explanation, some future theoretical physics is fancied, from which all observable phenomena of the entire universe (including organic life and mind) would be derivable. This thesis is, of course, not only problematic but also inevitably vague in that such a theoretical physics may have to be very different from its current stage. All that can be said at the moment is that the 'style' of explanation might be somewhat similar to that used in the present stage of the theories of relativity, quantum mechanics, and quantum electrodynamics. If this program is at all successful, it would resolve the much-discussed difficulties of emergent evolution.

It was Schlick, perhaps more distinctly and promisingly than anyone else among the logical positivists, who provided helpful suggestions regarding a coherent conception of the relation of the mental and the physical, or, as the widely used phrase expresses it, 'the place of mind in nature.' The traditional mind-body problem has indeed remained one of the most recalcitrant difficulties of a scientific empiricism. As many philosophers have seen it, this perplexing problem arises when we ask about the place of the apparently 'homeless qualities' of immediate experience in a world conceived of by means of the highly abstract concepts and postulates of physical theories. It was Schlick's contention that the 'physical' be conceived of not as a kind or aspect of reality but as a type of conceptual system radically different from the conceptual system by which we describe the phenomenal content of our direct experience. Anticipating again the conclusion of the later Russell (of 1948), Schlick proposed a new solution of the mind-body problem. It is the very nature of the concepts and theories of physics that they give no more than a *purely structural* account of the world. The content or qualitative nature of the events thus structurally described is largely unknown by acquaintance. But, wherever the physical description deals with the events in living brains, these contents are given, are indeed known 'by acquaintance', and hence are codesignated by the mentalistic descriptions. The main obstacle for appreciating this solution of the mind-body problem lies in our almost ineradicable habit of confusing the abstract, purely structural physical concepts with the intuitive images we have of physical bodies and processes. Once this 'introjection' of the pictorial elements into the physical conception of the world is abandoned, the vexation of the mind-body problem disappears. Whatever kind or intensities of direct experience ('sentience') we wish to ascribe to infants or animals must be guided by analogical reasoning. Once the analogies are as tenuous as they are in the case of lower animals (and, I am inclined to add, of electronic computers and robots), there is hardly any justification for ascribing human-like sentience (qualities of immediate experience) to them.

Schlick, of course, admitted that his solution of the mind-body problem depends on the feasibility of a universal physical explanation, that is, on the success of the program of the *second* thesis of the unity of science. And this success in turn is predicated upon a logically contingent ('brute fact'), yet fundamental and pervasive, feature of our world. Schlick realized that his view was based on inductive extrapolation as well as on logical analysis. But he was sure that it did not involve a metaphysics in the objectional sense.

I have attempted to review in brief outline some of the more important and scientifically relevant doctrines of the early phases of logical positivism. Most of these doctrines have been transformed and strongly modified under the impact of trenchant criticisms. Practically every one of the early tenets has been attacked, both from outside the Vienna and Berlin schools of thought as well as in the later developments resulting from revisions made by members of these schools. The distinction of analytic from synthetic propositions, the various formulations of the empiricist meaning criterion, the interpretation of theories, the very possibility of an inductive logic, the two theses of physicalism, and many other pivotal points have remained subjects of lively controversy. Looking back at the impact and the effects of logical positivism, I think it is fair to say that it has at least provided a powerful stimulus for much that has occurred in philosophical reflection during the past forty years.

NOTES

* [Publications mentioned in this article are listed in the bibliography to this volume. – Ed.]

1 Popper was not a member of the Vienna Circle but had intensive interchanges with, and offered incisive criticisms to, Schlick, Carnap, and myself in the early thirties.

2 This book appeared in its first edition in 1918 as the first volume of a distinguished series of monographs and textbooks of the natural sciences.

3 In 1948, contrary to his earlier empiricist views, Russell found synthetic *a priori* postulates indispensable foı the justification of inductive and analogical inference, such as he deemed necessary also for beliefs, for instance, in the existence of the external world and of other minds.

4 Or its modernized versions as in Mach, Avenarius, in one phase of William James, and in the early phase of Russell's doctrine of neutral monism.

5 See L. S. Stebbing [1937]. This is a deceptively clever positivistic critique of Eddington.

6 Carnap was the first to see this quite clearly, and consequently distinguished 'proper' concepts from 'quasi' concepts in his early article, 'Eigentliche und uneigentliche Begriffe'.

7 Space does not allow more than a brief reference to the philosophy of mathematics prevalent in the Vienna Circle: We were all duly impressed with Hilbert's formalism; we found Brouwer's intuitionism intriguing and challenging, but not necessarily acceptable. Schlick, Carnap, and Hahn were clearly logicists along the lines of Frege, Russell, and Whitehead. Kurt Gödel's famous proof of his undecidability theorem [1931] was perhaps the most exciting achievement in the Viennese philosophy of mathematics. It resulted from an arithmetization of the syntactical metalanguage of mathematics and thus utilized Carnap's important studies [1937]. Gödel's findings were discussed in that early work of Carnap's.

8 This view was anticipated in an early form by N. R. Campbell in his [1921]; it was independently and briefly expounded by Carnap in his essay [1923], by H. Reichenbach in his [1924], and has been presented also by the American philosophers of science H. Margenau, in [1950], and F. S. C. Northrop, in [1947], who speak of 'epistemic correlations' or 'corresponding rules' as the connecting links between the formal system and its empirical foundation. In Carnap's most recent publication, [1966], as well as in some of his earlier essays, he too uses the term 'correspondence rule'; see also C. G. Hempel's [1965] and [1966].

9 I think this is (among other things) what Richard von Mises had in mind when he suggested 'connectibility' as a meaning criterion. It is even more explicit in Schlick's essay [1936], in which he stipulated that any factually meaningful (descriptive) terms in a language must be connected by a chain of definitional steps with some terms of the language of direct experience.

SUGGESTED READINGS

A lucid and concise account of the development and the major doctrines of logical positivism is presented as 'Editor's Introduction' in A. J. Ayer (ed.), *Logical Positivism* [1959] (also available in paperback). This book also contains a collection of some of the more important and typical essays by members of this and some related schools of thought. The bibliography of books and articles is remarkably complete. (Among the historical accounts of the origin and development of logical positivism listed there, I recommend especially those by Victor Kraft and Jorgen Jorgensen.)

Of the several books (by logical empiricists) which have appeared since the publication of Ayer's anthology, and which are of distinctive relevance for the philosophy of science, I refer especially to the following:

Rudolf Carnap [1966]; Adolf Grünbaum [1963]; Ernest Nagel [1961]; Arthur Pap [1962]; P. A. Schilpp [1963] (a complete list of Carnap's publications is appended to this book); Moritz Schlick [1918] (Eng. tr. 1974).

Books containing searching criticisms by outstanding philosophers of science include:

Henry Mehlberg [1958]; K. R. Popper [1935] (Eng. tr. 1959) and [1962]; Israel Scheffler [1963].

Some important articles not reprinted in Ayer's collection are included in:

Feigl and Brodbeck [1953a]; and Feigl and Sellars [1949b].

I have tried to expound, defend, as well as comment on the difficulties of a new synthesis (based on both Schlick's and Carnap's ideas) of a monistic solution of the mind-body problem in 'Physicalism, Unity of Science and the Foundations of Psychology' contained in Schilpp's [1963] and in a long essay, [Feigl, 1958b].

My own critique of logical positivism may be gathered from my presidential address [1963b].

3. THE POWER OF POSITIVISTIC THINKING

*An Essay on the Quandaries of Transcendence**

[1963b]

PREFATORY PERSONAL REMARKS

(Originally not intended for publication)

Please permit me a few introductory personal remarks. Your expectations will be fulfilled. True to stereotype I shall of course talk about positivism. But let me reassure you immediately. Had it not been for the attractive title (which I owe to a suggestion by my friend Calvin Rollins), I might have chosen an alternative such as 'The Pernicious Proclivities of Positivistic Perversion and Prohibitionism'! The power of positivistic thinking is, as I see it, severely limited: I shall be emphatic on the need for incisive revisions and ample liberalizations. As the older members of our association may well remember, I was an ardent propagandist for the outlook of logical positivism in the early thirties. Somewhat like Alfred Ayer in England, I was an *enfant terrible* (and consequently a *bête noire*) on the American philosophical scene. I had arrived, in the autumn of 1930, deeply imbued with the spirit of the Vienna Circle. Along with my father figures, Schlick, Carnap, and the early Wittgenstein, I had come to think of ours as a 'philosophy to end all philosophies'. In other words, I behaved myself like a philosophical prohibitionist. However, this iconoclastic phase soon gave way to a more moderate, widely appreciative, and constructive outlook. To my dismay I had come to realize that Occam's razor mobilizes the castration complex of the metaphysicians. Yet, having stereotyped myself (in the notorious fanfare article written in colloboration with A. E. Blumberg and published in the *Journal of Philosophy* in the spring of 1931), as a 'logical positivist', the label has stuck to me ever since. As early as 1935, however, I abandoned the label (at least as far as I was concerned) and availed myself of the alias 'logical empiricist'. This was triggered by a remark of a French philosopher at the International Congress for the Unity of Science in Paris (1935). He burst out at me: "*Les positivistes, ce sont des idiots*!" He had in mind that small surviving group of the followers of Auguste Comte who still carried on the rituals of the '*èglise de raison*' or '*èglise de l'humanitè*'. Much more important, however, were my disagreements with the remnants of some of Mach's ideas, and, in general, of the phenomenalistic and the behavioristic reductionism – that is, the narrow verificationism of the Vienna Circle. I had opposed these trends vigorously but unsuccessfully already during most of my Vienna years, especially from 1924 to 1930. Dirty names, such as 'metaphysician' were used to stigmatize me during that period. Temporarily I was overwhelmed by the arguments of my great friend Rudolf Carnap, and of my revered teacher Moritz Schlick. These two men were extremely forceful, resourceful, and persuasive. But my original position, long before I decided to study at the University of Vienna, had been close to a critical realism which I had first formulated for myself in a rather unsophisticated manner when during my adolescence I was deeply impressed with the achievements of astronomy, chemistry, and theoretical physics. I began reading the positivists, but also the Neo-Kantian Alois Riehl, the German critical realists Külpe, Becher, and Freytag during my last year in secondary school; and finally on the suggestion of a distant relative, a prominent professor of medicine (at the German University of Prague), I read two books by Schlick which had been highly recommended to him by no less a person than his old friend Albert Einstein. Schlick's *Allgemeine Erkenntislehre* (first edition 1918, second edition 1925) struck me like a thunderbolt. In the beautifully lucid and magnificently penetrating book Schlick argued essentially for a critical empirical realism, presenting trenchant objections to what he called the philosophies of immanence – that is, mainly the positions of Mach, Avenarius, and the early Russell. This, together with his views on the analytic nature of mathematical truth, his empiricist critique of Kant and the Neo-Kantians, and his profound understanding of modern science motivated me to become his student at the University of Vienna in 1922. But I was acutely distressed to witness Schlick's conversion to positivism in the late twenties. This conversion was largely due to the influence of Carnap and Wittgenstein. As to whether the early Wittgenstein was to be interpreted as a positivist may be debatable. But that was the way in which Schlick, Carnap, Hahn, and Neurath did understand the *Tractatus*. My own emancipation began in the middle thirties and was stabilized in the forties. Studies and teaching in the field of the philosophy of science helped me regain, refine, and buttress my earlier realistic position. I was also greatly encouraged by the scientific realism of Hans Reichenbach and the realistic epistemologies of my steadfast dear friends Roy W. Sellars and Wilfrid Sellars.

It has been with a sense of triumphant vindication that I observed during the last three decades Carnap's successive steps toward the abandonment of the reductionist phases, first of phenomenalism, and later of his reductive physicalism and logical behaviorism. Ayer's book *The Problem of Knowledge* manifests a similar transition – and quite a reform this is! – an epistemology practically tantamount to that of critical realism. Hence, far from feeling like a repentant sinner, I could say to my erstwhile positivistic friends 'I had told you so' – already way back in the heyday of the Vienna Circle.

In choosing the subject for this address I considered such topics as: 'The Rise and Decline of Logical Positivism', or: 'From Logical Positivism to Hypercritical Realism', or: 'The Troubles of Transcendence and a Theory of the Epistemic Predicaments', or: 'Empiricism at Bay?' Each of these headings indicates in part the theme of my address. It is, in a sense, a condensed survey of my own epistemological struggles, *per noctem ad lucem*, as I like to think about them in my more optimistic moments. Although my other endeavors throughout the years were in large part devoted to the philosophy of the empirical sciences (and to a much smaller extent to moral philosophy and value theory), I wish to concentrate here on some of the basic *epistemological* and *metaphysical* issues. These issues have kept vexing me deeply. I am far from confident that I have found a tenable settlement. The unusual opportunity of an after-dinner address (with the added boon that I shan't have to defend myself in discussion) encourages me to take the liberty of submitting to you in highly sketchy outline a few ideas which seem to me to offer some hope.

If you find that I talk too much in terms of '-isms', (such as 'phenomenalism', 'realism', 'subjective idealism', *et cetera*) I regret this as much as you might. But if I did not allow myself this device of compression, my speech would take five hours instead of about the fifty-five minutes which I expect to impose on your kind patience!

The main object of this essay is an attempt first to describe, and later to resolve, the logical tensions that exist between the empiricist, and especially the positivistic, doctrines about meaning and justification, and the realistic ontology of common-sense and of science. There is, I am sure, hardly any need to review the historical development of this issue. Empiricists and positivists ever since Berkeley and Hume have claimed to give us *the* 'natural' view of the world. Ernst Mach, and especially Richard Avenarius, those stalwart 'empiriocriticists' of the nineteenth century, were quite emphatic on this. They regarded their epistemologies as a restoration of common-sense, and as a bulwark against the temptations of transcendent metaphysics. But a storm of protest arose against positivism. We can perhaps bypass Lenin's embittered (and often confused) attacks. After all they stemmed more from socio-political motives than from epistemological considerations. Much more serious were the criticisms of the German, English, and American critical realists. Among them we find outstanding scientists, such as Boltzmann and Planck, and in his later years even Einstein. Nor should it be forgotten that, anticipating all of them, H. von Helmholtz had formulated a brilliant and eminently sound representative physical realism. Some of the most incisive philosophical objections to phenomenalistic positivism are those of the early (pre-Vienna) Schlick and the trenchant arguments in Arthur O. Lovejoy's *The Revolt against Dualism.*

The motivations of the positivists were obvious and explicit from the beginning. Comte, Mill, Mach, Ostwald, Avenarius, and the flock of their disciples, in varying degrees, perpetuated the Enlightenment of the eighteenth century. Radically opposed to the obscurantism of transempirical theology and metaphysics, they insisted upon the restriction of knowledge claims to the positive facts of experience. As I read the nineteenth century positivists, their main weapon in this battle was the principle of parsimony. The use of the empiricist meaning criterion, in a psychologistic form quite explicit already in Hume, was not fully revived until the advent of the Viennese Logical Positivists in the nineteen twenties. The other component in the motivation of the positivist outlook was of course their high regard for the methods and the results of the sciences. The advocacy of a '*Wissenschaftliche Weltaufassung*' [see O. Neurath, 1973, p. 299f] was wisely tempered with their cautiously open-minded attitude toward scientific truth. Scientific hypotheses and theories were regarded as mere tools for the organization and prediction of the data of observation. Hence, the 'truth' of scientific knowledge could be claimed only tentatively (or 'until further notice'). And 'truth' was of course understood not as correspondence, but rather as heuristic and practical experience. The close parallels of this point of view with American Pragmatism, experimentalism, and instrumentalism have often been noted. James and Mach, it is wellknown, were tremendously impressed with one another. [Later, when (in 1904) Wilhelm Ostwald came to this country for a visit, one journalist described him as a man "one year ahead of his time." Ostwald at first found this a bit shabby, but soon rationalized that it made him a very fortunate man: he could harvest the fruits of his labors in rapid succession! – What could be more typically pragmatist? I often think of America's most brilliantly and consistently positivistic psychologist, B. F. Skinner, as a latter-day homologue of Ostwald's. Thoroughly instrumentalistic in his attitude toward scientific laws and theoretical assumptions, in a sense even anti-theoretical, Skinner exemplifies beautifully the pragmatist ideology; "By their fruits ye shall know them", "*Savoir pour prévoir*", and of course: "*Prévoir pour pouvoir.*"]

On the whole the sort of *Aufklärung* pursued by the nineteenth and twentieth century positivists was more in the tradition of Hume than in that of the French materialists of the eighteenth century, as for example, of the remarkable Baron d'Holbach. The antitheological and antimetaphysical tendencies of the positivists were supported primarily by epistemological rather than by cosmological arguments. Thus it was news to the theologians and metaphysicians, and it took their breath away (at least temporarily), when they were told that they were not making sense rather than that their beliefs were utterly unwarranted.

This intransigent antimetaphysical attitude of the positivists is historically understandable, but as nearly everyone nowadays realizes – it led to excesses which are inexcusable. Before I turn to my own critique of positivist reductionism, let me first explain and defend whatever may be defensible in the antimetaphysics of the positivists. It was perhaps the doctrine of the unknowable *Ding an sich* which more than anything else aroused the destructive impulses of the positivists. Not only could there be no use for such an idea, it was even highly questionable whether it made any sense whatever. The distinction of appearance and reality, in all its varieties, from Parmenides and Plato down to Kant, Spencer, and Bradley had indeed become almost the paradigmatic source of pseudo-problems. These problems were guaranteed one hundred percent insoluble owing to the manner in which, by their very conception, they were made immune to attack by any sort of test or argument. Inscrutability came to be built into the very concepton of 'Absolute Reality'. Perhaps Auguste Comte was not so far wrong when he considered metaphysics as a depersonalized theology. If anywhere, then here the meaning criterion of the logical positivists found its most justifiable and salutary application. Do we really still know what we are talking about when we *give* empirical specifications with one hand, and *take them away* with the other? Demythologization in theology, if carried out completely, leaves us indeed with nothing but pictorial, emotional, or motivative expressions and appeals. The cognitive meaning, except for the formal-logical structure, has then completely been lost. This *is* the bitter end of the *via negativa*, as well as of the process of 'whittling away'. Turning to the case of modern metaphysics, even in recent sophisticated discussions of ontology, the logical striptease that ends up with 'bare particulars' seems to confirm the positivistic suspicions regarding absolute substance. Need I review the wisdom of Leibniz in his critique of the Newtonian ideas of absolute space and time? To be sure, Newton had an excellent argument in regard to dynamics (only to be criticized on grounds of parsimony, not of meaningfulness, by Mach and Einstein: though this is currently problematic again); but in regard to *kinematics* Newton (or his disciple Samuel Clarke) really had no defense against Leibniz's searching criticisms. And no matter what he may have said later, Einstein did use the positivist criterion of meaningfulness in his repudiation of the ether hypothesis of Lorentz and Fitzgerald. In that last ditch stand of defense of the ether hypothesis, with its special stipulations of contraction and local time, it had become impossible to ascertain the state of motion, that is, the velocity of a body with respect to the ether. The theory had been made proof against disproof. To be sure, this happened unwittingly, without malice aforethought. A similar fate befell the metaphysical conceptions of vital forces or entelechies as used in the explanation of the marvels of organic life. The positivists were extremely keen in spotting the verbal sedatives that were offered (unintentionally) in place of genuine scientific explanations.

My defense of the meaning criterion comes to this. First, even psychologically speaking I maintain that we cannot connect any cognitive, factual meaning with sentences that contain terms which are radically severed from any observation basis. Second, it should be clear also that *logically* speaking, no factual meaning remains, much as it may be deceptively believed to be present because of the accompanying pictorial, emotional, or motivative appeals. And as a last resort logical empiricists can always retreat to the position that the very formulation of the meaning criterion is a *proposal* and not a proposition; and that as a proposal it is to be judged in view of its fruitfulness. Its fruitfulness then clearly consists in the fact that it is in this way, and perhaps only in this way, that we can avoid the notorious vexations with absolutely insoluble problems. In any case, my own formulation of the meaning criterion differs sharply from the notorious one of the logical positivists. I think that the meaning of a statement has precious little to do with the method of its verification. Nor do I admit that the meaning of statements consists in the use we make of them. In keeping with the semantic explication of the conception of truth, I suggest that the cognitive meaning of a factual statement consists in its truth conditions. And factual meaningfulness is to be defined in terms of 'in principle at least indirect and incomplete confirmability or disconfirmability'.

Of late, some of my empiricist friends, partly under the influence of Karl Popper's incisive and powerful ideas, keep urging me to abandon the meaning criterion altogether, and to replace it by Popper's criterion of the demarcation of scientific from metaphysical (or non-empirical, non-scientific) statements. In view of the notorious difficulties of an exact formal reconstruction of the meaning criterion (compare, for example, Hempel's well-known essay [1950a]), this sounds like good advice. But, much as I agree with the basic intent of Popper's demarcation criterion, I find it defective in that it puts together as 'metaphysical' three very different sorts of things. On the one hand, there is the type of hypotheses which by special logical devices or stipulations are removed from any possibility of test (that is, the hypotheses just referred to, for example, of absolute space and time, substance, the ether, vital forces, *et cetera*.) Secondly, there are such hypotheses as those of ancient (Democritean) atomistics which are not open to experimental refutation because of the indefiniteness of the specifications (for example of the size or mass of the atoms). And thirdly, there are the assumptions of commonsense or scientific realism regarding the existence of the physical world or of other people's mental states, which according to Popper's way of construing them are as immune to refutation as is the doctrine of solipsism. Now, surely this is a most unsatisfactory conception of the domain of the metaphysical. First of all I would urge that hypotheses like the Democritean be regarded as inchoate stages of scientific theorizing; deficient, it is true, in their specification of meaning, but nevertheless in principle testable – that is, at least very weakly confirmable or disconfirmable. Even the ancient common life observations of the states of matter (solid, liquid, gaseous) and their transformations into one another, may well be considered as at least tenuous evidence in favor of the atomic hypothesis. Quite generally, the sort of 'promissory note' type of hypotheses that so frequently functions as place-holders for more fully worked out theories, should certainly be classified as scientific rather than as non-scientific, let alone metaphysical. (Of course they may turn out to be *un*scientific, that is, disconfirmed in the sense in which we repudiate superstitions like astrology or phrenology.) Finally, I find it most unhelpful and utterly misleading to put in the same category, for example, Hegel's metaphysics of the Absolute and the assertions of ordinary realism. If ordinary realism contains metaphysical elements, then surely it behooves us to distinguish between such 'harmless' metaphysics on the one hand and 'pernicious' metaphysics on the other. Pernicious metaphysics, I submit, arises typically out of the mistaking of pictorial, emotional, or motivative appeals for factual meanings: and by (at least implicitly) precluding any sort of empirical testing. It seems, then, it will be helpful to distinguish between three kinds of transcendence: (1) The *innocuous* transcendence involved in ordinary assumptions about the past, the future, physical reality, and other minds. (2) The *precarious* transcendence of 'far out' hypotheses in the natural and social sciences. Some hypotheses in current cosmology, speculations regarding the origin of life, or theories in depth-psychology may serve as examples. There is, of course, no sharp line of demarcation between precarious and non-precarious hypotheses. This is the familiar scale of degrees that extends from the highly problematic to the extremely well confirmed. (3) The *pernicious* transcendence which results in unanswerable questions, unsolvable problems, or inscrutable mysteries.

With their notorious aversion against pernicious transcendence, the old line positivists unfortunately threw out the other two 'babies' with the 'dirty bathwater'. Mach and Ostwald seem to have had a real phobia of the invisible and the intangible – hence their repudiation of the atomic theory. Similarly, some of the more tough-minded behaviorists opposed the psychoanalytic theories regarding the unconscious components of the human mind. (Others even excluded mental experience altogether from the domain of psychology). It was rather late in the day when Ostwald, in 1912, declared that the evidence for the corpuscular structure of matter was so strong that he recanted and abandoned publicly this opposition to the atomic theory.

The greater sophistication of the logical positivists of the Vienna Circle enabled them to regard with equanimity both the harmless transcendence of common sense and the precarious transcendence of scientific theories. According to Carnap's point of view (as of 1928) *all* of our empirical knowledge was to be analyzed ('reconstructed') as a logical construction erected on the basis of a neutral, 'subjectless', immediately given. Hence there was, in a logical sense, no transcendence at all involved in the knowledge claims of everyday life and of legitimate science. Pernicious transcendence of course was ruled out. Carnap showed how it led to pseudo-problems. And Schlick exuberantly spoke of a decisive turning point in the history of philosophical thought when he advocated his '*Consistent Empiricism*' and repudiated the pursuit of unanswerable questions. All this would have been most gratifying and exhilarating if it had not been for the ambiguous and ambivalent state in which ordinary realism was left. This became especially and embarassingly obvious once the doctrines of phenomenalistic reduction and hence of neutral monism were abandoned. A neutral, subjectless, immediate experience is a chimera, or at best a highly artificial logical construct – or should I say 'destruct?' I think it was (more or less deliberately) designed to avoid (or evade) the problems of transcendence and the mind-body problem. Yet, the question '*whose* immediate experience?' could no longer be repressed. Carnap's portentous switch from experiential phenomenalism to physicalism (and logical behaviorism) of course tended to undercut the problems of privacy and solipsism. Knowledge could, after all, be reconstructed on an intersubjective basis, and the physicalistic thing-language was offered as the ultimate ground-level of epistemic reduction. This amazing turn can only rather misleadingly be assimilated to a transition from subjective idealism (roughly Berkeley's epistemology without his theology) to a materialism of commonsense.

But the traditional perplexities of epistemology and metaphysics cannot so easily be made to disappear. The issues which divide Berkeley and Locke, Hume and Kant, Mach and Helmholtz, the phenomenalists, neo-realists and the critical realists, cannot be resolved by positivistic fiat. They do not vanish by deciding upon a certain basis and a certain mode of reconstruction. The repression of philosophical problems produces symptoms not unlike those of a psychoneurosis: Anxiety, vacillation, ambivalence, inconsistency and even a certain unconscious dishonesty with one's self.

And yet, the power of positivistic thinking impresses itself upon us when we reflect upon the curious and peculiar nature of the problems of ordinary, innocuous transcendence. *Prima facie* (and, I maintain, even *ultima facie*) it appears preposterous to assimilate harmless to pernicious transcendence. But this is the skeptic's gambit. Somehow one cannot escape the feeling that it is only through surreptitious and self-deceptive steps that he arrives at the sort of philosophical doubt that questions our right to believe in the existence of such quite ordinary things as physical objects, other persons' mental states, the occurrence of past or future events, *et cetera*. The positivists did recognize that doubts of this sort are somehow different from the usual kind of doubt that may pertain to empirical knowledge claims. Carnap, in his essay on the *Pseudo-Problems of Philosophy* (published 1928) clearly anticipated the Oxford philosophers in pointing out by his examples about the mountain in Africa, or the emotions of another person, that it is one thing to ask the usual factual questions about existence or occurrence, and another the perniciously transcendent question regarding *reality überhaupt*.

Difficult as it may be sometimes for geographers, or for psychologists, to arrive at agreement, it seems altogether impossible for these same men to settle their differences once they entrench themselves in the metaphysical armchairs of representative realism or of subjective idealism. In a way similar to Carnap's, Schlick in his well-known essay on 'Positivism and Realism', advocated a neutralist position; and quite recently, Ernest Nagel in his superb book on *The Structure of Science* (in the chapter on the Cognitive Status of Theories) maintains that the conflict between the positivistic-instrumentalistic view and the realistic view of scientific theories is merely a matter of 'preferred modes of speech' and 'the question as to which of them is the 'correct position' has only terminological interest.' One is indeed tempted to treat the entire issue with scorn or ridicule. Just imagine if the morning paper headlines were to declare one day: 'Reality of External World at last assured by Philosophical Demonstration.' In a striking article, [Barrett, 1939], a long time ago, William Barrett complained: "So great, in fact, is the gap between this problem and any practice that one must stand convicted of shame when one tries to communicate this as a serious cognitive problem to anyone innocent of philosophy."

The trouble of course is that neither we in the teaching profession, nor many of our students, are innocent of philosophy, and that most of us, at least at one time to another, have suffered from the quandaries and perplexities of transcendence. In my own teaching career of thirty-six years I would say that I have been pestered much more frequently by questions from students who agonized over the reality problems rather than over any other problem in the whole range of philosophy. Therapy is clearly called for, and various therapeutic devices have been proposed. I have already referred to the well-known therapies offered by the neutral monists, the logical positivists, and by the linguistic analysts of the Neo-Wittgensteinian persuasion.

It must be acknowledged that the distinction between philosophical doubt and plain empirical doubt is suggestive and helpful. Philosophical doubt clearly and invariably makes things too difficult for itself: Might not the world have sprung into existence a minute ago, with everything that we usually call 'remnants', 'traces', 'indicators' and 'memories of the past' all ready made? Might not the qualities of private immediate experience be systematically interchanged from person to person (the problem of the 'inverted spectrum')? Might not all so-called human beings except myself be merely inanimate robots, lacking in sentience or experienced mental life altogether? Might not all so-called physical objects simply fade into non-existence in the intervals during which, as we normally say, they are not perceived?

The gyrations of the philosophers who have tried to remove these philosophical doubts are well known. In recent years they have been helpfully reviewed in differing ways by A. J. Ayer [1956] and Stephen E. Toulmin [1958]. The peculiar 'ballet' or 'minuet' that most of us perform in our courses in the theory of knowledge includes the steps from Naive Realism to Representative Realism to Skepticism to Subjective Idealism (or else, Phenomenalism) to Direct Realism (sophisticated rather than naive), to Hypothetico-deductive (or 'scientific') realism, and finally to the dissolution of the problem by descriptive analysis of the actual uses of language or the introduction of a logic of reasoning from 'criteria', 'warrants', and 'backings.' This last one is presented as a type of reasoning that uses for its guiding principles rules that go far beyond those of the ordinary deductive logic of analytic arguments by embracing 'substantial' arguments. Applied to the problems of ordinary, innocuous transcendence, these 'substantial' arguments and their logic turn out to be simple endorsements of common-sense reasoning. But that's where we started in our 'minuet' – it is merely naive realism refurbished a bit by reassuring terminology such as 'warrant' and 'backing.' Hence it is not much better for the refutation of skepticism than G. E. Moore's tough-minded way of confronting us with the obvious facts of existence. (Moore's technique reminds me, distantly, of the 'slap-in-the-face' replies Zen-Buddhists offer to metaphysically puzzled inquiries.)

There are of course still other ways of undercutting the problems of transcendence. Paul K. Feyerabend (see especially [1970a]), extending the epistemological views of Karl Popper, allows almost for no difference whatever between observation statements and theoretical statements. He insists that even what we are inclined to regard as the "simplest statements of direct observation" are already suffused with a great deal of theory. Despite his severely critical attitude toward Oxbridge type analyses, he uses something very like the Neo-Wittgensteinian reasons for the repudiation of sense-data as the testing ground of knowledge. I cannot here undertake to criticize these interesting and radical views in any detail. Suffice it to say that I think they repress rather than resolve the problems of transcendence. From the point of view of a rational or logical reconstruction there is, I think, an ineluctable and undeniable surplus of meaning over and above the content of observation or 'criteria' statements in our assertions about physical objects or other minds. For the sake of clarity, and in order to do justice to the arguments from illusion, and quite generally, if we wish to accommodate what is obviously correct in the causal account of perception, we must retain in our epistemology the logical difference between the confirming evidence and the truth conditions of our knowledge-claims. This corresponds of course to the classical distinctions of impression and object, no matter how they may be refined in the modern psychophysics of perception.

There is a severe shortcoming already clearly discernible in Mach's epistemology. When in his *Analysis of Sensations* he reduced physical objects to complexes of elements (neutral sense data) "A, B, C . . . ", he prevented himself, no doubt unwittingly, from formulating some of the most obvious facts of perception. In common language it makes perfectly good sense and is clearly true that the tree must be there if I am to have the corresponding percept or sensory impression. This is clearly a synthetic proposition. More fully explained, what I mean to assert is the common-sensical truism that under normal perceptual conditions the removal of the object (for example, uprooting the tree and carting it off) will result in impressions different from those experienced when the tree was still in its original place. (I formulated it carefully in order to steer clear of the tautology that would result if I were to use such achievement words as 'see' or 'perceive', that is, if you insist with 'Oxbridge' that these are the paradigmatic uses of those words.) Now Mach precluded himself from formulating that obviously synthetic statement of commonsense or the scientific theory of perception. On his own analysis, all he could afford to say is, "If the elements A, B, C. . . occur, they occur." But that's a pathetic tautology, and does not begin to render the intended meaning.

True enough, nineteenth century positivism was logically rather uncouth, but even twentieth century positivism, despite its logical sophistication, inherited some of these deplorable defects. Are the (possibly infinite) sets of subjunctive or counterfactual conditionals about sense-data really equivalent to the assertions of the independent existence of physical objects during the interperceptual intervals? Does not the question '*whose* sense data?' obtrude tenaciously? And if so, must we not in some fashion establish congruity between the epistemological reconstruction and the factual structure of the world as we have come to know it? But how can all this be achieved without vicious circle or *petitio principii*? The quandaries of transcendence, the basic tensions between the outlook of Locke on one side, and that of Berkeley and Hume on the other, are still with us. But, as I have hinted before, let us try to have a sense of humor about these philosophical headaches. They *are* academic; they are *our* troubles exclusively; commonsense and science will *never* be bothered with them; or '*hardly ever*' – there are after all the curious recrudescences of Berkeleyan subjectivism in the Copenhagen interpretation of quantum mechanics. Disregarding these scruples for the moment, however, let us be clear about what we philosophers can do, and what we cannot do in the matter of these perplexities.

As those of us who have pursued the problems of justification to the bitter end know only too well, proof and demonstration are after all of extremely limited value in the traditional issues of philosophy. The *malaise* that I am tempted to call 'demonstrationism' seems plausible (that is, psychologically understandable) only for the typically Aristotelian or the rationalist outlook, according to which intuitive self-evidence is the mark of the truth of the first principles which form the premises of its demonstrations. But we have learned to distrust and to discredit intuition and self-evidence. It is familiarity that breeds intuition, and familiarity can be woefully misleading. Intuitively self-evident insights may serve to tranquilize our skepsis-engendered anxieties, but in the end they are no better as a justification of first principles than are the verbal sedatives offered by some theologians. Must we resign ourselves to the conclusion (welcome in some metaphysical and theological quarters) that the realism of commonsense is a matter of unjustifiable *faith*? Indeed, what else *can* we offer when we are asked to state good reasons for the first principles to which we hold so firmly anyway? In the tragicomedy of wisdom that we dignify with the label 'philosophy', we are apt to run like bewildered chickens in the customary circle from self-evidence to correspondence to coherence to pragmatic utility. Correspondence, while for purposes of formalization most illuminatingly explicated in the pure semantics of Tarski, Carnap, and Martin, will bake no very nourishing bread for our particular needs. Correspondence, in this respect, remains after all the '*elende Diallele*' as Kant recognized and so branded it. The semantic conception of truth helps in making clear what we mean, or (*pace* Oxbridge!) what we *should* mean by 'truth' if we want to allot to this concept its rightful and indispensable place in the schemes of our rational reconstruction. But the semantic notion of truth as is generally admitted, does not by itself provide any criteria that would enable us to tell whether we have got hold of a truth, or even as to whether we have got nearer to (or farther from) the truth. Coherence, for well-known reasons, is no better – except that it includes consistency, a condition notoriously difficult to fulfill in complex philosophical doctrines, but of course absolutely indispensable if we are to communicate responsibly even only with ourselves. Since, a little later, I shall use a sort of indirect proof or *reductio ad absurdum*, I do want to go on record, even if at the risk of sounding both reactionary and dogmatic, that I consider the two-valued logic as unique, irreplaceable and thus as the absolute presupposition of all responsible discourse.[1] Justification of the harmless transcendence of ordinary realism is then not feasible by appeal to either the self-evidence, the correspondence, or the coherence view of truth. Characteristically, latter-day sophistication tends toward the pragmatic conception. Carnap's distinction [1950a] between questions concerning the very frame of cognition, and questions within such a frame, leads him to answer the first in terms of pragmatic convenience or utility, and only the second in terms of either deductive and/or inductive validation. Given his positivistic past, it is understandable that Carnap resorts to a pragmatist move when it comes to the justification of first principles (or, if you will, ultimate presuppositions). In a moment I shall venture to show that we can do just a little better than that. But before I do this, I think it will be helpful to examine Karl Popper's firmly entrenched opposition to any kind of justificational reconstruction in epistemology. (In recent years he has been even more emphatic on this than before.)

Popper, as is well known, accepts as definitive and unassailable, Hume's arguments as to the impossibility of any justification of inductive inference. In his own positive contribution he avoids Hume's psychologism, repudiates the Neo-Wittgensteinian dissolution of the problem of induction, and maintains steadfastly that there is no such thing as inductive inference, let alone a logic of induction. (In view of this it seems strange that the title of his *magnum opus*, in its English edition is *The Logic of Scientific Discovery* [Popper, 1959]. His recently published *Conjectures and Refutations* [1962] bears a much more characteristic title.) Popper maintains not merely that the only defensible rational procedure, but also the one that is naturally and normally employed, is the hypothetico-deductive method. The *modus tollendo tollens* is presented as the only way in which we can attempt to attain the truth or at least get nearer to the truth. The best we can do is to conjecture freely and boldly, to deduce vigorously, and to test severely. Whatever hypotheses withstand his incisive treatment are considered 'corroborated', at least until further notice. The growth of knowledge, Popper declares, is achieved as though by natural selection through the elimination of the "unfit" hypotheses, and the survival of others "nearer to the truth." But does Popper really have a criterion for 'being nearer to the truth?' I think his radical anti-inductivism precludes this. When he recommends holding on (tentatively) to well-corroborated hypotheses, he considers this a matter of mere practical advice, and not in the least as a justification of any predictive or extrapolative inference. Now, *ultimately*, I too am a practicalist (though certainly not a pragmatist) in all matters of justification. But as I have been at pains to point out repeatedly [Feigl, 1950a, 1952a, 1956c, and 1961c. Furthermore, cf. Salmon, 1957.] cognitive justification ('validation') can go a long way before we come to the practical justification ('vindication', as I call it) of the adoption of the very principles that we appeal to in validations. I am pretty well convinced that Reichenbach's vindication of the rule of induction is essentially sound, and in any case indispensable if we are to be empiricists and refuse the easy consolations of the synthetic *a priori*. Popper, his protestations to the contrary notwithstanding, does utilize induction in that he puts his trust in well corroborated theories, and in that he assumes that a well-refuted hypothesis will not begin to 'stand up' again from now on out. When challenged on this point, he replies that our belief in the basic invariance of the laws of nature is a piece of metaphysics, that is, essentially untestable and irrefutable, and he adds that this is a 'good' kind of metaphysics. How does he argue for this connection? Given his general outlook, I doubt that he can offer anything but – and really nothing better – than a practical justification. So as regards ultimate questions perhaps Popper's outlook is not so radically different from Reichenbach's and mine as he suspects. Be that as it may, I think that the hypothetico-deductive procedure which Popper extolls, can be utilized, up to a point, in the justification of a critical (or should I call it 'hypercritical?') realism. This I shall now attempt to sketch. In short, I am going to argue that the transcendence involved in the realism of ordinary life and of science is quite innocuous, only rarely precarious, and never pernicious!

Most of us logical empiricists (after our liberation from the narrower logical positivism) used to put it this way: The existence of physical objects during the interperceptual intervals; the occurrence of events in the (distant) past or future; the presence of immediately experienced mental states in other persons; the existence of the micro-entities of modern physics; *et cetera, et cetera*, is assertible in a factually meaningful way because statements of this sort are (at least) indirectly and incompletely testable. They are in principle confirmable or disconfirmable. It might then be asked how we know this. Since we are not phenomenalists any longer, the relation between the to-be-confirmed hypothesis and the description of the confirmatory evidence is not a purely logical or analytic one. Only if we start from theoretical premises which in a genuinely synthetic manner formulate the relation between the objectively existing situation and the actual and/or possible observations (which could serve as evidence pro or con), can we *deduce* the occurrence of some specifiable observations under specified test conditions. And here the modern Berkeleyans insist on asking how we can possibly know about these synthetic connections. Well, in one sense – the horribly restricted and the phenomenalist sense which the Berkeleyans have in mind – we don't and we couldn't. But here I agree completely with Popper: Nothing ventured, nothing gained.

All our knowledge claims, except possibly those about present moment first person immediate experience, transcend the evidence that might confirm them. Even W. T. Stace in his famous 'Refutation of Realism'[2] allows for the transcendence of the past and the future of one's own experience, as well as of other minds. This curious monadology à la Berkeley and Leibniz is of course open to the question: 'If I may infer other persons' mental states, why may I not infer with the same right, the persistence of material objects when not perceived?' It is hardly necessary to dwell upon this well-worn dialectical argument. But perhaps it is nevertheless advisable to stress the power of the indirect proof, the *reductio ad absurdum* that the dialectic contains. The very same reasoning that is used in behalf of subjective idealism, finally and inevitably forces you into the corner of the solipsism of the present moment. Strictly speaking, you are doomed to end up with a radical Pyrrhonism and the complete silence of Cratylus. This *is* the end of *that* road. As soon as you begin to talk in terms of a merely methodological solipsism (as did Carnap in the early days of the *Aufbau* ideology) you leave the door open for all the ordinary realistic interpretations, involving the differences between evidence and reference, that is, between confirmatory conditions and truth-conditions.

It is of the essence of the hypothetico-deductive method of cognition to make assumptions, to grant oneself the latitude of postulates. It is the typical phobia of positivism and radical empiricism to shy away from such latitude. But the redeeming feature of the hypothetico-deductive method is precisely that its conclusions are permanently open to criticism and refutation. Contrary to the logical positivists, I think that solipsism is not meaningless but so outrageously improbable as to be safely dismissed as false. Can't I be absolutely sure that it is false? Of course not. I can think of very fantastic hypotheses (somewhat akin to solipsism) which I could not possibly refute with finality. Consider, for example, this bit of science fiction: all my ordinary immediate experience is simply the subjective aspect of neuro-physiological processes in a cerebral cortex that is being kept alive and functioning in a glass case filled with a nutrient solution, and that by some ingenious and elaborate automation and programming the appropriate afferent impulses are fed into it, and efferent ones conducted out. Under such circumstances I (am I still 'I?') might mistakenly believe that I have a human body, that I have a lovely wife and children, take trips, make speeches to assembled philosophers, watch sunrises and sunsets, and so on and on for a 'lifetime' of ordinary experience. I grant that this is merely common error and illusion magnified to the n-th power, and deliberately and artificially protected from detection or correction. But the philosophical lesson should be clear: Though in principle dubitable, there are practical certainties which form – until further notice – the very frame-presuppositions of all our cognitive activities. Though they are in principle open to revision, we would require extremely strong reasons actually to revise them.

The much maligned causal theory of perception is one such basic frame assumption. I submit that the reluctance and suspicion with which it has been treated in some philosophical quarters is due to the exclusively *egocentric* perspective from which perception has been viewed in the positivistic tradition. This central view must clearly be supplemented by the *lateral perspective* with which we are so familiar in everyday life. We can observe other people observing things and events. We can see that the tree does not fade out of existence when the perceiver closes his eyes; or the bell does not stop ringing when he stuffs his ears with wax. We can leave a movie-camera clicking away in front of our fireplace when we are out of the room, and later find out from the film that our absence made no difference to the continued burning of the logs. Arguments such as these are usually dismissed as begging the question of transcendence. After all *someone* has got to observe the observer and the tree; someone has got to look at the film, *et cetera*. So ultimately we are back to immediate experience after all. Of course we are; I don't deny it. But the regularities of experience that make us accept the causal theory of perception, require an explanation, and the most parsimonious and plausible explanation is of course the realistic account of perception. Translate, if you wish (and if you can?) all those statements about sense organs, movie cameras, fireplaces, *et cetera*, into sense-data propositions. These conditionals are then the explananda for which the ordinary realistic assumptions furnish the explanantia. No matter how you formulate these regularities pertaining to the causal structure of the perceptual processes, it is these regularities which make the realistic account all but inevitable. Arguments of this sort seem to me so much stronger than the ostensive gestures of G. E. Moore. (I cannot enter here into a critique of the fashionable Copenhagen and complementarity views of measurement and perception in micro-physics. For all I know the Berkeleyan and positivistic arguments offered are not called for. Acceptance of a probabilistic theory of the interaction of, for example, electrons and photons goes a long way toward restoring 'objectivism' in quantum physics.)

It does not matter for our present purposes what sort of solution we accept in regard to the theory of perception for the mind-body problem. Dualistic parallelism or even interactionism will do; although for an all around synoptically adequate account I still favor some sort of synthetic identity theory.[3]

The important point to focus upon is the proper interpretation of the *epistemic predicaments.* As Ralph Barton Perry pointed out most helpfully, some fifty years ago [1910. Cf also 1916, esp. pp. 128–132], subjective idealists and phenomenalists exploit the egocentric predicament in an illegitimate way. Properly interpreted, the egocentric predicament constitutes no more than a natural limitation of direct verifiability. But whosoever demands direct verifiability for factual meaningfulness, and direct verification for validity, is doomed to the silence of a metaphysical solipsim of the present moment. He has confused the innocuous transcendence of ordinary realism with the pernicious transcendence of bad metaphysics. The egocentric predicament and the predicament of the present moment have their ontological-cosmological counterparts in the natural embeddedness of the knowing subject in the to-be-known world. Only through such a 'Copernican revolution' that assigns the knower its proper place in nature, only by thus repudiating Berkeley's *esse est percipi*, can we arrive at the desired congruence of epistemology and cosmology. The idealistic and phenomenalistic arguments collapse as soon as we realize that they demanded from the realists a proof which is logically impossible. Once you decide upon the inevitable distinction of the *given* from the *non-given* reality, it is not only preposterous, but downright inconsistent to ask for a proof of the existence of the non-given that would involve its being given.

The epistemic predicaments, I repeat, consist in a logically in-principle *contingent, natural,* limitation of direct verifiability. The finitude of human direct experience may just as well be accepted with equanimity. But this does not establish a limitation of conceptual thinking. We can transcend the limits of immediate experience, because of the power of the human intellect, the power that enables us to frame hypotheses and theories. And although it is not customary to regard the assumptions of ordinary realism as a theory, it is illuminating to do so for the purposes of epistemological reconstruction. In such a reconstruction we find a great deal of convergent or 'consilient' evidence, that can be used in support of our realistic assumptions regarding the physical world as well as other minds.[4] For once I am tempted to say, in true Oxbridge style, "you can view ordinary realism in analogy to scientific theories, but be careful in doing so!" The analogies are limited, in that we are dealing here with extreme or degenerate cases of 'theories.' Perhaps a more adequate way of putting the matter is that we are here dealing with metascientific analyses which reveal some partial parallels between the adoption of a realistic conceptual frame and the confirmation of scientific theories.

My conclusion then is this: The meaning criterion of the logical positivists, if appropriately liberalized, permits us within the frame of the hypothetico-deductive, inductive, and analogical procedures of cognition to countenance as harmless the transcendencies involved in ordinary realism. This implies, for example, that such puzzles as that of the inverted spectrum are to be treated as factually meaningful, and that they are to be resolved according to the rules of analogical inference. To spell this out; if the behavioral evidence indicates that my friend and I have the same powers of color discrimination, then it makes perfectly good sense and is very likely true that he experiences the color qualities very much in the way I do. The same sort of reasoning applies legitimately to the existence of physical objects during interperceptual intervals. If during those intervals all observed effects are 'as if' they were produced by the unperceived object, then we are justified to assume it as extremely likely that the object is there, even if not perceived. In viewing the epistemic predicaments as aspects of the place of minds in nature, we incorporate what is epistemologically a limitation of direct verifiability in a cosmology that is practically forced upon us by convergent evidence.

The primary aim of the positivists was to restore to philosophy a natural and reasonable view of the world. But in their zeal to purge philosophy of all pernicious metaphysics, they overshot their goal. They used Occam's razor so radically that they nearly cut away the very core of science itself. This could have been avoided. Science does not rest on faith in the sense in which the theologically anchored religions do. Scientific beliefs are confirmable. Religious beliefs may be either utterly unconfirmable; or if they are testable, then it turns out that there are no rationally defensible grounds for accepting such beliefs.

The power of positivistic thinking at its best was demonstrated in the detailed analyses of the experimental and theoretical procedures which contributed so greatly to our understanding of the scientific method. But the tremendous influence of the persuasive prohibitionism of the positivists proved in the end detrimental. They did not achieve a tenable epistemology. I have tried to outline the directions in which what is sound in positivism must be supplemented, and what is unsound supplanted, by a critical realism. But with the Wittgenstein of the *Tractatus* I am willing to concede how very little is achieved even if these problems were definitively resolved. The problems were of our own making. We had inherited them from Locke, Berkeley, and Hume. If my suggested solution proves workable and acceptable, perhaps we can rid ourselves permanently of what has been a persistent vexation. My constructive ideas will no doubt be called 'metaphysical.' To which I can only reply: If *this* be metaphysics make the least of it!

NOTES

[1] For a succinct defense of this point of view, cf. my essay 'De Principiis non disputandum. . .?' in [1950a] and in my 'Some Major Issues of the Philosophy of Science of Logical Empiricism, [1955c] esp. pp. 6–14.
[2] [Stace 1934]. (Professor Stace told me privately that he intended this brilliant *tour de force* as a joke!)
[3] For my fullest exposition of this theory, its merits as well as its difficulties, see my essay 'The Mental and the Physical' in [1958b]. Cf. also my essay 'Physicalism, Unity of Science and the Foundations of Psychology' in [1963d].
[4] I have dealt with these matters more fully in 'Existential Hypotheses: Realistic versus Phenomenalistic Interpretations' [1950b]; 'Logical Reconstruction, Realism and Pure Semiotic' [1950c]; and 'Other Minds and the Egocentric Predicament' [1958c].

4. THE *WIENER KREIS* IN AMERICA

[1969d]

I. ORIGIN AND DEVELOPMENT IN VIENNA

The migration of a philosophical movement from Central Europe to America is the topic of this essay. Since I am not a trained historian, and since the data available to me are rather incomplete and uneven, I decided to write the story in a somewhat impressionistic manner. This, of course, involves in part an autobiographical perspective. Although I realize that I am a minor figure in the development of Logical Empiricism, I have known fairly intimately many of the major figures, and have been in more or less continual contact with most of them. I trust that my procedure of presentation – necessitated by unavoidable circumstances – will not be regarded as presumptuous. In extenuation I can only mention that I was (after Schlick's brief visits) the first 'propagandist' of our outlook in the United States. I also happened to be, in 1930, the first of the group to enter the United States with an immigrant visa, and the first to acquire U. S. citizenship by naturalization, in 1937. Moreover, my friend Albert E. Blumberg and I were the ones who provided, in 1931, our philosophical movement with its international trade name, 'Logical Positivism.' For reasons to be sketched briefly later in this essay, most of us have preferred the label 'Logical Empiricism' or 'Scientific Empiricism' ever since about 1936.

The movement quickly aroused a great deal of discussion, criticism, and dispute. In retrospect I feel that we were most hospitably received in the United States, even by our most fervid opponents. It took me a while to realize how much we were indebted to our hosts for the generous and friendly treatment extended to every one of us. Several of our group arrived in the United States in a spirit of 'conquest.' We were deeply imbued with the conviction that we had found a 'philosophy to end all philosophies.' Naturally, we offended – especially the more tradition-bound thinkers in the new country. Nevertheless we met almost everywhere with very amiable and amicable philosophers. Our views were sharply criticized, in countless discussions and publications. Gradually we learned – or shall I say, we mellowed and matured. We realized that it is intellectually fruitful to 'compare notes' with the various kinds of opposition. Lines of communication that we had never even dreamed of in Europe opened up in the New World. Among our early contacts were also a good many scientists and scientifically-oriented philosophers who greated us as comrades-in-arms.

Before I review these varied reactions in detail, I must first relate, in outline, the origin, spirit, and organization of our movement in Europe.

Logical Positivism, or Scientific Empiricism, can be traced in its genesis to a number of factors and influences in European thought. The Vienna Circle (of which I was a member from its beginning in 1924 until 1930) and the Berlin Society for Scientific Philosophy had their roots in the view of the great scientist-philosophers of the second half of the nineteenth century and of the early years of the twentieth century. The most important influences came from the physicist-philosophers Hermann von Helmholtz, Ernst Mach, and Albert Einstein, and from the mathematician-philosophers Henri Poincaré, David Hilbert, and Bertrand Russell. In the period before World War I, which we in the Vienna Circle called with affection and respect (and not at all with derision) 'prehistoric', a group of young doctors of philosophy, with 'major fields' in physics, mathematics, and the social sciences, and including most notably Philipp Frank, Hans Hahn, Richard von Mises, and Otto Neurath, met in an old Vienna coffee house on Thursday evenings for discussions, mainly of issues in the philosophy of science. At that time it was primarily the positivism of Ernst Mach that inspired this small group of scholars. Mach's historical and critical studies, as in his *Science of Mechanics, Wärmelehre* (Thermodynamics), *Optik*, and his epistemological and psychological investigations in *Analysis of Sensations* and *Erkenntnis und Irrtum* (Knowledge and Error)* had engendered in a completely renewed and highly enriched form a positivistic philosophy that went far beyond the doctrines of Auguste Comte and John Stuart Mill. In its destructive aspects it was – after Kant's *Critique* – the most radically antimetaphysical philosophy of the nineteenth century. In its constructive aspects it contained many important, though controversial, contributions to the philosophy of the empirical sciences. Philipp Frank and Richard von Mises were certainly the ones most deeply imbued with Mach's positivism. This did not prevent them, however, from disagreeing with Mach's opposition to the atomic theory. (Mach's doubts regarding the theory of relativity belong to his old age and a long period of illness; they played no role in those early days of Viennese Positivism.)

As far as I can ascertain, the ideas of Franz Brentano and his foremost disciples, Alexius Meinong and Edmund Husserl, though not entirely uncongenial, had at best only a weak and indirect effect upon both the early as well as the later Vienna Positivists. Sigmund Freud's theories were on the whole more fully appreciated, though the more radical among the Viennese thinkers insisted that the psychoanalytic doctrines should be scrutinized very carefully. They felt that much in these doctrines was 'metaphysical', i.e., unverifiable. I shall return to this theme in connection with the later developments.

In 1924 under the direction of Professor Moritz Schlick, the Thursday evening gatherings of the 'prehistoric' Vienna Circle evolved into the Vienna Circle of Logical Positivists that achieved world-wide attention, acclaim, and opposition. Schlick had come to the University of Vienna in 1922, having held teaching positions previously in the Universities of Rostock and Kiel in North Germany. The Viennese mathematician, Hans Hahn (mentioned above as a member of the 'prehistoric' Circle) was influential in bringing Schlick to the famous 'Chair in the Philosophy of the Inductive Sciences', which had been occupied by the great physicist-philosophers Ernst Mach and Ludwig Boltzmann and had been vacant for many years. Schlick was clearly a most worthy successor. He had earned his doctorate with a thesis in theoretical optics under Max Planck in Berlin, but his life work was in philosophy. Schlick was among the first academic philosophers (one should group with him C. D. Broad and Ernst Cassirer) who had an adequate understanding of Einstein's theory of relativity and who recognized its profound philosophical significance. His slender book *Space and Time in Contemporary Physics* was first published in Berlin in 1917. This was followed by his magnificent *Allgemeine Erkenntnislehre* (General Theory of Knowledge, [Schlick, 1918]. This book appeared as the first in a series of outstanding monographs in the natural sciences published by Springer, Berlin. Not since the publicatons of Mach, Helmholtz, Boltzmann, and Poincaré had any book in the field of epistemology been as close to the outlook of modern natural science as Schlick's epoch-making work. The theory of knowledge and the philosophy of science were his major concerns. '*Naturphilosophie*', as it was then still called in German universities, had nothing in common with the obscure speculations of Schelling, Hegel, and others in the first half of the nineteenth century. '*Wissenschaftstheorie*', a designation that came to be used more frequently later, is a much better expression of the actual intent of these endeavors. S. S. Stevens, the Harvard psychologist, called it the 'Science of Science', correctly indicating thereby that the aim of this reflective enterprise was the clarification of the basic assumptions, concepts, and methods of the sciences.

Schlick was a man of considerable erudition. In contrast to many other members of the Circle, he was well informed in the history of philosophy and the history of science. He was an extremely lucid thinker and writer, but not a brilliant lecturer. His students had the impression that he was slightly bored with lecturing to his usually large audiences. But he gave more of himself and went to much greater depths in his seminars. There he often proved immensely stimulating. His was a warm, kindly personality. He seemed extremely calm and unassuming in his self-effacing modesty.

As I recall, it was in 1924 that F. Waismann and I – we were favored students of Schlick's – approached him with the idea of forming a discussion group. Schlick consented, and the result was a Thursday evening colloquium – the beginning of the Vienna Circle. Among its initial members were the mathematician Hans Hahn, the sociologist-economist Otto Neurath and his wife Olga (she was Hahn's sister and a mathematician and logician), Felix Kaufmann, then a lecturer in the philosophy of jurisprudence, and Victor Kraft, a philosopher well versed in history and much concerned with epistemology and the methodology of science. During that early period we were also joined by the German mathematician Kurt Reidemeister, who had just come from the University of Königsberg to an appointment at the University of Vienna. There were several other scholars, all highly competent in their respective fields but less influential in the development of the ideas of the Circle. Friedrich Waismann and I, together with a handful of other advanced students, were among the junior members.[1]

It was Kurt Reidemeister who in 1924, or perhaps 1925, suggested to us a project that was to become decisive in the development of the Circle's philosophical outlook. We read and discussed at length Ludwig Wittgenstein's *Logisch-Philosophische Abhandlung* which later became famous in English under the title *Tractatus Logico-Philosophicus*. This earlier publication of Wittgenstein's appeared in the last volume (1921) of Wilhelm Ostwald's *Annalen der Naturphilosophie*. I had chanced to read this essay in the Nationalbibliothek of Vienna in 1922. I must confess that although I was struck with whatever I could understand of this aphoristic and cryptic work, I dismissed it as the product of an eccentric, though incisively brilliant, mind. In the Circle we began to penetrate Wittgenstein's ideas on the nature of language and its relation to the world, his repudiation of metaphysics (notwithstanding a few aphorisms toward the end of the *Tractatus* that had a mystical flavor), and his conception of logical and mathematical truth. We had been well prepared for this venture, especially by Hans Hahn, who in an extracurricular evening course had introduced us to the major ideas of the great work of Alfred North Whitehead and Bertrand Russell, *Principia Mathematica*. Hahn had done for us what would otherwise have been a most arduous task: he extracted the philosophical message from that veritable 'cemetery of formulae.' I still remember how Hahn, equipped with a long stick, pointed to the formulae, beautifully displayed on the many blackboards of the Mathematical Institute.

In 1925–1926 Schlick and Hahn were considering two candidates for the position of lecturer (Privatdozent) in philosophy. These were Hans Reichenbach and Rudolf Carnap. These two young but already outstanding scholars had partly similar backgrounds in mathematics, physics, logic, and epistemology. Both had been much concerned with the philosophy of space, time, and relativity. They also had a tinge of Neo-Kantianism due to their early training. Carnap, having been a student of Gottlob Frege's at the University of Jena, was especially interested in the formal-logical problems and techniques, whereas Reichenbach was at that time primarily a philosopher of physics. (In some of his earliest work he was also concerned with the theory of probability, a subject to which both he, and later Carnap, were to make immensely important contributions.) Schlick, who became personally acquainted with Carnap as well as with Reichenbach, was equally impressed by both of these young and productive thinkers. The fact that Carnap was finally offered the position at the University of Vienna can perhaps be explained by Hahn's strong influence, for Hahn, a great admirer not only of Mach but more especially of Russell, was convinced that Carnap would carry out in detail what was presented merely as a program in some of Russell's epistemological writings (notably in *Our Knowledge of the External World*). In fact, several of us in Vienna were reading a large typescript of Carnap's which was then entitled *Konstitutionssystem der Begriffe* (A System of the Constitution of [Empirical] Concepts). In this great work, later published under the title suggested by Schlick *Der logische Aufbau der Welt* (now available in English as *The Logical Structure of the World*), Carnap attempted a logical reconstruction of the concepts of empirical knowledge. The logical form of this reconstruction was essentially the symbolic logic of Whitehead and Russell. In *Principia Mathematica* they had tried to show that all concepts of pure mathematics could be introduced by step-wise definitions on the basis of the concepts of a modernized logic. Similarly, Carnap sketched in considerable detail how the concepts of empirical knowledge could be defined on the basis of concepts pertaining to immediate experience. This seemed indeed the fulfilment of the original intentions of Mach's positivism, as well as a brilliant application of the tools of modern logic to some of the perennial issues of epistemology.

Carnap paid us a visit in Vienna in 1925. He presented a paper on the logic of space-time structure to the Circle. Everyone was impressed with his logical proficiency and scientific competence. Several of us felt that here was a very atypical (at the time!) philosopher who explained his logical reconstructions in the manner of an engineer describing the workings of a machine. Carnap – if psychological typing be permitted – was strongly introvert, Reichenbach strongly extrovert; but this character difference was much more noticeable in their attitudes and behavior than in their published work. Carnap and Reichenbach had become friends in the early twenties and kept up a lively correspondence on logical and philosophical issues long before they became co-editors in 1929 of *Erkenntnis*, the periodical which for ten years was to be the main organ of publication for the new outlook in philosophy. Reichenbach, after a spell of teaching physics at Stuttgart, had become in 1928 a lecturer in philosophy at the University of Berlin, where he formed his own Society for Scientific Philosophy. Outstanding members of that group led by Reichenbach and the mathematician Richard von Mises (originally a Viennese) were Kurt Grelling, Walter Dubislav, Alexander Herzberg, and, later Reichenbach's students Carl G. Hempel and Olaf Helmer. The aims and activities of this Berlin group were very similar to those of the Vienna Circle. Strong connections between the two groups existed from the very beginning, partly because of the personal friendships between Carnap and Reichenbach, and between von Mises and Philipp Frank, then professor of physics at the University of Prague. Frank, never an actual member of the Vienna Circle, was a frequent visitor, and a friend of Hahn's and Schlick's. Frank and Schlick soon became editors of a series of notable books called *Schriften zur wissenschaftlichen Weltauffassung* (Writings toward a Scientific World View). This series contained such important items as Frank's own book *Causality*; Carnap's *Logistics* and his *Logical Syntax of Language*; Neurath's *Empirische Soziologie*; and later, in 1934, K. R. Popper's *Logik der Forschung* (Logic of Scientific Discovery).[2]

Carnap's participation in the Circle discussions (1926–1931) brought about rapid developments. Even outside the regular Circle sessions, Carnap engaged in a continual interchange of ideas with Neurath, Waismann, and myself. Carnap and Neurath also had a great deal in common in that they were somewhat utopian social reformers – Neurath quite actively, Carnap more 'philosophically.' Neurath, a man of great erudition, especially in history and the social disciplines, was also a powerful propagandist of the Viennese positivist outlook. He was of powerful physical stature, extremely energetic, full of 'enterprise', with great talents for organization. He was also a very witty man, using sarcastic dialectics most effectively in discussions and controversies. I owe him a special debt of gratitude for sending me (I think as the first 'emissary' of the Vienna Circle) to Bauhaus Dessau, then, in 1929, a highly progressive school of art and architecture. It was there in a week's sojourn of lectures and discussions that I became acquainted with Kandinsky and Klee. Neurath and

Carnap felt that the Circle's philosophy was an expression of the *neue Sachlichkeit* which was part of the ideology of the Bauhaus. I don't know of any exact synonym in English for that German word *Sachlichkeit*. Perhaps the closest would be 'fact-minded, sober attitude.' This was indeed the basic mood of the Vienna Circle. In a pamphlet (*Überflüssige Wesenheiten*) in which Hans Hahn extolled the virtues of 'Occam's razor' (*entia non sunt multiplicanda praeter necessitatem*), he contrasted sharply the *weltabgewandte* with the *weltzugewandte* orientation in philosophy. This is essentially the distinction between 'other worldly', transcendent speculation or mysticism, and the 'worldly', secular scientifically enlightened types of philosophical attitude.

It was especially Neurath and Frank who envisioned and worked for a new era of enlightenment, propagating the Viennese form of positivism. In consonance with this idea, Neurath began the planning of the *International Encyclopedia of Unified Science*, modeled, though only distantly, on the ideas of the French *Encyclopédistes* of the eighteenth century. (Later an Institute for the Unity of Science would flourish in Boston, under the presidency and wise leadership of Philipp Frank.)

During Carnap's first year in Vienna (1926) the Circle took up a second reading of Wittgenstein's treatise. Wittgenstein himself, though he lived in Vienna from 1927 to 1929, never joined the Circle. He emphatically told the few of us (Schlick, Waismann, Carnap, and myself) with whom he occasionally met (either in cafés, at Schlick's apartment, or that of my fiancée, Maria Kasper, then a student of philosophy) that he was no longer interested in philosophy. He felt that he had said all he could in the *Tractatus*. Moreover, only on relatively rare occasions could we get him to clarify one or another of the puzzling or obscure passages in his work. He seemed himself rather unclear on the ideas he had developed during the First World War. During those Vienna years he was mainly preoccupied with architecture; he designed an almost palatial mansion for his sister in one of Vienna's most aristocratic and secluded districts. On occasion, he would read poetry to us (e.g., that of Rabindranath Tagore). We encountered him also at symphony concerts, and I remember that he shared my enthusiasm for Anton Bruckner's music. Beethoven, Schubert, and even Johann Strauss (which surprised me then, but now no longer) were very dear to his heart. It was quite clear that the genius of Wittgenstein might well have expressed itself in art or music had the circumstances of his life been slightly different. All of us were deeply impressed with his fascinating personality. Schlick adored him and so did Waismann, who, like others of Wittgenstein's disciples, even came to imitate his gestures and manner of speech. Schlick ascribed to Wittgenstein profound philosophical insights that in my opinion were in fact formulated much more clearly in Schlick's own early work.

I recall Wittgenstein, on one occasion, precipitating a quarrel with Carnap, which, as Waismann and I interpreted it, was mainly an expression of diametrically opposite personalities. Carnap was always the tenacious, exact, thorough, painstaking logician, Wittgenstein a man of profound intuition and, at least in conversation, not very articulate. Hence Wittgenstein became impatient to the point of exasperation when Carnap, with the best intentions, asked him to 'explain a little more fully' this or that point. "If he doesn't smell it, I can't help him. He just has got no nose!" Wittgenstein complained to me about Carnap. Indeed, for a time I had to suffer long harangues against Carnap before I was able to get some philosophical conversation going with Wittgenstein, and finally, sensing that I was closer to Carnap, Wittgenstein became inaccessible even to me. But I cherish the privilege of having known him, and of meeting him fairly frequently in 1927–1929.

I mention these matters because they reflect, at a personal level, the beginnings of a profound schism in modern analytic philosophy. Altough I believe that Carnap's and Wittgenstein's basic substantive positions were fairly similar in the twenties, their manners of approach were radically different. Later, this led to the sharp divergence between the method of Carnap's rational reconstruction and the procedure of informal analysis of the 'ordinary language philosophy', in England as well as in America, and that was inspired by Wittgenstein.

One more incident stands out in my memory. When the Dutch mathematician Luitzen Egbertus Jan Brouwer was scheduled to lecture on intuitionism in mathematics in Vienna, Waismann and I managed to coax Wittgenstein, after much resistance, to join us in attending the lecture. When, afterwards, Wittgenstein went to a café with us, a great event took place. Suddenly and very volubly Wittgenstein began talking philosophy – at great length. Perhaps this was the turning point, for ever since that time, 1929, when he moved to Cambridge University Wittgenstein was a philosopher again, and began to exert a tremendous influence. A veritable *bouleversement* took place in English philosophy, soon to spread to the United States and Australia as well. Bertrand Russell, dismayed at the new turn Wittgenstein's philosophy had taken, first was icily silent and later outspokenly opposed to it. Carnap, never very much disturbed by the fashions of the times, continued his constructive work in Prague and later in America.

I think it was by 1927 that we began to feel we had developed into a 'movement' in philosophy. Despite some disagreements and of course great differences in emphasis, we had attained a predominantly common 'platform.' Visitors from other countries began to join the Circle for various periods of time. Among them were Eino Kaila, a brilliant Finnish philosopher, and Dickinson S. Miller, an American philosopher who, though elderly and rather silent, was a keen analytic thinker. One of the most outstanding members of the Polish group of logicians, Alfred Tarski, came for a short visit in 1930, attracted primarily by Carnap's work. Carnap, a few months later, reciprocated with a visit to Warsaw where he continued discussions on logic and metamathematics with Tarski, and also with Lesniewski and Kotarbinski. This was the time when Carnap developed his ideas on the logical syntax of language, published in the Frank-Schlick series in 1934.

In the late twenties the Circle was greatly enriched by the attendance and participation of several advanced students, three of them primarily mathematicians: Gustav Bergmann, Kurt Gödel, and Karl Menger. Gödel, whose great abilities were quickly appreciated, had a lively exchange of ideas with Carnap (much of it outside the Circle) and began to develop extremely original ideas, culminating in 1931 in his famous proof of the essential incompleteness of a large class of mathematical systems. The proof itself was most ingenious, utilizing syntactical methods, i.e. an arithmetization of the syntax of mathematics. Gödel was a very unassuming, diligent worker, but his was clearly the mind of a genius of the very first order. Carnap immediately realized the importance of Gödel's discoveries and dealt with them later in various publications. On the personal side, I should mention that Gödel, together with another student member of the Circle, Marcel Natkin (originally from Lodz, Poland) and myself became close friends. We met frequently for walks through the parks of Vienna, and of course in cafés had endless discussions about logical, mathematical, epistemological, and philosophy-of-science issues – sometimes deep into the hours of the night. Natkin, whose doctoral dissertation concerned causality, induction, and simplicity, was a brilliant thinker – thoroughly Humean in his positivism, but able to apply it to some of the most difficult and controversial issues in the logic of the empirical sciences. For personal and practical reasons he decided on a career in business, and left Vienna after his doctorate and settled in Paris. There he became one of the world's outstanding experts on photography.

I visited Natkin in Paris several times, and in 1957 he, Gödel, and I had a most enjoyable reunion in New York. The first of my Paris visits was in 1929, after I spent a few days with Neurath in Geneva, serving as his French interpreter at a conference. It was also in 1929 that I visited Reichenbach in Berlin for the first time. He had read my doctoral thesis, as well as my first book (*Theorie und Erfahrung in der Physik*), and generously gave me the benefit of his very helpful comments and criticisms.

In the early autumn of the same year, 1929, the Vienna Circle together with the Berlin Group held its first conference in Prague. I shall always remember the excitement of that meeting, as well as that of the second meeting in Königsberg of the year after. A fairly large group of highly competent thinkers gathered on both occasions, and despite some divergencies, we felt that we all had basically a common orientation. This contrasted sharply with the situation among the more tradition-bound philosophers; they did not even seem to communicate effectively with one another. The Königsberg conference included among its personnel John von Neumann (who represented Hilbert's formalist philosophy of mathematics); the Dutch mathematician Arend Heyting (representing intuitionism); and Werner Heisenberg, the physicist and founder of quantum mechanics. Von Neumann's intellectual agility and genius was already quite noticeable in the public discussions and in the many private conversations that took place at Königsberg. At the preceding conference in Prague, Reichenbach, von Mises, Waismann, and I presented papers pertaining to the problems of probability and induction, Waismann, building upon some of Wittgenstein's ideas, formulated a logical concept of probability. This, much later, became the basis of Carnap's extensive work in this field.

There were two outstandingly brilliant minds in Vienna who, though close to us in philosophical orientation, never joined the Circle: Edgar Zilsel and Karl R. Popper. Both were convinced of their intellectual independence from us, and tried to preserve that independence by remaining outside the Circle. Indeed, I felt that both these men, each in his own way, were among our most valuable and helpful critics. Several of us met with them separately and privately – with Zilsel in the small discussion group that gathered at the houses of Victor Kraft, in which Professor Heinrich Gomperz (son of Theodore) was a prominent member. Gomperz was a most stimulating lecturer, a highly erudite philosopher, and like his father an authority in the history of philosophy, especially of the classical Greek period. He was a man of great wit – and despite his occasional sarcasm, a kind and helpful teacher. (He moved to the University of Southern California in 1938 where he enjoyed a happy life and continued his highly effective teaching.) Gomperz, in his skeptical wisdom, viewed the Logical Positivist movement with severe reservations, but nevertheless with considerable understanding and sympathy.

Edgar Zilsel, who had been a student of Gomperz, had published several books – his earliest and perhaps most original being *Das Anwendungsproblem* (1916) (The Problem of Application, i.e., problems of induction and probability, and more generally on the application of mathematics in the empirical sciences). This book, while severely defective in some respects, was extremely original, and certainly fascinatingly written. Zilsel, too, migrated to the United States – first settling in New York, working with the Horkheimer-Adorno group on social philosophy, later teaching at Mills College in Oakland, California. I met him only on a few occasions in the United States. It seems his was an unhappy and isolated existence in this country. We were deeply shocked to learn that his life ended in suicide.

Karl Popper's story is a very different one. He was a school teacher in Vienna. He was close to Gomperz and Kraft during his student years. I came to know him only shortly before my emigration. It must have been in 1929 that we spent a whole day and a large part of the night in discussion (in the apartment of my parents). Popper had read my doctoral thesis which dealt with problems to which he had already given much independent thought. His disagreements with my views, and also those of the Circle – especially with Carnap's epistemology (as in the *Aufbau*) – stimulated the further development of his thought. Schlick finally invited him to publish a monograph in the Frank-Schlick series. Many discussions with Schlick ensued; and as I know from Schlick's remarks (in Italy in 1935), Popper impressed him with his high originality and independence, and with all his characteristic 'intensity' of discussion. To make matters worse, Reichenbach had criticized Popper's ideas severely (and perhaps unjustly), so that Popper remained on the whole quite critical, if not antagonistic, to our movement. In any case, his book, *The Logic of Scientific Discovery* (1934), was outstandingly successful. Carnap, despite some basic disagreements, appreciated deeply Popper's highly original approach to many issues of common concern. Thus it came about that I arranged for a few weeks of summer vacation (1932) in the Tyrol where Popper and I joined Carnap for many fruitful discussions and conversations. Popper later moved to New Zealand, and finally to England where he has been professor of philosophy at the University of London's School of Economics since 1946, and a knight since 1965. He also came to the United States quite a few times on visiting appointments. But more of this later.

II. EARLY CONTACTS WITH AMERICAN PHILOSOPHERS, AND THE BEGINNINGS OF OUR MIGRATION

Several members of the Circle had a reading knowledge of English, but Schlick, whose wife was American, spoke English perfectly. Some of the conversations at Schlick's house were in English, notably with such visitors as Roger Money-Kyrle but occasionally even with Wittgenstein who also was fluent in English. Schlick was the first of our group to be invited to the United States. In 1929 he was visiting professor for a semester at Stanford University. At that time Paul A. Schilpp, also a German, who would later attain world-wide recognition with his Library of Living Philosophers, served as his assistant.[3] Schlick enjoyed his sojourn at Stanford, made many friends, and was promptly invited to another visiting professorship, this time (in 1931) to the University of California at Berkeley. Thus it came about that Schlick was the first to spread the Vienna 'gospel' (with a strong emphasis on Wittgenstein's ideas) in America. My own first journey to the United States occurred in September 1930, when I was fortunate to obtain an International Rockefeller Research Fellowship. This allowed me to work at Harvard University for about nine months.

My interest in visiting and quite possibly even settling in the New World was motivated by various circumstances. To start with, there were my early contacts with two splendid American philosophers, Dickinson S. Miller and through him Charles A. Strong. Miller, as I have already mentioned, spent a year in Vienna attending the Circle sessions. About a year later I received from him a most amicable invitation to visit him and his good friend Charles Augustus Strong in Fiesole near Florence. He offered to cover all my expenses and to be his house guest. (It occurred to me later that it was Strong, a son-in-law of John D. Rockefeller, who must have financed my trip.) Both Strong and Miller were retired from their respective positions in American universities, Strong from his chair in philosophy and psychology at Columbia University, Miller from a succession of posts, notably at Bryn Mawr, Smith, Harvard, and Columbia. Both of these gentlemen were wonderful hosts. Miller showed me the great works of art in Florence, and I enjoyed the Italian spring in the hills of Fiesole – it was indescribably beautiful. Both Miller and Strong knew German much better than I knew English at the time, so our conversations – mostly on epistemological issues – were in German. I reported and explained the views of the Vienna Circle to them. Nearly a year later they invited me to Fiesole once more – mainly to help Strong toward a better understanding of Einstein's theory of relativity. Strong was working on a metaphysical theory of space and time and wanted to find out to what extent his views were compatible with those of Einstein. This again proved a most enjoyable experience for me. I was impressed with the erudition and the generosity and open-mindedness of these American scholars.

During the summer of 1929 I met a young American student in Paris,

Albert E. Blumberg, who had come from Baltimore (Johns Hopkins University) and was interested in coming to Vienna. I encouraged him to do his thesis (on Emile Meyerson's philosophy of science) under Schlick's supervision. Blumberg and I soon became close friends, and I helped him, philosophically and linguistically, with the composition of his thesis. It was through Blumberg that I learned a great deal about the United States, its philosophers, its system of education, and its ways of life.

Most of us in the Vienna Circle were largely ignorant of American philosophy. We had, of course, read some of the work of William James and of John Dewey. But we had only a very vague idea of Charles S. Peirce.[4] We knew that James and Mach had some affinities and that they respected each other enormously. But for the rest, our ignorance was vast. We hardly knew anything about the American philosophical movements of Neo-realism and Critical Realism, though I personally had learned a bit about them from Miller and Strong, and later from Blumberg. I think it was through Bertrand Russell's books that our attention was called to American Behaviorism, particularly as represented in the work of John B. Watson and his followers.

In any case, what I found out through the personal contacts mentioned above and through my reading of some of the American philosophical literature attracted me strongly. 'Over there' I felt was a *Zeitgeist* thoroughly congenial to our Viennese position. It was also in 1929 that, I think through Blumberg's suggestion, we became acquainted with Percy W. Bridgman's *Logic of Modern Physics* (1927). Bridgman's operational analysis of the meaning of physical concepts was especially close to the positivistic view of Carnap, Frank, and von Mises, and even to certain strands of Wittgenstein's thought.

Thus encouraged, I applied to the Rockefeller Foundation in New York for a fellowship. I was 'looked over' by one of their officials in Vienna in the spring of 1930, and about two months thereafter I was awarded a fellowship for research in the logic of scientific theories at Harvard. I arrived at Harvard in September 1930, shortly after the conference in Königsberg.

I was tremendously impressed by many things in the United States. New York was the first overwhelming experience. Soon afterwards there were the scholars at Harvard with whom I was fortunate to become acquainted: in short order, P. W. Bridgman, C. I. Lewis, Henry Sheffer, and of course A. N. Whitehead. In addition there was Susanne K. Langer, of German parents, the wife of the Harvard historian, William Langer. We had already known in Vienna her fine first book, *The Practice of Philosophy*. At her house a small group of scholars met occasionally for an evening's discussion. Then there were the 'soirées' at Whitehead's house; most of them were long, rambling, but thoroughly captivating soliloquies – on all sorts of subjects – by the amiable, thoroughly British, philosopher. Whitehead had turned increasingly metaphysical (or 'cosmological') after coming to the United States a few years earlier. His lectures were rather poetic ("the flickering, perishing flame of life . . ."), though when his students handed him term papers, he was dismayed with their imitations of his flowery style. (Returning the papers in class, he said, with that slight characteristic stammer of his: "You – you can't (cawn't) do that – *I* – *I* do that!") Paul Weiss, then an instructor at Harvard and at that time still more interested in logic and epistemology than in metaphysics, was a most kind and helpful friend. George Morgan, who in 1927 had been secretary to Dickinson Miller in Fiesole, also aided me in my efforts of adaptation to the manners and intellectual climate of this new environment. (This gifted man later entered upon a career as diplomat; but there is a good book on Nietzsche by him.)

Among the graduate students at Harvard I was especially impressed with W. V. O. Quine, who later became one of the world's greatest logicians. Sheffer, an outstanding logician in his own right, remarked about Quine's first book, "Quinine logic – a bitter pill."

It so happened that Karl Menger was also at Harvard University for part of that year. He had by then achieved international recognition as a mathematician, especially through his work in topology and the theory of dimensions which he had begun in Vienna under Hans Hahn (stimulated by some of Poincaré's work). I was glad to have this reunion with a Circle member in the New World.

At Christmas, 1930, Albert Blumberg and I met in New York for a week's vacation, and for work on an article, 'Logical Positivism; a New Movement in European Philosophy', that was published in the spring of 1931 in the *Journal of Philosophy*. It was this article, I believe, that affixed this internationally accepted label to our Viennese outlook. I should mention that previously, in 1929, a slender pamphlet, *Wissenschaftliche Weltauffassung: der Wiener Kreis*, had been composed by Carnap, Neurath, and Hahn, aided by Waismann and myself. This was, as it were, our declaration of independence from traditional philosophy. We presented this pamphlet to Schlick upon his return from Stanford. Schlick was moved by our amicable intentions; but as I could tell from his facial expression, and from what he told me later, he was actually appalled and dismayed by the thought that we were propagandizing our views as a 'system' or 'movement.' He was deeply committed to an individualistic conception of philosophizing, and while he considered group discussion and mutual criticism to be greatly helpful and intellectually profitable, he believed that everyone should think creatively for himself. A 'movement', like large scale meetings or conferences, was something he loathed. Yet the expansive spirit of Neurath and Reichenbach, and to some extent also of Carnap, had taken hold of us. Blumberg and I felt we had a 'mission' in America, and the response to our efforts seemed to support us in this. We had, indeed, 'started the ball rolling', and for at least twenty years Logical Positivism was one of the major subjects of discussion, dispute, and controversy in United States philosophy. Among the early reactions were important articles by C. I. Lewis (Schlick replied to one of them), W. H. Werkmeister, V. C. Aldrich, and P. A. Schilpp; and a notable book by Julius Weinberg.

My own presentations at various meetings of the American Philosophical Association (especially those at Ann Arbor, Michigan, 1932; Chicago, 1933; and St. Louis, 1934), though concerned with specific issues such as induction, mind-body, and transcendence, were largely iconoclastic in regard to traditional metaphysics, and advocated the Viennese type of logical analysis. To some extent it was a '*succès de scandale*', perhaps somewhat similar to that of the early A. J. Ayer in England, whose aggressive and extremely well-written book, *Language, Truth, and Logic* (1936) contributed greatly to the propagation of the Viennese views, especially those of Carnap, Wittgenstein, and Schlick, in the English-speaking world.

I remember well the national meeting of the American Philosophical Association in Chicago in which I was given the special privilege of an hour and a half for my lecture and subsequent discussion. My topic was 'A Logical Analysis of the Psychophysical Problem.' I tried to show that the notorious vexations of the mind-body puzzles in traditional philosophy could be solved (or dissolved!) by conceptual clarification. Fairly soon I came to realize that my ideas were rather half-baked, and that the problems involved were much more complex and intricate than I had thought at the time. My paper was published in the first volume of *Philosophy of Science*, a periodical for whose initiation I was in small part responsible. The reactions to my paper ranged from enthusiastic approval to devastating criticism. The California philosopher S. C. Pepper repudiated my entire approach, shrewdly spotting some inconsistencies of which I was then unaware. At the Chicago meeting I encountered, for the first time, Professor Charles Morris, who very generously supported me in the discussion.

Morris and I became better acquainted later. He invited me to give a colloquium lecture in Chicago in 1934, where I was also his house guest. Morris had grown up in the pragmatist movement and had high regard for the philosophies of John Dewey and George H. Mead. He had published a book, *Six*

Theories of Mind, and was becoming increasingly interested in the theory of language. He had already formulated his general program of research. The broad discipline of *semiotic* was to combine studies in the syntax, semantics, and pragmatics of language. I encouraged Morris to visit Carnap in Prague, which he did in the summer of 1934. Morris, realizing what tremendous contributions Carnap could make on the American scene, mobilized those in power at the University of Chicago to invite Carnap. Thus it came about that Carnap started his American teaching career in the fall of 1936. He arranged also for assistantships for his young friends and disciples, Carl G. Hempel and Olaf Helmer.

Charles Morris was also instrumental in bringing Hans Reichenbach to the United States. Reichenbach had to leave Germany because of the increasing menace of the Nazi regime. He went first to Istanbul (Richard von Mises was also there at that time) where he taught first in French, and later even in Turkish. In 1938, the year in which Reichenbach's important book *Experience and Prediction* was published by the University of Chicago Press (Morris had arranged for this), he began teaching at the University of California at Los Angeles, where he taught until his untimely death in 1953. He soon established himself as one of the leading philosophers of science in America. He was a most productive scholar; he worked in many areas, and published several important books. Reichenbach was also a brilliant speaker and lecturer, beloved by his students. He and Carnap and I met occasionally at specially arranged conferences, or – as in 1951 – at a combined meeting of the American Philosophical Association and the American Association for the Advancement of Science at Bryn Mawr and Philadelphia.

Kurt Gödel accepted a research position at the Institute for Advanced Studies at Princeton in 1938. There he collaborated during several periods with Einstein on problems of cosmology and continued his work on the foundations of mathematics, notably on the Continuum Problem.

C. G. Hempel, one of the most clear-headed thinkers and most effective teachers of the group, held positions at Queens College, New York, later at Yale University, and finally at Princeton University. His studies in the logic of scientific explanation were begun in collaboration with Paul Oppenheim, also an émigré from Germany, who was a close personal friend of Einstein's, a private scholar who, in his quiet way, stimulated a great deal of interesting work in the philosophy of science. In Oppenheim's magnificent house at Princeton there were many social gatherings of distinguished scientists and philosophers.

Also in 1938, Gustav Bergmann moved from Vienna – first to New York and later to the University of Iowa. When, in 1940, I was called from Iowa to a professorship at the University of Minnesota, Bergmann became my successor at Iowa and began a highly influential teaching career and a long period of scholarly production there. Olaf Helmer, after his brief spell with Carnap in Chicago, joined the research staff of the Rand Corporation in Santa Monica, California. Alfred Tarski, who proved to be one of the greatest logicians and mathematicians of our time, settled at the University of California in Berkeley. And Henryk Mehlberg, another outstanding Polish philosopher of science, taught first at the University of Toronto (1949–1956) but settled in 1957 at the University of Chicago.

Philipp Frank left Prague a year before the Nazis occupied Czechoslovakia and Austria. He was offered a position in the Physics Department of Harvard University, where he taught primarily philosophy of science courses and seminars. The teaching and the books of this wise and humorous man contributed greatly toward the empiricist outlook on the American scene. Frank's book *Philosophy of Science* (1957) bore the subtitle 'The Link Between Science and Philosophy.' Thus, shortly before C. P. Snow's ideas on the 'two cultures' emerged in the limelight of public discussion, Frank had already pointed to one important bridge over the chasm.

Very much in the same spirit, and perhaps even more incisively written, was Richard von Mises' book *Positivism – A Study in Human Understanding* (1951). Von Mises, too, had found a haven at Harvard. His main work there was, however, in the fields of mathematical physics and aerodynamics.

Felix Kaufmann, until his premature death (in 1940), was a professor of philosophy in the New School for Social Research in New York City. Although he had been an active and cherished member of the Vienna Circle, his philosophical allegiance during those years was distinctly to Edmund Husserl's phenomenological approach. He had published books on the philosophy of law and the foundations of mathematics. During his American years he devoted a good deal of his work to the philosophy and methodology of the social sciences, and in this connection he was fairly close to the outlook of the Viennese positivists.

During the spring of 1931, while I was at Harvard University, it became clear to me that my chances for a teaching position in an Austrian or German University were extremely slim. True, the ever so optimistic and kindly Schlick was convinced that I would obtain a Privatdozentur (position as a lecturer) at the University of Vienna. But though I was Austrian by birth, I had become a Czechoslovakian citizen after the revolution in 1918. My home was then in Reichenberg (Liberec), in the Sudetenland, where I was born and grew up, and had attended primary and secondary schools. My parents, though thoroughly 'assimilated', were of Jewish descent. More realistic than Schlick, I abandoned the idea of a teaching career in Europe, and began applying for a position in a number of American universities. I had been given generous recommendations by Einstein, Whitehead, C. I. Lewis, and Bridgman. I received very courteous replies from about thirty American universities (I had sent out forty-five applications), but most of them then had 'no opening.' Three universities, Rutgers, New York University, and the State University of Iowa, were the only ones that wanted to 'look me over', and toward the end of May 1931 I visited all three places. Iowa seemed most strongly interested. The late Dean George Kay, a prominent geologist of Canadian origin, telephoned Professor Lewis long distance. As Lewis later related to me, Dean Kay asked him in detail about my qualifications, character, and personality. At the end of that (about twenty minutes!) telephone conversation, Kay finally asked: "Is he a Jew?" To this, Lewis, the noble New Englander, gave the – to me unforgettable – reply: "I am sure I don't know, but if he is, there is nothing disturbing about it." Thus it came about that a few days after my visit to Iowa, I received a telephone call from Dean Kay offering me the position of a lecturer. That telephone call reached me just fifteen minutes before my departure for Europe. It was one of the most exciting moments of my life. I returned to Vienna and married Maria Kasper (to whom I had been engaged for three years). We both obtained immigration visas for the United States and settled, in September 1931, at Iowa City.

The transition from cosmopolitan Vienna to the small town, then with a population of about fifteen thousand in the 'Tall Corn State', required some effort. But we were young and eagerly interested in starting a '*vita nuova*.' Moreover, we soon acquired a number of wonderful and helpful friends – some of them typically midwestern, others of German, Austrian, French, or South African origin – all on the Iowa faculty. I had improved my fluency in the English language considerably at Harvard during the previous year. Within a few weeks, I managed to communicate effectively with my students. I shall relate my further experiences at Iowa, and return to my 'propagandistic' activities, after the following section devoted to the 'ideology' of Logical Positivism.

III. LOGICAL POSITIVISM; LOGICAL EMPIRICISM – A BRIEF SKETCH FOR NON-PHILOSOPHERS

In order to make more intelligible the impressions, experiences, and influences of the European positivists in the United States, it is necessary to sketch, in brief outline, some of our basic doctrines and their early modifications. Logical Positivism (the position of the Vienna Circle) and Scientific Empiricism (the position of the Berlin Society), despite important differences, had much in common. Both ideologies were conceived as philosophies for our age of science. As a culmination of the empiricist tendencies in philosophy since the Renaissance, positivism attempted to retain from traditional philosophies the spirit of enlightenment and clarification. The opposition to obscurantist metaphysics and mysticism became the most conspicuous, and to some the most offensive, feature of positivism. The critique of metaphysics was directed essentially against two major traditional philosophical tendencies: first, the conviction that truths regarding matters of fact could be established by pure reason alone ('*a priori*'); the empiricist opposition to this position had already been clearly enunciated by David Hume in the eighteenth century. The second was the conviction that knowledge regarding 'transcendent' reality could be justified by speculation, intuition, dialectics, or other non-empirical procedures. Thus limiting the scope of factual meaning to what is testable by observation and experience, many positivists drew a line of demarcation between the knowable and the unknowable even more narrowly than some of the earlier empiricists.

The designation *logical* positivism seemed required in order to mark an important difference between our position and that of such empiricists as Comte and Mill. Logical and mathematical truths were considered, for example by Mill, as a most general kind of empirical knowledge. In this regard, we favored the different attitude already adumbrated by Leibniz and developed much more fully by Frege and Russell. The truths of *pure* mathematics (i.e., not including physical geometry or other branches of the factual sciences) are *a priori* indeed. But they are *a priori* precisely because they are *analytic*, i.e., because they are validated on the basis of the very *meaning* of the concepts involved in the propositions of mathematics. Empirical certification of mathematical truths is neither required, nor indeed possible. The Vienna Circle regarded, for example, the identities of arithmetic as necessary truths, based on the definition of the number concepts – and thus analogous to the tautologies of logic (such as 'what will be, will be'; 'the weather will either change or remain the same'; 'you can't eat your cake and not eat it at the same time'). With the formalists (e.g., Hilbert) we would consider mathematical proofs as procedures that start with a given set of sign combinations (premises, postulates) and according to rules of inference (transformation rules) lead to the derivation of a conclusion (theorem). Russell (who was not a formalist himself) formulated this doctrine in his famous quip "Mathematics is the science in which we never know what we are talking about, nor whether what we say is true." With the logicists (e.g., Frege and Russell) we regarded mathematics as a branch of logic. As we saw it, there were no insurmountable difficulties connected with the mathematical problems of the infinite (or its Cantorian levels). Our view appeared derogatory to the mathematicians. They understood us to say that mathematics is no more than a huge tautology. But this is derogatory only if 'tautology' is taken as synonymous with 'triviality.' Far from maintaining anything so silly, we insisted that the proof of the very tautologicality of the implication that holds between the postulates (and definitions) and the theorems of mathematics often requires the intellectual powers of a genius.

In any case, we were convinced that there are only two kinds of genuine knowledge: the purely formal (logico-mathematical) and the factual (empirical). This distinction, already clearly drawn by Hume, but greatly refined and elaborated later, is also inherent in the two meanings of the word 'proof' as used in common language. It is one thing to 'prove' (by deductive derivation) a theorem in mathematics, and quite another thing to 'prove' the truth (or rather the likelihood or credibility) of a factual assertion (hypothesis, prediction, theory) by inductive or hypothetico-deductive procedures and observational-experimental (or statistical) confirmation.

These are in essence the guiding principles of the epistemology of Logical Positivism. Schlick preferred the label *Konsequenter Empirismus* (Consistent Empiricism), but our designation, emphasizing as it did our view of logical truth, gained world-wide acceptance.

It became necessary to guard against misunderstandings of our main thesis. Our concern was not with the psychological origins or the social conditions of the cognitive enterprise. Our distinction was based on differences in the method of validation. In Reichenbach's terminology, we were analyzing knowledge claims in the 'context of justification' and not in the 'context of discovery.' We always admitted that all sorts of intuitive processes (e.g., 'hunches', consciously or subconsciously discerned analogies), may well be extremely instrumental (heuristically) in the *genesis* of hypotheses and theories. We never claimed that great scientific theories could be constructed by a logic-machine. (This was long before the age of computers – and who knows what may yet be achieved by them?) We firmly held (with Einstein, and with Popper) that there is no straight logical path that leads from the data of observation to an explanatory theory. Great ingenuity, with all the risks of 'guessing wrong', is the order of the day in theory construction. Many of us agreed with Popper that the procedure here is: bold conjectures, and their criticism through persistent and perspicacious attempts at refutation. If the theory in question survives such searching experimental challenges, it may 'until further notice', be considered as corroborated, and as such be included in the corpus of science – as long as no refuting evidence turns up.

The Logical Positivists' most controversial tenet was their 'principle of verifiability.' This was intended as a criterion of factual meaningfulness. Actually, as formulated by Carnap as early as 1928, it was not verifiability but *testability-in-principle* (*Prüfbarkeit*) which we used as the distinctive mark of factually (not just scientifically) meaningful assertions. We used this criterion in repudiating any *transcendent*-metaphysical propositions. We thus continued and sharpened the critique of the *absolutes* (such as substance, space, time, vital force, soul, the deity) that had already been characteristic of the empiricist tradition.

Telling metaphysicians and theologians, not that they make false or unjustifiable statements, but that their verbal utterances make no sense, this was clearly the most iconoclastic and offensive challenge that could be put to them. Judging by the tremendous flood of polemical reactions, we had touched a sensitive spot here. To be sure, we had to modify and to mollify our attack, for we were soon confronted with pertinent and provocative questions regarding the very meaning, and the justification of our criterion. We explained that we were distinguishing the various functions of language, or (what amounts to the same thing) the diverse types of significance. While admitting that the cognitive and the non-cognitive uses of language are almost always combined, and psychologically fused, it seemed imperative to warn against confusing one with the other. The pictorial, emotional, or motivative functions of language must be sharply distinguished from the informative function. In metaphysical and theological assertions the emotive (i.e., pictorial, emotional, and/or motivative) type of significance is often erroneously taken to be a genuinely cognitive-factual meaning.

Despite the many changes that have been made in the formulation of the criterion of factual meaningfulness, its basic intent has remained the same. Essentially it amounts to eliminating from cognitive discourse questions which can neither logically nor empirically be answered. Problems which are guaranteed unsolvable are *cognitively* meaningless. But as existentialist torments indicate, questions such as 'the meaning of life', or 'why is there something rather than nothing?' may be highly significant emotionally or motivationally; cognitively, however, they are absolutely unanswerable. The proffered answers or assertions are proof against disproof. They are beyond the limits of rational discourse. To put it yet another way: we tried to establish a clear line of demarcation between genuine problems and pseudo-problems.

It is true that some of the early formulations were too drastic in that they eliminated difficult questions along with nonsensical ones. This was remedied by later more circumspect and more liberal formulations. Discussion and dispute concerning the very feasibility of an adequate formulation continues. It should also be noted that we never accused the metaphysicians or theologians of deliberately talking nonsense. Our diagnosis was rather that they were unwittingly mistaking pictorial or emotive significance for factual meaning.

Since the word 'metaphysics' is used in a variety of ways, and thus designates several quite different philosophical endeavors, it is important to note that by no means all of them were 'taboo' from our point of view. If 'metaphysics' designates speculative extrapolations based on scientifically obtainable evidence, then such an endeavor, though risky, is not necessarily meaningless. Examples in various domains would be cosmology (essentially a branch of astrophysics) and psychoanalytic theory (an attempt to explain behavior and experience on the basis of speculative assumptions regarding unconscious motivation). Such types of 'inductive' metaphysics are thus continuous with science, but its theories are more precarious because they are only very indirectly and incompletely testable. Another time-honored meaning of 'metaphysics' may be characterized as categorial analysis, i.e., an investigation of the basic concepts and conceptual frames used in our knowledge of reality. This is not fundamentally different from the sort of logical analysis pursued by the positivists themselves.

What we did repudiate as illegitimate was rather the use of 'verbal sedatives', that is, high sounding phrases that may tranquilize scientific curiosity and thus impede the progress of research. We were convinced that the emotional needs of man can be fulfilled much more adequately in poetry, music, and the arts in general, if not in life itself. We pointed out that the word 'belief' is quite ambiguous in that it may mean: (1) a cognitive attitude connected with testable propositions, or (2) the acceptance of transempirical (untestable) assertions, or (3) a commitment to certain values or ideals. We were convinced that transempirical assertions were devoid of cognitive meaning, and that their utterance (as in sermons, prayers, rituals) merely enhanced the motivative power of commitments. Thus our attitude toward theology and religion was that of the naturalistic or scientific humanists. Indeed, several of us found in the general position of the American Humanist Association an ideology that seemed very similar to our basic philosophical attitude. If, as most humanists prefer, 'religion' is not connected with any theology whatever, then a deep commitment to such human values as basic and equal rights, the civil liberties, the ideal of a peaceful and harmonious world community, may well be said to be the religion of the humanists – and of the positivists.

We encountered strong opposition to some of our early provocative pronouncements in regard to the significance of moral judgments. We agreed (for example with G. E. Moore – though this important insight had already been explicit in Hume's work) that it is logically impossible to derive moral norms or imperatives from purely factual premises about human nature. But we disagreed with Moore's idea (and with many other philosophers, past and present) that moral judgments could be justified by 'intuition.' It is *familiarity* that breeds intuition (in many areas) – but this does not in any way validate the intuited judgment. What appears as intuitively cogent may well vary from person to person, or from culture to culture. But going beyond these observations, some of the positivists assimilated moral judgments to judgments of taste, and thus not only offended well-intentioned moral philosophers but actually did not do justice to the nature of moral reasoning. Schlick's largely psychological approach to the problems of ethics found little attention. But A. J. Ayer's chapter on ethics in his *Language, Truth and Logic* aroused strong opposition. The important work of the American philosophers Charles L. Stevenson and Charles Morris in moral philosophy and value theory helped greatly in paving the way toward a less offensive (by no means purely 'emotivist') position in ethical theory. The issues are logically delicate and intricate. Here, just as in epistemology and philosophy of science, our outlook has undergone considerable change and development.

At this point I should explain the reasons for the change in label from 'Logical Positivism' to 'Logical Empiricism.' The original positivism of Auguste Comte, as well as that of the late nineteenth-century exponents, especially Mach and Avenarius, were often misunderstood as being yet another version of subjective idealism. Careful reading of all these authors reveals that this is indeed a misunderstanding. In any case, Carnap's work (especially in the *Logische Aufbau*) was similarly misinterpreted. Carnap had quite emphatically explained that his reconstruction of the empirical knowledge was phenomenalistic or solipsistic only in a methodological sense. That is to say, he chose as the basis for his particular reconstruction the data of immediate experience. But he pointed out that a reconstruction on a physicalistic basis was equally possible. Actually he later came to prefer this latter kind of reconstruction. Carnap was equally explicit on the 'metaphysical neutrality' of either (or any other) type of reconstruction. He considered metaphysical interpretations as illegitimate intrusions of *Begleitvorstellungen* or of *Begleitgefühle* (accompanying imagery or emotions): an irrelevant surplus beyond the genuinely cognitive content.

Perhaps the most important and constructive aspect in the transition to Logical Empiricism was the element of empirical or scientific realism that became increasingly prominent in our views. Reichenbach and I had already opposed the phenomenalistic reduction during the twenties. In this regard we were closer to the views of Zilsel and Popper. We regretted that Schlick had abandoned his early critical realism, and we tried to reinstate it in a more defensible form. This was achieved through the liberalization of the empiricist criterion of meaning. Verifiability was replaced by (at least indirect and incomplete) testability; or, as we now usually put it, by confirmability or disconfirmability. On this basis it makes perfectly good sense to speak of the existence of theoretical entities – be they the particles or fields of modern physics or the unconscious mental processes in psychoanalytic theory.

The most important positive contributions by the individual members of the movement pertain to the philosophy of science. The foundations of mathematics and the structure and confirmation of the theories in the natural and social sciences have been among our major concerns. Carnap's work in syntax and semantics and in the logic of probability and induction, and Hempel's contributions to the logic of concept formation and scientific explanation, are outstanding examples. Most of these more technical endeavors are far too intricate to permit brief description. It should also be noted that even more than during the initial stages of the movement (in Vienna and Berlin) there have developed considerable divergencies in point of view as well as in method among the individual members of the group. In addition to our conspicuous preoccupation with the philosophy of science, there is one aspect of method that the movement as a whole still shares. It is the pursuit of formal logical reconstruction, and in this regard it is quite different from the neo-Wittgensteinian informal linguistic analysis. Some of us accept the informal procedures as a useful first step, i.e., the sorting out of explicanda, and thus paving the way for exact explication. Others find the procedures of Oxford-type linguistic analysis completely uninteresting and fruitless. The need for formal reconstruction, it should be remembered, was already recognized by Aristotle. He could never have succeeded in codifying syllogistic inference had he not transformed the phrasings of ordinary-language sentences into standard patterns. Though this was only a first simple beginning, it is the prototype of all the reconstruction in terms of 'ideal languages' pursued in our century. That some of these reconstructions deviate considerably from the patterns of the natural languages does not matter. Any sort of reconstruction is bound to appear 'procrustean' from the point of view of ordinary language. Our answer here, as well as in all related issues, is the pragmatic one: 'By their fruits ye shall know them.'

IV. THE TERMINATION OF THE VIENNA CIRCLE AND THE BERLIN SOCIETY: INTERNATIONAL MEETINGS, EXPANSION AND MIGRATION

The rise of the Third Reich, and the invasion of Austria in 1938 and of Czechoslovakia and Poland in 1939, put an end to the Continental phase of the Logical Positivist movement. As mentioned before, two prominent members of the Berlin group, Reichenbach and Richard von Mises, found a temporary haven in Turkey before they came to the United States. The activities of the Vienna Circle had continued until 1936, the year of Schlick's tragic death. He was murdered by one of his former students, in all probability not a political assassin but a paranoid personality. All members of our now partly dispersed group were deeply shocked and grieved. Schlick had been extremely kind to all his students, even to the man who later committed that horrible crime.

There had been several international meetings of the Unity of Science movement – in Prague, 1934, Paris, 1935, Copenhagen, 1937, and Cambridge, Massachusetts, 1939. I participated (coming from the University of Iowa) in the Paris congress where I had a happy reunion with my erstwhile Vienna friends (Carnap, Neurath, *et al.*) but also met for the first time Bertrand Russell and Alfred Ayer. Ayer, who had not known Russell before this occasion either, and I introduced ourselves to Russell. I said, "In a manner of speaking, we are your [Russell's] intellectual grandsons." In characteristic fashion Russell instantly asked, "And who is your father?" "We have three of them", I replied, "Schlick, Carnap, and Wittgenstein." (Laughter on all sides.) Charles Morris was one of the American participants at the Paris meeting. Tarski (from Warsaw), Jorgenson (from Copenhagen), and Ajdukiewicz (from Poznan) were also present. For Carnap, the Paris meeting marked the beginning of his work in semantics. Although Neurath was skeptical if not outright opposed, Carnap was persuaded of the importance of semantics by Tarski.

A year later, in the summer of 1936, Carnap was invited to participate in the tercentenary celebrations of Harvard University. I joined Carnap there for a few weeks. This was also the first occasion on which we met Ernest Nagel, who was to become a good friend of both of us. Nagel had by then been teaching at Columbia University for several years. He had previously been a student of John Dewey at Columbia and Morris R. Cohen at City College. Cohen was an outstanding thinker and teacher, and perhaps the only prominent American philosopher of science after C. S. Peirce. Cohen's keen critical mind greatly influenced many of his students who were later to become notable philosophers in their own right. His own position was on the whole more rationalistic than empiricist. He opposed and criticized the positivists in many of his publications and lectures. Nevertheless Cohen, more than anyone else, brought Bertrand Russell's thought to the attention of the American scholars. When I came to America Cohen and Abram Cornelius Benjamin were the only really distinct representatives and teachers of the philosophy of the empirical sciences. Of course, there were also the philosophizing scientists, notably the physicist Bridgman and the psychologists E. G. Boring and S. S. Stevens (all three at Harvard) who were concerned with the logical and methodological aspects of their respective disciplines. The physicist Victor F. Lenzen, at the University of California at Berkeley, was another outstanding scholar in the philosophy of physics. The American intellectual climate, then, was not entirely unprepared for the tremendously accelerated upswing of interest in the philosophy of science that came about through the influx of the European positivists and empiricists. In 1939, just at the time of the outbreak of the Second World War, a Unity of Science conference took place at Harvard. This had been organized by Neurath (coming from Holland) and by Charles Morris (University of Chicago). A number of American sympathizers, and also friendly critics, participated in this conference. But the *Journal of Unified Science*, which was originated as a successor of *Erkenntnis*, was doomed to expire after publication of its first two numbers. The war and the dispersion of our groups temporarily impeded our collaborative efforts. Still, there was a sort of revival of the Vienna Circle in 1940. This was at the time when Bertrand Russell gave his William James Lectures (later published as *Inquiry into Meaning and Truth*) at Harvard. Carnap and Tarski were at Harvard as visiting professors. I spent part of my second Rockefeller research fellowship year at Harvard and thus was able to participate in the fascinating regular discussions that took place in the fall of 1940 and included, beside Russell, Philipp Frank, Richard von Mises, W. V. O. Quine, E. G. Boring, S. S. Stevens, P. W. Bridgman, and I. A. Richards, among the more active members. Issues in the foundations of logic and semantics, as well as of the theory of probability, were extensively discussed. Carnap offered a seminar on semantics at that time. Although he and Tarski agreed on the basic principles of the (semantical) metalinguistic approach, there remained some differences (also with Quine) on the nature of logical truth. Tarski and Quine refused to draw a sharp line of distinction between analytic and synthetic propositions, and hence between logical and empirical truth. Carnap, except for some recent modifications, upheld the sharp distinction that was one of the main tenets (later called 'dogmas' by Quine) of Logical Empiricism. Russell also was critical in other respects (especially in regard to the principles of inductive inference) of the original Vienna doctrine.

The closest allies our movement acquired in the United States were undoubtedly the operationalists, the pragmatists, and the behaviorists. The leader among the physicists was P. W. Bridgman. His tough-minded outlook, his generosity, and to some extent even his (as I think, misguided) subjectivism made him sympathetic to our ways of thinking. Bridgman was famous for his work in experimental physics; he was awarded the Nobel Prize for his investigations of the effects of very high pressures on the properties of metals. But he was also highly competent in theoretical physics. Unassumingly, he always protested against being considered a philosopher, and he did not like the label of 'operationalist.' All he claimed was that in order to understand the concept of physics, the operational approach proves most useful. By that he meant a reflection upon, and analysis of, the procedures employed in the use of concepts in the various contexts of measurement, experiment, and theory. This is very close to one aspect of Carnap's and Wittgenstein's views. 'Don't ask me for the "meaning" of a concept, ask me about the *rules* according to which the concept is used.' Although this is not an exact quotation, it reflects the attitude shared by all three (otherwise quite different) thinkers. Paraphrasing some of Bridgman's own formulations, he advised not to 'fall into mystical bewilderment' about the nature of space, time, matter, or energy. Let us rather see according to what rules of 'physical' (observational, mensurational, experimental) and 'mental' or 'paper and pencil' (logico-mathematical) operations we use the symbols (representing concepts) in the actual procedures of the sciences. Operationalism in this broad sense would indeed cover the entire range of cognitive 'meaning.' Bridgman himself was not a formalist, and I suppose he felt that Carnap and other logicians were doing a rather farfetched and artificial job of all-too-exact reconstruction. Bridgman, who was philosophically not too well informed, had perhaps not even known to what extent his ideas had been adumbrated by C. S. Peirce, the great American philosopher, about fifty years before him. Indeed, reading Peirce's famous essay on 'How to Make Our Ideas Clear' (1878!) one is impressed with how close Peirce came to anticipating the basic positivistic and operationalist outlook. To be sure, empiricists view Peirce as a 'split personality.' We were prone to disregard his profound metaphysical essays, but I don't think we ever managed to understand them.

The operational orientation soon became prominent in the methodological work, or pronouncements, of several distinguished American psychologists. E. G. Boring, S. S. Stevens, E. C. Tolman, C. L. Hull, and B. F. Skinner, anxious to make of psychology an 'honest natural science' of the behavior of organisms, endeavored to provide operational definitions of the basic concepts of that discipline. This harmonized well with John Dewey's outlook and that of his disciples in their instrumentalistic view of science. According to that philosophy, concepts, hypotheses, theories are *tools* in the organization of empirical knowledge.

The quick response of these physicists and psychologists was, I believe, responsible for the friendly reception accorded to us, and our remarkable success in the early years of our American existence. We soon struck a responsive chord too with some of the American sociologists, notably G. A. Lundberg. Other scientists with whom we had a large measure of agreement were the physicists Victor Lenzen, Henry Margenau, and R. B. Lindsay.

Philipp Frank established very cordial relationships not only with Bridgman but also with a number of scholars outside the field of physics. His genial temperament and his interest in the socio-cultural-historical settings of the scientific enterprise attracted historians, linguists, social scientists, and philosophers. Similarly, Neurath – who had found a haven (after his escape in a small boat from Holland and his temporary internment on the Isle of Man) in England – visited the United States several times, establishing relations between our movement and leading representatives of the social sciences.

It was my good fortune to teach summer session courses in the University of California at Berkeley in 1946 and 1953 and also to spend a part of a sabbatical leave in Berkeley in 1948. There I had close contacts with the psychologists E. C. Tolman and Egon Brunswik and Brunswik's wife, Else Frenkel-Brunswik. The Brunswiks already were friends of mine (and of my wife) during our Vienna years. They were students of the great psychologists Karl Bühler and Charlotte Bühler; but they were also students of Schlick's. It was Tolman, truly a 'prince of a man,' who facilitated the immigration of the Brunswiks and who appointed Egon Brunswik to a position in the University of California Psychology Department. Brunswik and I organized a Unity of Science meeting at Berkeley in the summer of 1953. I also had very pleasant relationships with several of the California philosophers, especially with W. R. Dennes, Paul Marhenke, and David Rynin. These three had known Schlick during his sojourn in Berkeley in 1931. Later Stephen C. Pepper and

I came to be good friends. His early (1936) sharp criticism of my first essay on the mind-body problem, though painful at the time, had stimulated my re-thinking of that baffling and perennial problem. When I presented (in 1954) a drastically revised paper on the same issues at the Meeting of the Pacific Division of the American Philosophical Association at Seattle, Pepper happened to be in the audience and expressed his enthusiastic agreement with my new outlook.

Most of the Continental scholars participated in various conferences of the American Philosophical Association, the American Physical Society, the American Psychological Association, and others. In the early thirties I defended Logical Positivism in discussions with such noted American metaphysicians as Charles Hartshorne, David Swenson, and Paul Weiss; and with the critical realist Roy W. Sellars. Later on, the entire atmosphere of the debate became different. Great changes had taken place in our own outlook, and, of course, American philosophy, too, underwent tremendous transformations, expansion, and diversification. Some of these developments can be gleaned from the contents of that remarkable book edited by P. A. Schilpp, *The Philosophy of Rudolf Carnap*. There twenty-six philosophers, many of them American, are represented by their critical essays. Carnap's replies, but also his intellectual autobiography in the first part of that book, bear testimony to the changes that have taken place in the thought of one of the leaders in our movement.

My own experiences at the University of Iowa, and later at the University of Minnesota, reflect the rapidly growing influence of our scientifically oriented outlook in philosophy. The Iowa philosophy department had only three members when I arrived there. And while a general course on 'Philosophy and Science' had been offered there, mine was the first course in the 'Philosophy of Science.' The situation was similar at Minnesota in 1941. On my suggestion the young and brilliant philosopher Wilfrid Sellars (son of Roy W. Sellars) was called to Iowa in 1938, and he joined the Minnesota department in 1946. Sellars was a most helpful collaborator. In 1949 we published the first anthology in analytic philosophy [Feigl and Sellars, 1949b]. In this volume we included a good many of the now 'classical' essays of the Logical Positivists, as well as those of their best critics. In the same year we began issuing, with May Brodbeck, John Hospers, and Paul Meehl as co-editors, the journal *Philosophical Studies*. This was, and still is, exclusively devoted to topics in analytic philosophy, and is thus the American counterpart of the British journal *Analysis*.

For a few years in the late forties and early fifties, Sellars and I, together with May Brodbeck, John Hospers, Paul Meehl, and D. B. Terrell, made up a discussion group in which occasionally visitors from other universities would participate. Gradually we came to think about organizing a more official center for research in the philosophy of science. Encouraged by the generous financial support of the Louis W. and Maud Hill Family Foundation in St. Paul, the Minnesota Center for Philosophy of Science was established in 1953. During the first few years the local staff members were Paul E. Meehl (an outstanding psychologist and philosopher), Wilfrid Sellars (who later left for Yale University and is now at the University of Pittsburgh), and Michael Scriven (originally from Australia and now at Berkeley). In the fourteen years of its activities, the Center has enjoyed visits of various durations by many outstanding American, European, and Australian and New Zealand scholars. Our major publications (thus far three volumes of *Minnesota Studies in the Philosophy of Science* and one volume entitled *Current Issues in the Philosophy of Science*) have aroused considerable interest. Several of the younger generation philosophers of science have been our visitors, among whom have been Scriven, Adolf Grünbaum (Pittsburgh), Hilary Putnam (Harvard), N. R. Hanson (Yale), Wesley Salmon (Indiana), Karl R. Popper (London), Paul Feyerabend (Berkeley), Bruce Aune (University of Massachusetts), Henryk Mehlberg (Chicago), George Schlesinger (Australia, now North Carolina), and Arthur Pap (Yale). For shorter visits we had C. D. Broad, Gilbert Ryle, Antony Flew, Peter Strawson, Gavin Alexander, and a number of scholars from various other countries.

Partly stimulated by the Minnesota Center's success, other centers or departments were established by men who had been our visitors, most notably at Indiana University and the University of Pittsburgh. Parallel but largely independent developments occurred in other places – the Boston Colloquium (with Robert S. Cohen and Marx Wartofsky leading), the Stanford University group (led by Patrick Suppes), and for a few years, the Delaware Seminar.

The Minnesota Center was established as a *research* department in the College of Liberal Arts of the University of Minnesota. Its staff members, in recent years primarily Grover Maxwell, Paul Meehl, and I, have continued to be active in the regular teaching departments of the University. A good many graduate students of philosophy, physics, and psychology, and a few from the biological and social sciences, have been regularly admitted to most of our Center colloquia. In view of the great demand for philosophers of the empirical sciences, this sort of training appears most helpful. There is hardly any university or college in the United States today in which philosophy of science is not represented. In fact, currently the demand for competent philosophers of science still exceeds the supply by far. The experience of my friends (Carnap, Reichenbach, Hempel, *et al.*) through the years has shown that many gifted young students prefer to become pure logicians concerned with relatively technical and advanced topics. On the other hand, anyone with strong interests and abilities in the empirical sciences is more likely to become a scientist, and thus pursue a career in pure science or in the various technologies. Nevertheless, there are now a number of outstanding philosophers of science in the younger generation. Some received their major inspiration from such scholars and teachers as Carnap, Reichenbach, Hempel, Nagel, Bergmann, Margenau, and Frank, or from the groups and centers mentioned above and their members.

Practically every one of the erstwhile Continental Logical Empiricists has been invited to visiting appointments in various universities inside and outside of the United States. Reichenbach gave a series of lectures at the Institut Henri Poincaré in Paris. Carnap and Hempel have been at Harvard, Carnap also at Princeton, Bergmann in Stockholm. I have been at Berkeley twice, once at Columbia University, and in 1964–1965 in Austria and Australia. It was most gratifying to note that after a long spell of neglect (if not of suppression) there was again a strong interest in the philosophy of science in Vienna. There, in the Institute for Advanced Study and Scientific Research (established by the Ford Foundation in 1963), I enjoyed a reunion with Karl Popper, Hilary Putnam, and Karl Menger. In the summer of 1964, Carnap, F. A. von Hayek, P. K. Feyerabend, and I joined forces in the European Forum (Austrian College) in the lovely Tyrolean village of Alpbach, where we were offering a seminar in the philosophy of science to an excellent group of advanced students, most of whom – like those in the Vienna Institute – had already obtained their Ph.D. degrees in Austrian or German universities.

I was also especially pleased with the interest I encountered in Australia. American philosophy of science was well known there, no doubt through the influence of several very active and brilliant scholars: J. J. C. Smart (Adelaide), John Passmore (Canberra), Brian Ellis (Melbourne), D. M. Armstrong, and David Stove (Sydney). The Australian psychologist William O'Neil, who had often been a visitor to American universities, is one of the increasing number of scholars in his field who is also intensely interested and competent in the philosophy and methodology of science. Under his guidance several younger scholars did worthwhile work closely related to American Logical Empiricism in the foundations of psychology.

On my trip from Austria to Australia, I encountered strong interest in and familiarity with American philosophy of science in general and Logical Empiricism in particular: in Athens, in India, in Hong Kong.

It is gratifying to know that the Continental Logical Empiricists contributed so greatly to the worldwide developments in the philosophy of science and more generally in a scientifically oriented philosophy.

V. SOME PERSONAL IMPRESSIONS AND REMINISCENCES

Among the many scientists who welcomed and supported the 'imported' outlook of the Logical Empiricists there is in my personal experience one man to whom much is owed: Carl E. Seashore. When I came to Iowa in 1931 he was the head of the Psychology Department. Having himself come from Sweden, he was a pioneer in American experimental psychology. At the time he was also Dean of the Graduate School of the University of Iowa. An energetic, forthright man, he was largely responsible for the important developments in experimental, theoretical, and child psychology. He was greatly loved and admired by faculty and students. But he was also somewhat feared for cracking down on questionable projects in the University. Thus it was with some trepidation that I asked him (I believe it was in 1932) whether he would approve of my offering a seminar in philosophical problems of psychology. I was afraid he would reply with his notorious and crushing "NO!", but to my pleasant surprise he slapped my shoulder and shouted "Good idea, Feigl!" Even since then I have enjoyed the support and encouragement that came to me from many outstanding American psychologists, notably E. G. Boring, S. S. Stevens, C. C. Pratt, B. F. Skinner, C. L. Hull, K. W. Spence, E. C. Tolman, and from the Continental immigrants Kurt Lewin, Wolfgang Köhler, and Egon Brunswik. In 1958 I was the invited guest speaker of the American Psychological Association at its national meeting in Washington, D. C., where I spoke to about 3500 listeners on 'Philosophical Embarrassments of Psychology.' One of my friends remarked: "Aristotle never had it so good!"

On the whole I have found greater interest in our work on the part of psychologists and social scientists than among physicists, chemists, or biologists. But I have been invited for lectures or contributions to publications quite frequently by their societies, departments, or individual scholars. My immigrant confrères had similar experiences. Perhaps because of their great interest in methodology, the psychologists and sociologists are, by and large, more appreciative of philosophy of science than are the physicists. Some thirty years ago the great English astronomer and cosmologist Sir Arthur

Eddington said that the physicists put up a big sign, "Reconstruction going on here; philosophers please keep out." This was understandable in view of the almost permanent revolution in theoretical physics, and because of the small number of philosophers properly trained and competent in that highly intricate domain. But the situation has changed greatly since the thirties. There are now a considerable number of well-equipped philosophers of science specializing in the logic and methodology of physcis. In addition to the older generation of American physicist-philosophers such as Henry Margenau, Victor Lenzen, and R. B. Lindsay, a younger group has come to the fore. Among them are Hilary Putnam and Wesley Salmon (both formerly students of Reichenbach), Adolf Grünbaum (influenced by Reichenbach's work), N. R. Hanson (a brilliant American philosopher-physicist who had spent many years in England and tragically perished in an airplane accident in 1967), Henryk Mehlberg, Grover Maxwell, Richard Schlegel, and P. K. Feyerabend (originally from Vienna where he was a student of Victor Kraft's but later became closely associated with K. R. Popper).

I met Feyerabend on my first visit to Vienna after the war (my last previous visit was in 1935). This was in the summer of 1954 when Arthur Pap was a visiting professor at the University of Vienna. Feyerabend had been working as an assistant to Pap. Immediately, during my first conversation with Feyerabend, I recognized his competence and brilliance. He is, perhaps, the most unorthodox philosopher of science I have ever known. We have often discussed our differences publicly. Although the audiences usually sided with my more conservative views, it may well be that Feyerabend is right, and I am wrong.

Arthur Pap and Paul Feyerabend, concurring with Professors Victor Kraft and Bela Juhos, the only two members of the Circle who had remained in Vienna, told me that the new spirit in the Philosophy Department of the University was quite hostile toward anything that even remotely reflected the ideas of the Logical Positivists. Peculiar alliances of Hegelianism and Existentialism, as well as Catholic Philosophy, were predominant in 1954. I found this still confirmed ten years later when I revisited Vienna (in 1964–1965) in connection with my appointment at the Institute for Advanced Study and Research. This Institute is administratively completely independent of the University. Its teaching staff, consisting mostly of visiting scholars, is truly international and thus includes a number of Americans.

As I reflect on my motivations in connection with the Minnesota Center for Philosophy of Science, it seems fairly clear to me that my formative experiences in the Vienna Circle, and again at Harvard in 1940, have encouraged me to endeavor collaborative teamwork in philosophical research. In this regard, I have found the intellectual atmosphere in American philosophy and science even more favorable than that of the Continent in the twenties and early thirties. On the whole, I have found that American scholars are remarkably open-minded, willing to accept criticism as much as to proffer it. Of course, there is occasionally a rare 'prima donna' who has to be handled with care, and to be exploited in 'cafeteria style' – you take from them whatever valuable ideas they offer, but you don't have to accept whatever seems dogmatically rigid or narrow; nor should you expect to convert them to your own point of view.

Among the American philosophers who have left with me a lasting and deep impression are – in the early years – John Dewey, Ralph Barton Perry, C. I. Lewis, C. H. Langford, R. W. Sellars, and Morris Cohen, John Dewey came to Harvard as a visiting professor in 1931, and he offered a rather tedious seminar in the philosophy of logic. While one could not help being impressed with Dewey's personality, he was clumsy and diffuse as a lecturer. I saw him again in 1940 in New York, when (at the New School for Social Research) I was an invited participant, together with Paul Tillich and Herbert Schneider, on issues in the theory of values. Tillich was almost unintelligible to me. During the public discussion John Dewey, with extended threatening forefinger, asked whether I had absolute values in mind when I spoke of 'terminal values.' I reassured him that I meant nothing of the sort.

Morris Cohen and I had a *rencontre* in 1938 at one of the meetings of the American Philosophical Association. We were both, together with Evander McGilvary, panel members on a symposium on Universals. Cohen, always brilliant and challenging, took sharp issue with me – and we even continued our conversation at breakfast on the following day. I remember the 'squelch perfect' he sometimes used in reply to students (or also his opponents): "As a matter of fact you are wrong; but supposing you were right, what of it?"

C. H. Langford and I for several years spent the first evening of association meetings together. It was on our first encounter (in 1932) that he challenged me with the problem of the logical analysis of contrary-to-fact conditionals, which we Viennese had neglected. Langford was certainly one of the keenest and personally most attractive and congenial American logicians and philosophers.

During my early years (in the forties) and later at Minnesota, I was very fortunate in having a number of interesting colleagues on the faculty, in as well as outside of the Philosophy Department. The brilliant psychologist B. F. Skinner and I became close friends. We disagreed sharply on philosophical issues of psychology, but this never disturbed our personal relations. In a small group of psychologists and philosophers, Skinner read to us, chapter by chapter, his utopian novel *Walden Two*.

During my first year at Minnesota I became acquainted with the man who probably was my best student, Paul E. Meehl. Although about twenty years younger than I, Meehl became one of my most cherished friends. He has been and still is, one of the two or three most helpful and contributive members of the Minnesota Center for Philosophy of Science. He is best known as a highly versatile psychologist, but there is no question in my mind that he is also one of the most penetrating philosophers and methodologists of science in our time. He has applied his tremendous intellectual powers not only to the psychology of learning, to clinical psychology and psychiatry, but also to the analysis of the basic concepts and methods of psychology. For several years (in the fifties) we gave seminars together on the philosophical problems of psychology, and were warmly rewarded by the response.

In 1962–1963 Professor Karl R. Popper was a visitor of the Center for a semester. He, Paul Meehl, Grover Maxwell, and I had many extremely fruitful private discussions, in addition to a regularly scheduled seminar on the philosophy of physics. In this memorable seminar we had five professors: Popper, Maxwell, Feyerabend, Edward L. Hill (University of Minnesota theoretical physicist), and myself, debating with each other before an audience of graduate students and some faculty members. One afternoon we were joined by the theoretical physicist Alfred Landé (then at Ohio State University, now retired). We discussed especially the philosophical issues of quantum mechanics and some of the fundamental problems of theory construction in physics. This was surely one of the occasions when I felt we had attained the high quality of discussion characteristic of the Vienna Circle or its equivalent at Harvard in 1940.

One of my most thrilling experiences was my visit with Albert Einstein at Princeton in 1954 (one year before his death). I had once previously visited Einstein – in Berlin in 1923, when my teacher Schlick had sent me to the great man. The visit at Princeton was made possible through the kind assistance of Dr. Paul Oppenheim. Oppenheim had told me that Einstein loathed 'journalistic' interviews and also merely 'socially pleasant' conversations. Einstein liked visitors who would come with questions for him. Naturally, I arrived with a long list of questions. Einstein, greeting me in the entrance hall of his home on Mercer Street, hardly indulging in any social amenities (though we had tea later), after a few minutes of ordinary conversation, asked, "And what shall we talk about?" At that point I pulled out my long list and began by asking him about the important changes that had taken place in his philosophical outlook throughout his long life. He admitted that in his early years he had been strongly influenced by the positivism of Hume and Mach, but that he had changed his views gradually toward a scientific realism. Of course, I had known about this from his publications. I wanted to hear some of the detailed reasons that motivated this transformation. I was immensely pleased to learn that there was a large measure of agreement in this respect with the development of Logical Positivism toward a more liberal Logical Empiricism. Einstein was in a very happy mood, and often laughed loudly when he said something slightly shocking or surprising. I remember that upon one of my questions regarding the mind-body problem he said (I translate from his forceful and somewhat coarse German phrase): "If it were not for the 'internal illumination' (that is, by consciousness) of the physical universe, the world would be no more than a mere pile of dirt!" Einstein also discussed with me his reasons for considering quantum mechanics (despite its admitted fruitfulness) to be incomplete. He was still hoping for some sort of deterministic and unified field theory.

In his autobiography Carnap relates some of his fascinating conversations with Einstein and his impressions of that genius. I think to many of us Europeans, the very presence of that reversed scientist-philosopher in this country was a source of inspiration and encouragement.

For me, personally, it was also and primarily my frequent visits with Carnap (in Chicago, Princeton, Los Angeles, and for a while at his summer home in the hills outside Santa Fe, New Mexico) that I always found instructive, encouraging, and helpful. After Schlick's death, Carnap and Reichenbach were in my estimation the leading exponents of our philosophical approach. Thus I had the privilege of learning first hand of Carnap's developing ideas, be it in the field of semantics or in the field of inductive logic and the foundations of probability. Carnap also was a willing listener and an acute critic when I submitted to him my own ideas on scientific realism or on the mind-body problem. As I reflect on my more than forty years of frequent contacts with Carnap, I feel he is perhaps the most ametaphysical philosopher I have ever known. The metaphysical neutrality, already explicitly formulated in his early work (1928), pertains to all important traditional philosophical issues. For Carnap, such disputes as those between Nominalism and Platonism, and Realism and Phenomenalism, are not substantive issues of an ontological sort. In his reconstruction he considers them as questions regarding the linguistic frame of knowledge (in more traditional terminology, the basic 'categorial' conceptual frame) whose acceptance or rejection is a pragmatic matter. By contrast, questions within the chosen frame are genuinely cognitive, and are susceptible of solution by deductive, inductive, or hypothetico-deductive methods.

All this was evident in our most recent meetings – a small-scale conference I had arranged at the University of California in Los Angeles in March 1966, and in a different context at the University of Hawaii, where both Carnap and I were invited for a week's seminars and colloquia in March 1967.

Both my sojourns at Hawaii – for a semester in 1958 (as Carnegie visiting professor) and the much briefer visit in 1967 – were among the most pleasurable experiences of my life. The peerless hospitality of the friendly and excellent faculty of the University of Hawaii, the marvelous climate and scenery, make this indeed the 'Paradise of the Pacific.' I shall never forget the amicable Professor Charles A. Moore who, to my great sorrow, passed away in April 1967. When I arrived in my office at the University of Hawaii's Philosophy Department in February 1958, Moore appeared with a big stack of books and put them down on my desk. He said: "These are the best books on Oriental Philosophy. You will want to read them." Somewhat astonished, I replied: "I thought you wanted me to bring modern Western Philosophy of Science to Hawaii!" "Of course", he replied, "that's your obligation, but here is your opportunity to learn something different!" I shall always be grateful, and remember Professor Moore fondly, for I really achieved at least a smattering of information about (especially) Indian philosophies and religions at that time.

Many members of the Vienna and Berlin groups have been honored with distinctive positions in various scholarly organizations of America. Reichenbach, Hempel, Bergmann, and I were elected (at various times) to the presidency of the American Philosophical Association (Pacific, Eastern, and Western Divisions, respectively). I was elected a vice-president of the American Association for the Advancement of Science and chairman of Section L (for History and Philosophy of Science) in 1959. Carnap received an honorary doctorate from Harvard University in 1936 and from the University of Michigan in 1966. Most of us have been visiting lecturers at several United States universities at various times. All of us are deeply grateful for the splendid opportunities that have been ours in the New World, for the magnanimous friends, and many brilliant students it was our good fortune to find here.

NOTES

* [Publications mentioned in this article are listed in the bibliography to this volume – Ed.]

1 At the University of Vienna there was no distinction between undergraduate and graduate students. Waismann, about six years older than I, was already a lecturer at the People's Institute (*Volkshochschule Wien*). I attended several of his brilliant courses, which were mainly in the foundations of mathematics. I was just then beginning to work on my Ph.D. thesis, 'Chance and Law', concerned with the problems of probability and induction in the natural sciences.

2 Schlick had invited me to publish my doctoral dissertation ('Chance and Law') in this series, but my thesis was completed in 1926, the very year of the revolution that quantum mechanics engendered in the determinism-indeterminism issue. And while in my work I had open-mindedly discussed that issue at length, I had not, and could not have foreseen, let alone digested, the specific form that indeterminism was to take in the theories of Heisenberg, Born, Jordan, and Bohr. Thus I decided not to publish my thesis at that time, and except for a few articles on probability, induction, and causality, this work in which I had invested an enormous effort will remain unread in the library of the University of Vienna.

3 Schilpp's general scheme, as realized in the Library's many volumes, was to have prominent philosophers (e.g. Dewey, Whitehead, Russell, Einstein, Carnap) write their intellectual autobiographies; to follow these writings with critical essays by about twenty-five writers; and then to conclude with the philosophers' replies.

4 This is partly excusable: Peirce's *Collected Papers* did not begin to appear until the early thirties.

5. SCIENTIFIC METHOD WITHOUT METAPHYSICAL PRESUPPOSITIONS

[1954]

As the title of this article indicates, I contend that there are *no philosophical postulates* of science, i.e., that the scientific method can be explicated and justified *without metaphysical presuppositions* about the order or structure of nature. My positivistic or logical empiricist background, I must admit, may have made me somewhat allergic to the term 'metaphysics.' I realize full well that there are uses of this word that cover quite respectable or at least semi-respectable endeavors. If 'metaphysics' designates the examination and explication of the basic concepts, methods, and assumptions of the quest for knowledge, then, surely, logical empiricists are metaphysicians – only they prefer to label this sort of study 'Logical Analysis', 'Theory of Knowledge', or 'Philosophy of Science.' The term 'metaphysics' is sometimes used also for the highly extrapolative – and in this sense precariously speculative – pursuits of the synthesis of a well-rounded world view. 'Inductive Metaphysics' is perhaps a good label for this type of endeavor. But metaphysics in this sense is continuous with science. It is merely the most venturesome part of scientific theorizing. Modern cosmological theories in physics and astronomy furnish a good example of this sort of speculation. Generally, it seems scientists are much more competent than professional philosophers to judge to what extent, if any, hypotheses of this more sweeping sort are warranted by the empirical evidence on hand.

I shall refrain from discussing other meanings of the badly ambiguous term 'metaphysics,' such as the alleged intuitive or dialectical methods of arriving at ultimate truths concerning the nature of existence.

The two senses in which the term 'metaphysics' covers enterprises that seem objectionable to the logical empiricist are of course (1) transcendent, i.e., in principle untestable assertions, and (2) the belief in factual truths that could be validated *a priori*, i.e., in complete independence of the data of observation.

The history of modern and recent physics provides a long series of devastating object lessons in both respects. From Galileo down to our days of the theory of relativity and of quantum mechanics, physicists have become poignantly aware of the futility of untestable assumptions and of the questionability of erstwhile unquestioned *a priori* postulates or presuppositions.

In an impressive sequence of incisive changes the very postulates of physical theory have undergone profound revisions. The impact of these developments upon philosophy has quite generally tended to diminish, if not to abolish, the traditional prepossessions of Rationalism. Little, if anything, is left of the *verités éternelles* of Leibniz or of the Kantian synthetic *a priori* and other principles of this sort, formerly considered as self-evident, indubitable, or as the pre-conditions for the very possibility of scientific knowledge.

The discovery of the non-Euclidean geometries more than a hundred years ago, and especially their utilization in present-day physical and astronomical theories, shattered the rationalistic and absolutistic conceptions of space and time. The electro-dynamic, quantum, and wave-mechanical theories of matter have transformed the time-honored idea of substance beyond recognition, if they have not indeed made it completely obsolete. In this connection the ancient dogma of continuity, one of Leibniz' basic truths of pure reason, had to give way to the conception of the discontinuous interaction between matter and radiation. Connected with the transition to discontinuity was the even more fundamental critique of the deterministic conception of causality. The principles of sufficient (or of insufficient!) reason, as well as the closely related symmetry principles used in many demonstrations of classical and statistical physical theory have long been exposed as only speciously *a priori*. Ever since the critique by Ernst Mach of Archimedes' proof of the law of the lever, and Richard von Mises' critique of the principle of indifference (e.g., in Keynes' theory of probability) we have recognized that the correspondence of physical with geometrical symmetries is a matter of empirical fact. Indeed, one would wish to know what the rationalists have to say on the geometrical relation of electric and magnetic field vectors – this basic asymmetry in our universe should certainly shatter anyone's faith in *a priori* discernible laws of nature.

But to continue with more recent revolutions in our concept of nature: Neither the idea of particles that remain self-identical and indestructible nor the idea of the strict predictability of micro-processes can be maintained any longer with confidence. A large mass of experimental evidence militates relentlessly against practically each and every one of the alleged truths of pure reason held inviolable for so long by countless philosophers. Adding insult to injury, recent physical theories (introducing advanced potentials) seem to upset completely all classical notions concerning the temporal structure of causal relations, and recent cosmological theories have called into doubt even the dictum *ex nihilo nihil fit* – and blithely assume the possibility of spontaneous generation of matter, along with assorted doctrines of expanding or oscillating finite but unbounded universes.

Even if some of these developments need not be taken too seriously, their total trend and significance is indisputable. The mere fact that some of the allegedly indubitable first principles have been called into doubt, indicates that they cannot be indispensable presuppositions of science. The great conservation principles (of energy and of momentum), for example, were for a while (around 1924) under suspicion of being merely statistical macro-laws – along with so many other such manifestations of the 'law of large numbers.' But the detailed studies of the Compton effect reassured the physicists – at least up to the discovery of energy discrepancies in radio-active β-disintegrations – that these basic laws still hold good in a strictly deterministic fashion, even for micro-processes. On the other hand the neutrino has not thus far been independently identified, and it is conceivable that the original suspicion expressed by Bohr, Kramers, and Slater may yet be justified on different grounds. I shall not tire my readers with the narration of further instances. The *a priori* in physical theory is either of the analytic, purely definitional sort or it is nonexistent. Of course, if one wishes to speak of a *relative* or *pragmatic a priori* – in the sense of C. I. Lewis, Victor Lenzen, Arthur Pap, or in the sense of Wilfrid Sellars – this is another matter. This sort of *a priori* is different from the classical rationalistic one (especially from the Kantian) in that it is ultimately under the jurisdiction of experience. This conception of the *a priori* connotes only universality and necessity, the latter in the sense in which even synthetic propositions can be true by virtue of the meaning of the terms they contain. I shall return to this point later.

The upshot of our discussion thus far is simply this: Any proposed assertion concerning the order and structure of our universe, no matter how fundamental its role or pervasive its scope, must be regarded as tentative and may be held only until further notice; such notice being given by data of experience which may conceivably motivate us to modify, if not to abandon, the assumption at issue. In other words, any assertion regarding nature, if it is to be scientifically meaningful, must in principle be confirmable or disconfirmable.

In a very interesting recent article [1953] A. W. Burks argues for the necessity of unconfirmable presuppositions for the very *confirmation* of any scientific laws (be they causal or statistical) and of any scientific hypothesis or theory. Along lines fairly familiar since J. S. Mill and Jevons, and elaborated in various ways in more recent decades by Edgar Zilsel, W. E. Johnson, C. D. Broad, J. M. Keynes, J. Nicod, Bertrand Russell, and others, he asks what assumptions regarding the uniformity of nature are required in order to account for the probability of inductive generalizations. Burks claims he does not intend to furnish a *justification* of the inductive leap. This, he recognizes [1951] is an insoluble problem precisely because – at least in its traditional setting – any proposed solution involves either a self-contradiction or a *petitio principii*. This has of course been known ever since Hume's incisive critique of causality and induction. If we attempt to transform induction into deduction we require premises whose validity cannot be anything but *inductive*. And if we try to demonstrate the (certain or even only the probable) success of inductive inference on the basis of its success to date, then we assume the very principle we propose to prove.

What, then, is the function of the sort of presuppositions that Burks deems indispensable? He tells us that they are required in order to *explain* induction. This seems to mean that these presuppositions are needed as premises from which the assumed or actual success of the inductive methods can be *deduced*. Unfortunately Burks gives no more than a hint as to precisely what presuppositions he has in mind. He suggests that they are of the type of Keynes' principle of limited variety, or, more generally, that they are assumptions concerning the uniformity of nature.

I fail to see the philosophical importance of any attempt in this direction. If it were the success of human adaptive learning and theorizing behavior that is to be accounted for, I would be the first to admit that this is a genuinely meaningful question – but surely a question of science, not of philosophy. This question can indeed be answered. And the answer is clearly along the lines of the biology and psychology of knowledge. It is the same sort of question that can be raised on a more lowly level in regard to the learning and generalizing behavior of that pet of our psychologists, the white rat. Given the rat's equipment of learning capacities, how complicated a maze will it be able to master, in how many trials, under what conditions of previous training, etc.? While it is a long way from the orientation of rats in a maze to the intellectual adaptations (if I may be forgiven the irreverent comparison) of the Newtons, Maxwells, and Einsteins in their theoretical constructions of the physical universe, the nature of the problem is the same: What type and degree of uniformity must the universe possess in order to be successfully predictable by means of the inductive and hypothetico-deductive procedures of modern science? I think the answer to this question is very obvious. The universe must have precisely the type and degree of uniformity which the successfully confirmed laws and theories ascribe to it (or rather, to some of its aspects).

Burks, however, does not propose this sort of conception of presuppositions or premises for the explanation of induction. He states explicitly and emphatically that the presuppositions of induction are in principle unconfirmable. They result in the choice of a definition of initial probabilities (or of a concept of degree of confirmation) which, once adopted, bestows certainty upon its own presuppositions, and a probability of zero to any alternative presupposition. This trivial and tautological confirmation of the presuppositions of all inductive inferences whose probabilities are determined by the respective choice of a definition of degree of confirmation can indeed not be regarded as a genuine appraisal on the basis of empirical evidence. The three inductive methods which Burks outlines are according to his claim altogether immune to the testimony of experience. Nevertheless Burks indicates that in a world of completely random character one method (Mr. 'Dagger's') would yield results superior to those of the standard or normal procedure of inductive extrapolation (Mr. 'Star's' method; best adapted to a uniform world). The third method, finally (i.e., Mr. 'Diamond's') would serve most effectively in a universe of such diabolically perverse structure that what for the normal method is the most probable, would in that universe be the least probable predictions or hypotheses (and vice versa).

Without wishing to go into technical details of Burks' comparison of the three methods, I should like to raise a more fundamental question: What is the point of the presuppositional analysis if only an omniscient being could decide which presupposition actually applies to this universe of ours in which limited human beings grope for reliable knowledge? If finite, limited empirical evidence has no relevance for the preference of one presupposition as against another and if limited empirical evidence is the only sort of evidence that human beings will ever be able to marshal, then what, in Burks' opinion, justifies us in clinging to the standard method and in considering the other two as 'perverse'?

Of the many traditionally proposed solutions of the problem of induction, which one is relevant or helpful in deciding this issue? The psychologistic, or 'animal faith' doctrine of Hume and Santayana merely restates what the human animal as a matter of stark propensities and of habituation believes so strongly anyway. It furnishes no objective reason for a preference. The Kantian theory of knowledge in its anthropological version yields an equally psychologistic, but implausible doctrine of forms inevitably impressed upon the contents of raw experience – and thus depends for its plausibility and effectiveness on the assumption of the constancy of the forms of pure reason. But this is clearly an inductive assumption and thus ineffective in resolving our problem. The presuppositional interpretation of Kant's critique is of course the one to which Burks (along with many other modern thinkers) appeals but it does not enable him to ground the preference for the standard inductive method. Probabilistic solutions of the problem of induction, such as that of D. C. Williams are equally unavailable because they too endorse the standard method by presupposing (D. C. Williams' protestations to the contrary notwithstanding) uniformity. The common-sense school of philosophy (mostly the disciples of Moore and Wittgenstein) reminds us that the very meaning of 'rationality' and 'reasonableness' includes besides formal consistency especially the normal method of profiting from experience, i.e., standard induction. But since no reasons are given why this standard method should be dignified with the honorific label 'reasonable', we are, despite the helpful reminder, left in the dark as far as our problem is concerned.

It should scarcely be necessary to explode the illusion that the presupposition of uniformity is an ultimate *postulate*. 'Postulate' in modern science, especially in mathematics, means an assumption which serves as a premise for deduction. In contradistinction to the term 'axiom', it does not carry the traditional connotation of self-evident truth or indubitability. In keeping with its etymology, a postulate is a demand, a requirement. But the mere fact that we require or demand uniformity does not make it sure or even only likely that nature will be good enough to conform with such demands.

I shall forego the discussion of intuitionistic, metaphysical, and other clearly ineffective approaches to the problem of induction. Instead I shall more fruitfully turn to a brief discussion and comparison of two outstanding contributions by logical empiricists, i.e., those of Carnap and Reichenbach. Since it would take much more space than is available here, I shall have to assume some familiarity with the basic features of these two theories. It is clear that the normal method of induction (or at least something very close to it) is here absorbed in Carnap's *definition* of a logical concept of probability. Like J. M. Keynes' concept of probability, Carnap's degree of confirmation is relational in that it determines the degree to which some given evidence supports an inductive conclusion. Reichenbach, though an avowed proponent of the frequency conception of probability, defines a concept of *weight* for singular predictive inferences as well as for the probability of hypotheses. Since the supporting evidence in Carnap's conception consists in observed frequency ratios, and since Reichenbach's concept of weight is similarly defined in terms of the relative frequency of a certain type of event in a properly chosen reference class, these two theories of probability are perhaps not as irreconcilably different as might appear at first glance. It is true, Carnap's concept of null-confirmation (i.e., a degree of probability on the basis of no evidence whatever) allows for a qualified use of a principle of indifference and thus seems objectionable to Reichenbach, the staunch empiricist. But Carnap's '*a priori*' probability (if I may label it in this perhaps somewhat misleading manner) is rendered fairly innocuous by the qualification that in his theory the reliability of assertions of degrees of confirmation on scant evidence is extremely low. This corresponds to the absence of appraised posits in the so-called 'primitive induction' according to Reichenbach's theory. And this means that numerically specified weights for predictions and hypotheses can be justified only after certain basic frequency ratios have been posited without assignable weight. Furthermore, Reichenbach, when faced with the notorious problem of how to confirm or disconfirm statements regarding limits of relative frequency has to resort either to a problematic finitization, or to the utilization of a probability-logic which bears a vague though suggestive resemblance to a theory of degree of confirmation. Add to this that Carnap's degree of confirmation according to his own theory is equivalent to an estimate of relative frequency; and that the two rival theories are, of course, isomorphic to the extent that they both yield the customary axioms and theorems of the calculus of probabilities – and one begins to wonder whether Carnap and Reichenbach are as far apart from one another as is generally believed.

Perhaps an even more convincing, and for our present purposes extremely important, common feature of the two theories is the justification of induction. According to Carnap there is a large class, in fact a continuum, of inductive rules (or what is tantamount: of definitions of degree of confirmation) of which Reichenbach's rule is an element, and which all share the following significant feature: If the world has some degree of order at all, predictions made according to any one of the inductive rules will in the long run (strictly speaking, in the limit) not only converge with the others, but can also be shown (deductively!) to be the only type of predictions that utilize evidence methodically and are capable of anticipating that order of nature. In some of my early papers [Feigl, 1930a, 1934a] I had been groping for this sort of solution of the problem of induction and I think I came fairly close to a tenable formulation in the paper of 1934. But with genuine appreciation I credit the late Hans Reichenbach (see especially [1949a]) with the independent discovery and the more elaborate presentation of this solution.

Let me now show in some detail that this solution implies that there are no ultimate factual presuppositions of science. The search for invariant first principles may be humanly understandable, but in the light of the history of science (as I briefly sketched in the beginning of this paper) it must be considered as misguided. The inductive and hypothetico-deductive methods of science do not logically presuppose any premises concerning the uniformity of nature. The invariant character of the basic rules of scientific method hinges exclusively upon the *aims* of scientific inquiry. In order to take the air of dogmatism out of my formulation I shall state it in the form of a doubly conditional proposition: If it is the goal of scientific investigation to discover lawful relationships and thereby to render the observed phenomena maximally predictable, *then* the methods of (normal) inductive generalization and of theory construction will attain this goal, *provided* it is attainable at all, i.e., *if* nature is at all uniform in some sense, some aspect, and to some degree.

Hume has shown once and for all that no matter what type or degree of order our observations may have confirmed thus far, it is logically conceivable that this type or degree of order may not prevail in the as yet unexplored domains of nature. These domains may be the future, or distant regions of space, or untested ranges of the magnitudes that figure in the functionally formulated quantitative laws of nature. A survey of the conceivable types and degrees of order (uniformity, dependence, independence) is a matter for purely formal analysis in modern logic and mathematics. A study of the criteria of confirmation for the empirical actuality of any one of these types and degrees of order is a task for philosophical analysis and for the theory of probability. Despite the present controversies in the theory of probability, it seems safe to say that there is one basic rule of procedure for the extrapolation from the known to the inferred portions of the universe: 'Generalize on the basis of as broad a background of experience as can be secured, and in accordance with the principles of factual simplicity.' If there is an order of nature, not too complicated, or too deeply hidden, to be discoverable by limited human beings, then this rule of procedure will yield successful predictions. This last statement is of course a tautology. But I think it is illuminating nevertheless. It shows that, while it is impossible to *validate* any *over-all* assumptions concerning the uniformity of nature, the *procedure* of induction which posits *specific* uniformities can be pragmatically justified or '*vindicated*' [Feigl, 1950a]. If the degree of uniformity in our world were much lower than it actually is, the rule of induction would still be a helpful tool in the anticipation of the as-yet-unknown. Even in a universe so completely chaotic that all specific generalizations were invariably fruitless, there would yet be one very modest generalization which would hold: namely, the expectation that the chaos will continue. And is this not also an induction?

The principle of induction when interpreted as a rule of procedure, and not as a postulate regarding the order of nature, is of course not subject to confirmation or disconfirmation. Its adoption depends merely on our purpose: to discover and generalize, tentatively and self-correctively, whatever type or degree of uniformity the accumulating observational evidence suggests. It was perhaps the dim recognition of the implicit functioning of this most basic rule of all empirical knowledge that led so many philosophers to mistake it for a genuinely synthetic truth concerning the universe that could be known on the basis of pure reason.

Waiving important technical questions which are still highly controversial, I venture to state rather succinctly the following tentative conclusions:

1. No matter how strong or how weak the uniformities are in our world, the procedure of normal induction is the only methodical procedure of which we can prove *deductively* that it *can* disclose such uniformities.

2. Our world, as we have come to appraise it through the guidance of the principle of induction, seems to contain a vast amount of independencies (nonuniformity, disorder) in addition to a limited amount of dependencies (uniformity, order of various types, such as the causal and the statistical). This *inductively* established very general assumption serves indeed as a presupposition – or if you will as a postulate – for all types of more special causal and statistical research. But this postulate is 'philosophical' only in the somewhat regrettable sense that it is terribly vague and abstract. It is certainly not metaphysical in the sense of being either untestable or *a priori* (i.e., it is not independent of the jurisdiction of experience).

3. While differing in technical points, as well as in the specific values of probabilities based on relatively small amounts of evidence, such inductive methods as R. A. Fisher's (of maximum likelihood), Reichenbach's (rule of induction), and Carnap's (definition of degree of confirmation C*) yield in the long run equal results and are perhaps merely different versions of one and the same basic idea.

4. This basic idea (again waiving controversial issues) may be formulated as follows: If we wish for a method of generalization or of individual predictive inference that utilizes evidence and yields (at least in the limit) unique results, then the 'normal' method (or methods) of induction are quite distinct from any 'perverse' ones. The 'perverse' methods are either insensitive to the testimony of accumulating evidence – and therefore *not* self-corrective – or else they lack the *uniqueness* that is characteristic of the methods of *simplest* generalization or of *maximum* likelihood.

5. Confronted with the specific problem of determining the most adequate value for the limiting frequency of some statistical phenomenon (as, for

example, in radioactive disintegration, quantum-transitions of various sorts, or in Mendelian inheritance, etc.), we may choose either straightforward generalization of the statistical ratios obtaining for specifiable aspects under operationally identifiable conditions; or else construct a theory, an essential part of which must be assumptions regarding statistical distributions, in such a fashion that the observed data will have a maximum likelihood (according to a non-perverse definition of likelihood) in the light of the assumed theoretical model. This suggests that the 'star' method or something very close to it could be given an objectively justifiable preference.

6. In testing the correctness of a specific estimate of a limit of relative frequency it would indeed be 'perverse' to assume that the available samples are not representative. They may of course not be representative for all that. We may at any time, or even all the time, be the victims of what in the light of further evidence would appear as a 'terribly improbable' run of bad luck. But to assume this to start with not only makes our estimates non-unique (it opens the floodgates to limitless possible hypotheses) but would amount to a complete defeatism. The method of science is that of a fallibilistic optimism. ('If at first you don't succeed, try, try again!') We must work with what we have and cautiously utilize every bit of uniformity that a broad domain of evidence suggests for generalization. This I accept as the most valuable core of the contribution to epistemology and the logic of science contained in the pragmatist philosophies of C. S. Peirce, John Dewey, and C. I. Lewis.

7. There is one very important issue which I wish to propose for further discussion (I have not yet attained a stable decision on it myself): Are scientific *laws*, and *theories with lawlike assumptions*, subject to appraisal in terms of any concept of inductive probability? It should be noted that unlimited generalizations from finite evidence receive invariably a degree of confirmation of *zero*, according to Carnap's theory of induction. Laws can therefore be appraised only on the basis of their instance-confirmation. This suggests that any assignment of inductive probabilities really presupposes a set of assumptions of lawfulness, certainly of the statistical type and, possibly, in addition, some of the deterministic type. These assumptions would then be *posits* in Reichenbach's sense, i.e., tentative assumptions to which we cannot assign any numerical or even comparative probabilities. This is in keeping with the prevalent view of theoretical physicists according to which the basic premises of scientific theories may be accepted or rejected as they are more or less fruitful, convenient, simple, or expedient.

Probabilities can be determined only relative to a chosen frame of those basic theoretical assumptions. Far from being metaphysical (again in the sense of being exempt from the jurisdiction of experience) these basic posits are held, until further notice, that is to say as long as they provide the frame from which specific testable conclusions can be derived and successfully confirmed by the data of observation. It seems indeed questionable as to whether theories which introduce new magnitudes, as, for example, the electro-magnetic field theory of Faraday and Maxwell, could be appraised as more or less probable in the light of experimental evidence. The introduction of new magnitudes semantically considered consists in an essential enrichment of the language of science; and it seems that probability estimates presuppose such a semantical frame but are not applicable to it, as it were, from the outside [Feigl, 1950b]. Perhaps here is a point of genuine agreement between Burks and myself. Moreover, these lines of thought seem in accordance with the conception of natural law in terms of modal logic mentioned earlier and labeled 'pragmatic *a priori*.' At the risk of making a bad joke, this sort of view could be said to maintain the existence of synthetic *a priori* truths that are indistinguishable from analytic *a posteriori* truths. [See W. Sellars, 1953a, 1953b.]

8. In short, the position I am inclined to favor is this: Instead of postulating any general principle of uniformity, I think it is philosophically more defensible to retain the sound core of that principle, either by absorbing it (à la Carnap) in a definition of *inductive probability* or even more explicitly by formulating it (à la Reichenbach) in terms of a rule of procedure, or as William Kneale very aptly put it, as a feature of the *policy* of induction. The tentative assumption of specific causal and statistical laws then *is* subject to confirmation in the sense of acceptance or rejection in the light of evidence. But these laws are not capable of probabilification – precisely because they function as the premises of all specific predictions and hypotheses whose probabilities can be ascertained only on the basis of those more fundamental posits.

By way of a *brief summary* I should like to say that there are postulates of science, but they are neither philosophical nor metaphysical in any fruitful sense of these words. The postulates of physics are the basic assumptions which make up the physical theories themselves.

The explication of the methods by which we confirm physical laws, hypotheses, or theories has been attempted in three ways: presuppositionally, definitionally, and procedurally. Burks, in agreement with Kant, Mill, Keynes, and others finds the presuppositional approach illuminating. He feels that without reference to untestable presuppositions the other two approaches remain incomplete, arbitrary, or unjustified. Carnap's definitional and Reichenbach's procedural reconstruction of inductive probability, despite other important divergencies, do agree in the only essential point for our concern: the pragmatic justification of the inductive method. This justification can be derived from the most central purpose of the scientific enterprise: the achievement through the utilization of empirical evidence, of a maximum of predictability among the observed phenomena. The rules of induction and of the hypothetico-deductive method of theoretical physics can be explicated as well as justified without reference to metaphysical assumptions.

SUPPLEMENTARY REMARKS

The foregoing article is a revised and considerably shortened version of a paper presented in a symposium on philosophical presuppositions of science at the joint meeting of the Eastern Division of the American Philosophical Association and the History and Philosophy of Science Section of the American Association for the Advancement of Science, held in Philadelphia, December 29, 1951.

It is gratifying to note a certain convergence of recent philosophical opinion in regard to the problem of the justification of inductive inference. From the point of view of the new common-sense (or common language) approach, Paul Edwards. Max Black, P. F. Strawson, Frederick Will, and others have in various ways exposed the confusions that underlie some attempts to furnish a justification of induction. If the standard of rationality is identified with the logical cogency which is characteristic exclusively of deductive inference, then indeed one is hopelessly entangled in a pseudo-problem. But it is not enough to point to the (undeniable) fact that 'rationality' as commonly understood includes besides deductive consistency and conclusiveness also the patterns of normal induction. It has to be shown that the normal inductive procedure is reasonable, not merely by definition in the light of common usage, but that this definition itself is not merely a result of linguistic habituation or of an arbitrary fiat. This can be achieved, as I have tried to show [1950a], largely in agreement with Reichenbach's ideas, if the adoption of the rule of induction is made a matter of practical justification (vindication) rather than a matter of theoretical justification (validation). But it is important to note that vindication in this case does not require factual premises. We are dealing here with an extreme case of practical justification by means of purely tautological reasoning. This view, long ago formulated by Reichenbach and myself, seems essentially shared by Carnap [1950b], Kneale [1949], C. I. Lewis [1946, p. 325], P. F. Strawson [1952], and J. O. Wisdom [1952a].

6. PROBABILITY AND EXPERIENCE

[1930]

Among the manifold problems that are connected with the concept of probability, we have to differentiate between those that are within the individual sciences and those that are a part of epistemology in general. For example, the problem of contemporary physics 'causal or statistical lawfulness of elementary processes' doubtless is very important and topical but the epistemologist, for the purpose of a deeper clarification, first has to examine what is really *meant by the concept of the law of probability*. To date, only a very few thinkers have noticed that there is a serious difficulty here at all (see P. Ehrenfest, *Mathem. Enzykl. IV*, §30; M. Schlick, *Allgemeine Erkenntnislehre* [1925], p. 359f; and H. Reichenbach in his recent lecture in Prague). As a clear solution for this problem has not yet been presented and as I am convinced on grounds of principle that logic cannot be replaced or supplemented by a logic of probability (and indeed there is no need for such), I take the liberty to discuss this problem in some more detail and to suggest a way to its solution which, to my mind, seems to be the only one feasible.

Mr. Waismann demonstrated in his [recent] lecture how the meaning of the *logical* concept of probability can be reduced to the foundations of propositional logic. In this sense, probability is a relationship between the measures of certain well-defined extensions of concepts. This definition of the concept of probability expresses exactly what von Kries meant by 'ranges of possibilities' (*Spielraumverhältnisse*) and all that which one traditionally wanted to express by the really imprecise concept of 'objective possibility.' In contrast to an axiomatic basis for the calculation of probability, this introduction of the concept of probability has the advantage that the mathematical principles (theorems of addition and multiplication) are directly included because here they can be deduced in a purely logical, in particular a set-theoretical, manner. We are dealing here only with a more precise, more general, and logically more satisfying formulation of the classic combinatory concept of probability, i.e., with the quotients: favorable relative to possible cases. But in doing so we do not intend to deny that an axiomatic, in which the concept of probability is implicitly defined by the axioms, has its value. At any rate, in both cases it is a fact that logic does not have to be expanded when the concept of probability is introduced; and accordingly, the concept 'probable' can not at all be coordinated with the fundamental concepts of 'true' and 'false.'

But beyond this sphere of the purely logical concept of probability the problem arises as to how to apply this concept and with it the general calculus of pure probability to reality, i.e., to the *statistical* states of affairs as they are given in experience. Here, we do not refer to the completely unproblematic application to *descriptive* statistics which limits itself to the description of certain finite, closed statistical areas with respect to their frequency relations, or to the determination of their combinatory probability. It is not until we get to the formulation of probability *laws* which are to assist us in deducing *predictions* about future statistical distributions that we are confronted with the question as to what can be done at all for experience by using the combinatory concept of probability. The system of the probability calculus as it was developed by R. von Mises certainly is an interesting effort to eliminate this problem by using the concept of relative frequency from the outset in order to define the concept of probability; but, as we shall see, the basic difficulty which exists for all statistical inquiries cannot be overcome in this way.

It is probably beyond dispute that the scientific value of the probability calculus for our knowledge about nature rests solely upon its application to various *statistical* problems. For, as long as we remain within the range of the purely formal concept of probability we say nothing about reality. (Of course, here we can ignore the subjective concept of probability as for example in the sense of intensity of expectation.) Our problem then aims at the relationship between statistics and the probability calculus.

When we formulate this question in a general epistemological way it goes like this: which factual state of experience is represented by a general probability statement referring to reality. If it should not only express logical relationships of possibility but also general statistical assumptions, then the question immediately arises as to under what conditions can we speak of the *confirmation* of such a statement. The basic difficulty here depends – in popular terminology – on the concept of 'large' in the 'law of large numbers.' The fact of experience – that with an increasing number of statistical events in dealing with chance phenomena the counted relative frequencies get closer and closer to a constant value – might suggest at first that one could give a precise meaning to the law of large numbers by expressing it in the form of a *postulate of convergence*. This can be efficiently illustrated in the following manner: as abcissa, we choose the ordinal numbers of the members of a sequence, and as ordinate, the relative frequency of the statistically counted elements which are present in the finite portion of the sequence up to each member. Then the curve of relative frequency (in a statistical sequence that is thought to continue until infinity) always has to remain within two limiting curves which themselves asymptotically approach a fixed straight line (which is parallel to the axis of the abcissa). It is clear that indeed within tolerant limits, a concrete statement has been made about the approximation of statistical frequency to a fixed value in a growing sequence of events; and naturally that is the case only when the convergence of the boundary curves is in fact given in the form of a certain function, such that for every arbitrarily small deviation ϵ of the frequency curve from the desired fixed value (in coin games, e.g., this is ½) a definite number for a member of the sequence can be given from which point onward the deviations will continually become smaller than ϵ and never larger. But if one talks in general about the existence of a limiting value of the sequence without concretely describing the convergence in this manner, then the statistical statement remains empirically without content because every arbitrarily formed finite empirical sequence (empirical sequences are always finite) has to agree with it. Even if, in the given portion of the sequence, the relative frequency might not at all agree with the expected value (e.g., based on other statistical expetiences), the possibility still remains open of interpreting this as a 'rare chance occurrence' and of saying that the expected value would still result if one were to arbitrarily continue the sequence. But for knowledge of nature, empty promises about the indeterminate and the infinite do not help us in the least. Sentences like that are not only useless but also they are *completely devoid of meaning* because they don't communicate anything in the least about the world.

At this point one perhaps could object that it is exactly by means of the probability calculus that a postulate of convergence of the desired kind could be proven. Bernoulli's theorem shows that with increasing length of the statistical sequence the probability of realization of the approximately most probable frequency distribution approaches the limit 1 with a strict, exactly describable convergence. Here, we must beware of a crude mistake in reasoning that frequently occurs; it consists in trying to draw conclusions about the behavior of reality from purely mathematical deductions. The concept of probability that is discussed in Bernoulli's theorem is *combinatory*, i.e., purely mathematical-logical. The entire content of this theorem is based on the fact that different conceivable arrangements of elements can be evaluated in a combinatory manner according to membership in definite classes of distribution. The kind of convergence in question here is the convergence of a *probability* and not of a relative frequency. If we try to make this probability observable as a frequency once more by going to even longer sequences, the problem of the meaning of the probability statements would thereby merely be postponed, and if we were to continue the procedure we would get into an infinite regress. The other mistake that is frequently made at this place is simply to equate a very high combinatory probability with truth, such that in the end one wants to turn the probability calculus into a calculus of certainty after all.

Nevertheless, exactly this thought process, which draws our attention to the assumption of a strict convergence of relative frequency which should apply to all statistical phenomena (in mathematical terms, uniform convergence for all statistical sequences), is extremely precarious. By this, I do not refer to the self-evident fact that a proof of (uniform) convergence for empirical sequences cannot be given; such a proof is only possible for mathematically formulated, i.e., lawfully defined, sequences. And I do not want to insist on the equally irrefutable argument that the concept of the limiting value applies only to infinite sequences and that in experience we can only deal with finite sequences; for here it must be said that theoretical description of nature always carries certain generalizations and idealizations; and the approximation of relative frequencies to fixed values which can everywhere be observed may certainly be 'idealized' to a convergence thesis – no special objections from the standpoint of logic can be raised against it. Our present objection is quite different: we ask whether such a generation for the purpose of achieving a concretely describable statement of convergence appears acceptable in the light of the *facts* and with regard to the normal *process of induction* in our knowledge of nature. Exactly because one now wants to make a *definite* statement one runs the risk of making a *false* statement. That would not be so serious in itself because we constantly run this risk when we make inductions, e.g., when we formulate laws of nature. But the structure which we impose upon nature when we accept convergent statistical sequences in the strict sense of convergence is just too peculiar to be accepted by any research worker who has at least some sort of healthy instinct concerning his field. Although in this formulation the worst absurdities have been avoided as, e.g., those risked by Marbe and Sterzinger in their 'theory of statistical balance'; there they claimed that especially improbable distributions do not occur at all and that consequently very small probabilities of arrangements and distributions were, without further argument, to be evaluated as having a frequency zero. But even when, for the convergence thesis, very small probabilities do not immediately mean impossibility, they nevertheless do *prevent* the appearance of certain very deviant arrangements (e.g., pure groups or something similar) *before* certain places within the statistical sequence because such arrangements would break through the convergence. (The frequency curve would break through the boundary curves.) The convergence thesis, therefore, would disrupt the characteristic principle of all of applied probability calculus, according to which, so to say, everything that is possible can appear at any arbitrary place, because the probability that at a certain place (interval) of a finite statistical sequence a certain (although maybe very deviant) distribution appears, can always be determined in a combinatory manner and, at any rate, it is always greater than zero. Although it would be imaginable from a purely logical point of view that the empirical sequences would show these qualities of convergence right up to eternity, we nevertheless cannot bring this assumption into agreement either with the facts or with the normal procedure of induction. For *experience shows us that even very improbable events can happen* and that *we can draw no sharp limits without being arbitrary*. (Also, in every thesis of convergence, the starting point of the statistical sequence where we begin to count would get a very peculiar distinction which would be difficult to make consistent in view of its merely subjective and accidental nature.)

By presenting these arguments I hope to make the principal difficulty clear that confronts us in the interpretation of the meaning of the law of large numbers. This law tries to do justice to two claims which after more exacting analysis turn out to contradict each other: on the one hand everything is, so to say, thought to be possible and every arrangement and distribution can appear at any time. On the other hand, these events, according to combinatory probability, are thought to appear with a certain frequency. In order to formulate this statement about frequency relations in a *definite* way, so that (although with a lot of freedom but yet within well-defined limits) an unambiguous statement is made, it is inevitable that we exclude certain 'possibilities.' If convergence is not defined precisely, one has said nothing; if it is defined precisely, one says something that is basically to be considered false (and *tertium non datur*!).

Of course, it is easily imaginable that our nature is built according to a very tolerant convergence thesis; or even that the doctrine of statistical equilibration is correct. But what would that mean about the behavior of nature? Doubtlessly it would mean that the imaginable possible distributions of statistical phenomena reveal a certain lawful restriction in their realization; therefore, if a far-reaching uncertainty exists in respect to individual aspects, this uncertainty has firm upper limits. When we interpret it in this way, statistical (*regularity*) lawfulness would be a kind of framework for lawlike regularity, a kind of partial causality.

From this argument, it becomes quite evident why the effort to formulate the law of large numbers in agreement with the unrestricted applicability of the probability calculus cannot possibly succeed: the demand of the traditional idea of probability that every complexion of events, as improbable as it may be, might be realized with a certain frequency – this claim that everything possible could happen at least once – is simply incompatible with the convergence thesis, however tolerantly it may be formulated, because such a convergence contains at least a certain measure of regularity. But the applications of the probability calculus refer to *accidental irregular* events and therefore it should have been evident from the start that the 'law of large numbers' simply cannot be a *law* at all. So, our principal dilemma dissolves itself into a pseudo-problem which merely resulted from the effort to combine contradictory elements.

It was not our problem to discover how nature is constituted and what kind of lawfulness or lawlessness it reveals in this or that respect. We wanted to present the contents of the laws of probability for experience, and we have obtained the surprising result that in a strictly logical sense this is not possible at all. The uncertainty which thereby unavoidably enters into the formulation of every statistical law is not an uncertainty that can be confined within certain boundaries. And this is the salient point. It is not possible peacefully to resign oneself to the standpoint of a mathematics of approximation because the limits that statistics deals with are themselves again *only probable* limits.

The deeper reason why we cannot provide a precise formulation for the laws of probability is based on the fact that general statements about unlimited quantities or sequences can be made only when they obey a law. But irregularity cannot be formulated in general concepts because a conceptual formulation means precisely a *lawful* description. Thus, the characteristic of statistical sequences can only be expressed negatively, or characterized by individual positive existence statements.

From this we derive our opinion about the probability calculus of von Mises. We must reject his particular interpretation according to which the calculus of 'collectives' represents the mathematical theory of the statistics of natural events just as much, for example, as Maxwell's electrodynamics represents the theory of electromagnetic phenomena. For, in this latter case, as in all other examples about the application of geometry or theoretical physics to reality, we can distinguish clearly between a closed structure of concepts (a system of axioms) and its correlation with experience. This would be valid for an axiomatic probability calculus only if the concept of probability were *implicitly* defined, but not von Mises' theory. For this theory as it is presently formulated cannot be understood as a purely mathematical theory because it uses concepts that are not acceptable in a mathematical formulation. Thus, the concept 'independence of the choice of partial sequences of the characteristic differences' which is used for the definition of irregularity cannot be expressed mathematically. In fact this is a concept which cannot be justified objectively – even if one tried to formulate it empirically.

Further, it is not acceptable to present the concept of the collective as an idealized abstraction from experience. No doubt this is what we are confronted with when we talk about the concept of velocity (the ratio of distance to time) in classical mechanics. For there admittedly it is imaginable without contradiction that by a series of measurements we can approximate the 'true' value as closely as we wish. And exactly this is impossible from the beginning in statistical investigations because of the quite particular preconditions of probability calculations (in other words, irregularity). The mistake can still be of any arbitrarily great magnitude even after an arbitrarily large quantity of observations; it is even questionable here whether one could talk at all about approximation in a meaningful way without already explicitly assuming convergence – and, therefore, it is again true that in the statement about convergence criteria something absurd is said, and without such a statement we have nothing but empty words.

Nevertheless, I would like to believe that one could give a logically satisfying setting to von Mises' brilliant and fruitful theory by reformulation.

All these remarks about the nature of probability laws must appear to be terribly destructive, especially so in our use of statistical physics. Modern physics uses probability laws to make predictions which are not essentially inferior in their precision and testability to the laws of classical causal physics. How is this compatible with the insights we have just gained?

A follower of Vaihinger's philosophy of the 'as if' would gladly consider this peculiar problem as a confirmation of the conception that the law of large numbers is strictly a fiction, contradictory in itself. But it is obvious that this does not help us much, just as it does not help us much to consider the law of large numbers as a convention or as a postulate.

When physics works with statistical laws and thereby even achieves very great success in the mastery of empirical facts, it is the undeniable task of epistemology [the logical analysis of knowledge] to clarify the meaning of these laws and their position in the general system of our knowledge about nature.

The solution to the puzzle, which cannot be proven here in sufficient detail, results from a general consideration of the characteristics of our knowledge of nature.

To us, as we seek knowledge, the world divides itself into lawful and lawless aspects. Doubtless, we can imagine other worlds which reveal a lawful order of much higher degree or, as it may be, a much lower degree, than ours. But our world has some characteristics which are very advantageous for cognition, and which we can best call the basic *simplicity characteristics* of our world. The most important of these simplicity characteristics is the fact that the lawful and the accidental can be distinguished from one another without an excessive number of observations. This quality, which was shown especially by Zilsel and Reichenbach in their penetrating works, is one of the most essential conditions for the fact that mankind with its limited abilities can nevertheless gain such far-reaching inductive knowledge about nature. This is the ultimate fact, and if we want to express this mythologically, we can say that it is a gift from nature. Further analysis shows that the kind of frequency distribution, as we observe it in statistical phenomena, has a certain uniqueness (in the sense of an extremal quality) – by which, to put it briefly but somewhat imprecisely, the *most probable* distributions are realized. And it is exactly in this kind of distribution that contingency, accidentalness (irregularity) in the sequence of statistical results can be discovered most easily.[1] Irregularity in the appearance of individual events will therefore be most pronounced with approximate constancy of frequencies in the large. In the realm of irregular phenomena, therefore, one will practically always use a convergence theory (although one that is rather tolerant) and make predictions along its lines. Nevertheless one cannot offer any guarantee whatsoever for the fact – even under the precondition of strict determinism (e.g., in the form of unchanged validity of the micro-laws for all future times) – that the statistical relations themselves cannot change drastically tomorrow. For as long as our observations deal only with the macroworld and as long as we can explain the initial states of the microworld only statistically but cannot understand them in detail, we shall continue to be basically uncertain about all our statistical predictions. And a fundamental intensification of this uncertainty is displayed in modern quantum mechanics.

But we make statistical inductions nevertheless, and we count on the realization of the most probable distributions for the future as well because of the uniqueness and simplicity of these distributions, as we mentioned above. Here, as in every induction, we are not dealing with a well-founded procedure for reaching conclusions, but with practical activities, with a decision. The discernibility of regular lawlike connections and the existence of certain accidental distributions are not merely two different sides of one and the same thing. The fact that, in our statistical inductions, we do not expect to discover particularly improbable distributions in the near future of human observations, merely expresses our hope that nature will remain knowable in the future in the same way that it has been up to now. I think there is no sense at all in using the probability concept once again for this most general induction (except in the subjective sense of an intense 'believing' confidence).

When we transfer this thought process from the epistemological to the physical-cosmological level of inquiry then our result, according to the state of contemporary scientific understanding, is as follows: organic life, and along with it the ascent of man and his ability to know, is embedded in gigantic world epochs, in which the statistical structure of the world (the existence of a 'normal' dispersion) is extraordinarily illuminated by the discovery of natural laws. But it is possible that at some point world epochs will arise in which, in spite of the continuing existence of today's causal laws, our knowledge of nature will become almost or completely impossible and in which higher forms of organic life will probably not develop at all, e.g., during periods of considerable reduction of entropy.

We generalize the lawfulness we find nowadays according to the rules of induction which themselves are merely a reflection of the simplicity characteristic of the nature of our time.

Unfortunately, I cannot discuss here the conclusions which can be drawn from this with respect to important material questions and with respect to problems of nature philosophy in a more limited sense from the applied probability calculus. The problem of so-called *a priori* probabilities, the relation of statistical and causal laws, as well as the problem of induction, all will appear in the right light only – so I believe – when seen from the standpoint of the insight we have just gained. (I shall present a more detailed description of the entire problem in my book, *Zufall und Gesetz*, in the series *Schriften zur wissenschaftlichen Weltauffassung*, ed. by Frank and Schlick.)[2] I thought it important to discuss here the rarely noticed problem of convergence in order to point out the insolubility of this question in a strictly theoretical sense and its solubility in a practical sense.

NOTES

[1] Of course, it is not our intention to make the completely meaningless claim that our nature has an especially 'probable' structure. The appearance of most probable distributions, as we know from experience, is tied to certain material conditions (criteria of chance), and these conditions are fulfilled to the highest degree of purity only in certain areas.

[2] [The book, derived from Feigl's doctoral dissertation, was not published, due, as Professor Feigl has explained, to the impact of the probabilistic character and perplexities of the 'new' quantum mechanics of the mid-twenties. – Ed.]

7. MEANING AND VALIDITY OF PHYSICAL THEORIES

[Chapter III of *Theory and Experience in Physics*]

[1929]

1. THE PROBLEM OF THE CONCRETE IN PHYSICAL THEORIES

We have sketched the most important theories of physics. All we said in the introductory chapter about the nature of theory and of physical laws in general can now be explored in more detail and more broadly. The group of problems which we are going to investigate more exhaustively in the present chapter deals with the meaning and validity of physical theories.

The conceptions of the nature of physical theories which emerged after studying them from the philosophical point of view differ very much according to the various philosophical attitudes. We will have to distinguish mainly between dogmatic and skeptical, rationalist and empirical, metaphysical (realist, idealist) and positivist outlooks, all of which naturally appear mainly in hybrid forms.

For the time being, we want to characterize the problems which are most frequently the subject of dispute for these opposing viewpoints (without making any claims as to their completeness) by asking the following tentative questions: Is theory a true picture [representation] of reality? Or is it only a fictitious model, a working hypothesis which helps us to discover new states of affairs? Can the truth of a theory somehow be guaranteed as a matter of principle or of reasonableness, or is theory as knowledge of experience always subject to a degree of uncertainty? Are theories arbitrary constructions which could be just as easily replaced by others, or is there an unambiguous way that leads from experience to theory? Do we obtain the knowledge of an objective, transcendent reality in physical theories, or do they merely signify an economical, simplified description of our immediate experience?

Not before we have reached an assured position on these and related problems will a clear insight into the nature of theory be possible for us. A clear point of reference for answering many of the questions mentioned can be found in the development of physics itself as we outlined it earlier. For example, this is the case with the problem of *concreteness* (*Anschaulichkeit*) of physical theories which is contained in the first question. It has often been said that physical theories are *images* of the world. In fact, Heinrich Hertz's view of this is well known:

We form for ourselves [illusory] images [*Scheinbilder*] or symbols of external objects; and the form which we give them is such that the necessary consequents of the images in thought are always the images of the necessary consequents in nature of the things pictured [*Principles of Mechanics* (Dover ed., New York, 1956), p. 1 – Ed.].

The word 'image' [*Bild*] and the demand for concreteness (*Anschaulichkeit*) of course can be interpreted in several ways. First, in the sense of a concrete representation which represents its object essentially as it was perceived. But when we are dealing with objects which cannot be perceived directly, such as electrical and magnetic states, these images will always contain something artificial, an arbitrary element; they can only be formed along the line of analogies, most of which are useful only here and there. In the mechanical view of nature, we have learned about the attempt to construct all of physics on a concretely representable basis. Motions and forces of masses are so accessible to our senses that it is only too tempting to think that all natural events are composed of these elements. This tendency was substantially strengthened by the successes which mechanics achieved in the mastery of acoustical phenomena and then especially in the theory of heat. But we know that the usefulness of mechanical theories became more and more doubtful with respect to optical and electromagnetic phenomena, until finally the mechanical view of nature itself was dropped because it failed to provide a truly comprehensive and fruitful explanation of nature in modern physics.

The fact that non-concrete theories could be formed and that these theories then proved to be extremely useful is a fact of the history of science which contradicts the demand for concrete representability to which physics was often subjected. – Furthermore, a more exacting examination of the principles teaches us that the demand for concreteness is completely dispensable, and even that, indeed, genuine knowledge as such *must* be non-concrete. Knowing means discovering relationships and the formulation of knowledge will always consider it its highest aim to strive toward representing these relationships in the most exact form, i.e., strictly logical or mathematical form. But in this form, all concreteness has disappeared, and at best, can be viewed as nothing more than a convenient and merely practically important representation and illustration. This becomes very clear in geometry which is usually presented as the model of a concrete science. As soon as the statements of geometry are formulated in all their generality and rigor, which after all is possible only with a system of axioms (e.g. Hilbert's), we notice that concreteness is entirely eliminated and also that it never could be as useful as the logical-mathematical formalism. Exactly the same applies to physics. Only representations in memories or photography have entirely pictorial characteristics. But knowledge is never a representation in this sense. If we want to preserve the ambiguous concept of representation at all, then it must be formulated in a much more abstract, formal way, e.g., as it is used in mathematics. There, representation means only a form of unambiguous correlation. But this idea that theories are images of reality in the sense of a merely formal correlation has its difficulties also, as we shall see. –

If nevertheless we speak of the representability of individual physical theories as e.g. of mechanical theories, then this representability does not reside in the fact that now the entire contents of a theory could be represented pictorially but solely in the fact that concrete representatives can be found rather easily for the mechanical basic concepts.

We might object that physical theories are not only purely mathematical systems but that they are above all meant to represent reality, and this could only be done by establishing a link to concreteness. If we interpret this state of affairs as concreteness, then all physical theories would have to be called concrete because whatever they say about reality can only be tested in fact by gradually translating the abstract propositions of the theory into judgments about immediate experiences. This demand for translatability is obviously unavoidably necessary (and its meaning and indispensability clearly emerge from Carnap's writings in his *Physikalische Begriffsbildung*). Although at times physicists or philosophers protested against the non-concreteness of some theories, they did so mostly not because their connection with experience did not seem to be sufficiently assured but because the concepts of the theories did not allow an immediate concretely pictorial interpretation. The geometric concepts of the general theory of relativity or of the phase space of statistical mechanics, e.g., simply have no direct concrete representation, and even if such a representation can be made possible by roundabout ways, this nevertheless serves only as an example for some very special case and not for the general concept.

Consequently, physical theories without doubt are not images of the world in the primitive sense of a representation by a portrait.

2. WORKING HYPOTHESES, FICTITIOUS MODELS, IDEALIZATIONS

To some extent, the other extreme of the conception of the nature of theory, which we just criticized, can be found in different, partly sceptical, partly rationalist views which stress the discrepancy between theory and reality just as one-sidedly as agreement had been stressed before. In both conceptions, theory is represented as a construction which is more or less independent of experience. In that way, the more sceptical direction places its emphasis upon the provisional and arbitrary character of theories; it interprets them as mere fictions, analogue models, idealizations, working hypotheses – similar to scaffolding which becomes superfluous after it has fulfilled its task. They are bridges to ideas, mental aids to help the mind to find its way in the world of experience and to enlarge the realm of empirical knowledge. Although this opinion carries the essentials, it is in no way exhaustive. Each of those elements play an important role in the genesis of physical theories and in the vital process of development of science. As in most cases, the establishment of a theory temporally precedes its experiential verification, so a theory *in statu nascendi* indeed must at first be viewed as a mere working hypothesis which always has something provisional about its establishment. The *working hypothesis* tentatively establishes a certain view about real relationships and thereby gives grounds for new observations or experiments about whose results it expresses a conjecture. The means of reaching the working hypothesis are, as a general rule, analogy considerations which often also have a purely fictitious character. Thus, in the mechanical model-constructions which Faraday, Maxwell and others used in their investigations, we can see such working hypotheses based on analogy whose success finally led to the discovery of exact general laws of electromagnetic phenomena. The conception which we hold since Hertz about these laws no longer has any place for mechanical analogies, and as a result they are dispensable even if they may have been heuristically necessary. The same seems to be true for Bohr's atomic model, whose model characteristics lie in the partial analogy with relations of classical mechanics (the planetary system), and which similarly should itself be an extremely fruitful working hypothesis but which nowadays becomes superfluous in Heisenberg's comprehensive and completely abstract theory.

Frequently, there are analogies between different areas of physics which, after more detailed observation, turn out to be merely partial agreements but which nevertheless in a more limited area allow for great methodical facility. Such analogies exist, e.g., between sound- and light-dispersion, between heat- and electrical conduction, etc. The conceptual forms by which one area can be represented in such cases can be carried over to the other. But the fact that thereby merely the first step is taken toward a theory about the area in question is shown by the fact that the analogy applies only to one part of the area for which a comprehensive theory is sought.

Furthermore, in this context it is an important fact that all measurements are of only limited precision because of disturbances which simply can never be completely eliminated; therefore, in a strict formulation of the laws, a certain range of arbitrariness always remains. The simplifying representation which is offered here [as in Ch. 6 of Part I of the entire book – Ed.] has often been called an *idealization*, and indeed whole parts of physics are based on such idealizations. The laws of mechanics, for example, can never be applied to observable phenomena with full rigor because mechanical energy as such is never entirely conserved but inevitably is transformed into heat as by friction or impact; and thus, the concept of the purely mechanical ('conservative') process is indeed an idealization. Such a process cannot be observed even as a limiting case. – As their names already indicate, we can find similar idealizations in the concepts of the ideal-rigid body, the ideal gas, the ideal fluid, etc. In all cases, we are dealing with certain stylizations of actual states of affairs by which we either make a simple description possible as a first approximation, or we identify certain limiting cases which we can gradually approach by experiments by choosing appropriate conditions. This is the case, e.g., when we reduce the pressure of a gas sufficiently by increasing the volume; then it shows in close approximation those simple laws which define the concept of the ideal gas.

If physical theories are understood as idealizations, then this often points to the more or less artificial, arbitrary character of the theories. The theory is seen as an arrangement of facts according to viewpoints which comply more with the abilities and wishes of the knowing mind than with the states of affairs in reality.

3. THE THEORY OF KNOWLEDGE *A PRIORI*. KANTIANISM

We find a similar idea in Kantianism and conventionalism but turned more toward the rationalist side. The strongest motive for these two philosophical tendencies lies in the fact that one did not regard it as satisfactory to accept experience as the final decisive basis of knowledge. Uncertainty and inexactness, which exist in all empirical analyses and predictions, stood in opposition to the ideal which had been formed of strict and absolutely valid sciences. The model of mathematics was the yardstick for everything else and it is understandable that this absolute certainty was especially expected of physics which took on a more and more mathematical form. The attempt to ascribe unconditional validity to physical propositions had to lead to the result that the ground for their validity was sought in something other than experience. Here, Kantianism and conventionalism take separate routes. The validity of the highest physical principles is *a priori* in both conceptions, i.e., independent of experience, but Kant and his school assumed that there exists true substantive knowledge of reality which originates from pure reason, while conventionalism denies this and sees the basic laws of physics as mere definitions based on agreements (conventions).

Kant himself wanted to prove the possibility of 'pure natural science', whose validity is independent of any experience, by trying to show that certain general principles are unavoidably necessary preconditions for all knowledge of nature (conditions of the possibility of experience as such). He presented these principles as *a priori* forms of conception and thought. Among them for example are the basic laws of geometry (conception of space), the basic laws of time order, the principles of substance and causality. In Kant's opinion, these forms are impressed upon all perceptions of the external world, and so experience and knowledge become possible only by the fact that the unknown 'thing in itself' appears to us in this way in the light of our mind. We reach knowledge of these forms which are imprinted everywhere by analyzing our reason. Thereby, it is possible to deduce from the outset (*a priori*, i.e., by pure reason) certain most general traits of reality which are expressed in the highest fundamental principles of the natural sciences.

But a healthy natural attitude rejects this conception according to which reason dictates the laws to nature. And indeed the more recent development of the natural sciences, and subsequently the development of philosophy in part as well, reveal more and more clearly that the forms of conception and thought enumerated by Kant are not at all unavoidably necessary conditions for knowledge of nature. In Kant's writings, the principles mentioned were understood, partly explicitly and partly implicitly, according to the situation of the natural sciences at that time. Geometry was Euclidean, space and time derived their more detailed specifications from Newtonian mechanics, mass was considered to be a substance, and causality was thought to be a law without exceptions which tied together consistently continuous natural processes. From our survey of the development of physical theories we know that all these presuppositions have been abandoned, or at least they are nowadays considered to be problematic. They were merely the most general propositions of the natural sciences *at that time* and as such they were the premises for deduction of all the more specialized laws of that system only. But far-reaching minds had stressed the possibility and acceptability of other kinds of foundations (Gauss, Riemann, Helmholtz and others) even before the great innovations in physics, and thereby had stressed their empirical character; a conception of space and time which approaches that of today had already been prepared by Leibniz.

In order to preserve Kant's basic idea, the modern Kantian had to resort to much more general, more abstract formulations of the *a priori* principles, because any more specific statement could be contradicted by the further development of natural science. Knowledge is simply not tied to such special conditions, and even if this were the case, true knowledge of the world based on pure reason would nevertheless be impossible. For the Kantian conclusion went like this: knowledge, science exists; consequently the necessary conditions for it must be fulfilled as well. – But it is in itself a fact of experience *that* knowledge and science are present and that they have a certain form. So, if therefore in the sense of those highly abstract and quite unspecific statements of neo-Kantianism, one wanted to decree, for example, that a certain degree of order and lawfulness of some kind had to be given in the world so that knowledge could be possible, then this is correct but it does not constitute an *a priori* judgment about reality but merely a definition of the concept of knowledge. By knowing, everywhere one understood and still understands the indication of some order, the discovery of lawfulness. But *whether* this is always possible, obviously depends on the material and not on the forms of our knowledge. As we have seen and as I want expressly to emphasize once more, any specific preconceived form would be too restrictive; the surprises which result from constantly growing experience could finally shatter it one day.

Therefore, it is an irrefutable fact for the present time that we cannot reach valid judgments about the world by analyzing the knowing reason.

4. THE CONVENTIONALIST CONCEPTION OF PHYSICS

The rationalistic tendency which so clearly manifests itself in Kantianism appears in conventionalism in another form. (The leading ideas of this school were represented by Poincaré in a somewhat moderated form, and later they were radically expanded and exaggerated by Dingler.)

This philosophical conception too tries to save the certainty of science from the everlasting fluctuations of experience. In this view, the process of thought, in short, is the following: the idea that physical theories are gotten from experience by generalization (induction) is rejected. Experience might inspire the establishment of certain laws but the unconditional validity of these laws could not be guaranteed by experience; this is to be achieved by the decision to uphold these laws and carry them through at any cost.

Therefore, one decides to set forth certain general judgments as axioms or postulates of research at the outset and only then to place individual facts of knowledge in this framework. Conventionalism starts from the assumption that the wealth of experience could be grasped theoretically and unambiguously in different ways, so that we have a certain freedom of choice for the foundations of theories. Each of these foundations is chosen arbitrarily, established by agreement. In doing so, we employ the freedom of choice in such a way that we prefer that theory which is characterized by the simplest, clearest mathematical form. Therefore, according to this conception, the principles of physics are *conventions*.

The essence of a convention consists in the fact that a concept (or several concepts) at first is defined by purely formal relationships, and then that those objects which fulfill these concepts and relationships are sought out in reality. This procedure can be illustrated in a very simple example that, for the sake of easier understanding, I shall not take from physics. The statement that all ravens are black is normally understood as a generalized statement of experience. As such it is never valid with absolute certainty because it could be disproved by later experiences (appearance of white ravens, for example). But one could endow this sentence with eternal validity by declaring it to be true *per conventionem*, i.e., by establishing that from now on only black birds are to be considered as ravens (naturally only when they also have the particular zoological characteristics of ravens). If a white bird is then found that agrees in all other characteristics with the raven, the statement can no longer be disproved because the black coloring is now part of the definition of the raven. But definitions are merely regulations about the way in which certain signs (names) are to be used; therefore regarding their validity they are completely independent of experience and can never be confirmed or disproved by experience.

Now, we want to choose two important examples from physics for our discussion. Several thinkers would present the law of the *conservation of energy* as a convention. Thus, the unrestricted general validity of this law is viewed as not proven by experience but said to lie in its definitional character: We define energy as a quantity which remains constant throughout all changes of the forces. Carrying this convention through in physics can in principle be done in two ways only. Either we define energy according to normal methods of measurement and attribute disagreements which might show up to hidden quantities of energy so that the law of the constancy of energy is upheld by introducing suitable auxiliary assumptions, or, we say in the case of such disagreements: The measuring procedure which directly proves the constancy of energy without any auxiliary assumptions still has to be discovered; then that which in experimental physics to date was understood as energy is not 'the' energy. (Even if we did not know any black ravens, the sentence that all ravens are black would be incontestably correct as a convention – i.e., for the ravens which still have to be found.)

The situation is very similar in another example, namely in *geometry*. We can interpret the entirety of the axioms of Euclidean geometry as definitions of a rigid body. These axioms represent the possible positions of rigid bodies. The question now arises as to whether a body which is defined in such a way can be found in reality. And again we have the same two possibilities: either we interpret possible actual deviations from the established relationships by using auxiliary assumptions about the presence of distorting forces; or, we try to find a Euclidean body by direct measurements some day.

The impossibility of disproval is no doubt assured by the description of the way of the convention. But it probably makes sense at this point already that, by doing so nothing has been achieved for physics as a science of reality and that the meaning of physical principles and theories must be something else. Merely establishing a way to use signs, by agreeing to describe reality by means of a wholly defined network of relationships, it might be possible to reach judgments of unconditional validity – but as definitions, they are merely a formal importance, they only express inner relationships within the system of concepts, they say nothing about reality. (We have not gained knowledge when we agree to apply the name 'raven' to black birds alone.) If we establish a system of axioms in the conventional manner, and draw deductions from it, then we are engaged in mathematics but not in physics. For such a system to gain physical meaning, it is imperative to add material definitions to the purely formal ('implicit') definitions which merely relate the concepts to each other; these material definitions would serve to *correlate the concepts to reality*. The material definition will be brought about by exact description of measuring instruments and methods of measurement, by which those quantities which at first were defined merely by their formal relationships can now be defined. Only then would these systems of concepts become physical theories because it is first through definitions of correlation [*Zuordnungsdefinitionen*] that we can talk about reality in the language of the conceptual system. Therefore, if we want to deal with reality we must rely on both kinds of definitions.

Here the problem immediately arises whether the materially defined concepts fulfill the relationships of the formally defined ones or not. This problem is nothing else but whether our physical judgments are true or false.

The stratagem of conventionalism will assist only pure conceptual systems to become absolutely valid, but their concrete applications, on the other hand, remain as doubtful as ever. However it is clear that physics as science of reality wants to make predictions and that it tries to make correct statements in fully concrete cases. For that reason, conventionalism is not at all helpful.

This reveals itself also in individual cases of physical research which at first sight appear to confirm conventionalism. Thus, in the energy principle it appears at first as if by the introduction of the concept of potential energy all loopholes had been artificially closed which might show up in measurements of energy. If the entire energy of a system is broken down into a directly measurable part and a potential part, then the latter seems to offer the possibility of balancing all discrepancies because, according to what the need might be, we can assume different quantities of potential energy. But on closer scrutiny this is not the case since the actual process of scientific research is different: potential energies are assumed only where we know from experience that a reconversion into actual energy is possible, and the significance of the energy principle rests exactly in the fact that it makes a well-defined verifiable statement for all kinds of force transformations; namely that in all these transformations – no matter how they might happen – a certain quantity which can be measured in an exactly specified way remains unchanged. That contemporary physics is completely convinced of the empirical character of the energy principle becomes perhaps most strikingly evident in the fact that a few years ago, in view of certain atomic [energy] relationships there was some doubt about the strict validity of this law and that these doubts were abandoned only after new and direct experimental determinations became known (the Bohr-Kramer statistical conception of the energy principle, and the experiments of Compton, Bothe and Geiger).

The example of the law of inertia, which is mostly presented as a convention, is interesting as well. As we know, it states that force-free motions are uniform and in a straight line. This law is considered to be a definition of force-free motion, and as such would be empirically without content. But this is not the case because, first, the presence of forces can be established in other ways than by accelerations, as for example by elastic tensions; and, second, we can use a completely different formulation for the law of inertia which certainly will not be mistaken for a definition: Bodies which are at sufficiently large distances from each other (fixed stars), and within definite but otherwise arbitrarily chosen time intervals, cover straight sections of paths, whose lengths are in definite relationships to each other and which apply to each of these bodies in the same way. We shall return later and in a different context to the example of conventions in geometry.

The motives which led to the conventionalist conception of physics appear to me, in addition to the rationalist tendency, to endow knowledge with absolute rigor and certainty, and to consist furthermore in the consideration of the following circumstances: often theories precede experience in time, and thus we sometimes get the impression that experiences have been hand-chosen for the theory just as if one could draw off (*'exhaurieren'* [pump out]) those facts from nature which confirm the theory and neglect those that oppose it. Even if the formation of the theory comes from experience, it nevertheless always has the form of an idealization, for it contains arbitrary elements. Furthermore, there are cases in which a certain freedom of choice exists for the establishment of whole theories. Experiments cannot always decide unequivocally in favor of or against a theory because frequently a theoretical interpretation of the experiments themselves is required if they are to have a verifying power.

Even though we had to reject the basic idea of extreme conventionalism as an error (physical theories are no conventions, and even if they were, they could not provide absolute validity for our knowledge of reality) but under the circumstances which I mentioned before there is a series of leads which were decisive for a more moderate form of conventionalism. This less demanding standpoint, which chiefly has the merit of having presented the problems of the relationships between theory and experience in a more fruitful way, in many ways already comes close to a satisfactory solution which is quite compatible with the concepts of a sensible empiricism. On the one hand, this standpoint interprets theories as constructions which – in principle – contain a certain arbitrariness when compared to facts of experience; but on the other hand, it points out that in practice this arbitrariness is considerably restricted by the fact that here the point of view of simplicity plays an important role. This idea is represented in various colorings by Poincaré, Mach, Duhem, Enriques, Schlick, Reichenbach, Carnap and others.

In discussing these matters we shall now return to the results which we had obtained before [in the third and sixth sections of Chapter I] and which we now want to discuss in more depth.

5. THE LOGICAL STRUCTURE OF THEORIES. THE RELATIONSHIP BETWEEN THEORY AND EXPERIENCE

In the preceding presentation and critique of different philosophical conceptions of the meaning of physical theories, we have partly gone back behind the insight gained in our introductory chapters. But the conviction about the formal and empirical character of theories which we had introduced there has been considerably strengthened by this discussion. In this and in the following sections we shall pursue the problems systematically without referring to specific philosophical doctrines.

Physical theories are hypothetico-deductive systems. As such, they fulfill the task of an explanation because they allow us to deduce conclusions in a purely logical-mathematical way from a system of assumptions, and these conclusions can be directly compared with experience. The goal of physical theory is the uniform and consistent explanation (logical deduction) of experimental laws. So that it may reach this goal, it must satisfy certain requirements of a logical and empirical nature.

The *logical* requirements that a theory has to fulfill are the same as those that are decisive for *systems of axioms* in general. For in the state of greatest perfection, the theory will take an axiomatic form. The only condition a theory necessarily has to fulfill is *that it cannot be contradicted*. In most cases this is guaranteed by proving that the theory can be presented in a closed mathematical form, but furthermore this can be proven by a special logical investigation (representation in a conceptual system which is generally recognized to be without contradictions). A further logical requirement whose fulfillment is often quite valuable although not necessary is the *independence* of axioms from each other. This requirement demands that axioms cannot be deduced from each other so that every single one can be true or false without changing the truth-value of the others. If this requirement of independence is fulfilled, we can test each individual axiom for its empirical admissibility which is of special importance in the case of disputed theories. As long as the theory does not have an axiomatic form, it is often difficult to separate empirical statements from logical conclusions. Only axiomatization dissolves the intertwined knots and permits us to decide about the logical situation and the empirical accuracy of the individual propositions. – And finally, there is a third requirement, namely that of *completeness* which is fulfilled when the axioms are sufficient to permit the deduction of all propositions of the field in question. If a system of axioms is incomplete, it can be completed by adding one or several further axioms – a step which often has to be taken in the elaboration of physical theories when there are facts which cannot be deduced from the original theory.

When these three logical requirements are fulfilled, then we have a closed system before us, which can give us extremely detailed information about the connection between theorems and axioms. Up to now, only several individual physical theories have been axiomatized in a strict form but, at any rate, there is a growing tendency to try – if possible – to bring all of physics into the form of an axiom system.

The deduction of experimental laws from the general principles of the theory is always achieved by inserting particular conditions. For example, if we want to deduce the experimental laws of free fall from Newton's theory of gravitation, then we insert the particular distances of the falling body from the center of the earth, and its mass and the mass of the earth, into the law of gravitation and thus get its acceleration.

From the criticism of conventionalism, we have learned to distinguish between the theory as pure structure of concepts as it is in the case of the system of axioms, and its application to reality. We know that this application is made possible by the correlation definitions and that only in this way will the real goal of physical research, namely knowledge of reality, be reached. But we would be greatly mistaken to believe that by pointing out the difference between axioms and correlation definitions, the relationship between theory and experience has been clarified. The circumstance that, in the course of the development of physics, axioms as well as correlation definitions are modified again and again draws our attention to the fact that in their relationship all problems referring to the connections between theory and experience are hidden. Mostly these problems are combined under the headline of 'problems of application' since they refer to the conditions of applicability of certain conceptual systems to the reality of experience. For our purposes we could express the same idea more simply by asking for the truth of physical theories.

Besides the logical requirement for internal absence of contradictions, the requirement for *truth* is the most important condition that a theory tries to satisfy. We can call it an *empirical* requirement because we understand by the truth of a theory its agreement with experience. As the most far-reaching fulfillment of the requirement for truth, we would have to consider the case in which all propositions of a theory, i.e., all axioms (and thereby all derived propositions) could be directly proven to be empirically correct. In principle, this is impossible for physical theories because they normally express *general* assertions, that is (as we explained in sections two and three already), they have an inductive character and therefore can only be understood as systems of more or less probable hypotheses. As especially high or especially low degrees of probability are practically equivalent to truth or falsehood (the fact, for example, that all people must die is merely enormously probable!), we now have above all to investigate the conditions of the probability of theories.

But before we do so it is necessary to secure the inductive character of theories against certain objections. Some thinkers (and especially the conventionalists) have tried to prove that physical theories never are simple inductive generalizations but that they are conceptual constructions which have to fulfill only one purpose, namely to bring the experimental laws into a deductive connection. In doing so they rely upon the historical state of research and present impressive examples for it.

Thus, Newton would never have come to his gravitational theory if he had tried merely to generalize Kepler's laws inductively. According to Kepler's laws, a planet moves in an elliptical orbit around a completely stationary sun. But according to Newton's theory of gravitation, the sun moves as well, namely around the center of gravity of the sun-planet system. Therefore, Newton could never have developed the theory of *mutual* attraction of masses by mere induction. – The example of an atomic theory is probably even more striking: It would have been impossible to develop the kinetic gas theory by a mere generalization of the gas laws. For the gas laws only connect the quantities pressure, volume and temperature so that by generalizations, one would merely have obtained statements about these quantities but not about their underlying molecular motions.

These examples are meant to show that theories are not mere generalizations which everybody could have found by applying the most basic recipes of logic. Theories do contain new ideas, they are comparable to works of art, and like these have to be viewed as creations of genius.

All this certainly is true but it is relevant only for the *genesis* of physical theories. Of course, not a single theoretical advance in physics resulted from simple induction. But in addition to the historical-psychological point of view, there is another one, namely the systematic-logical. (If modern philosophy has any merit at all, it lies in the fact that it has learned to distinguish clearly between the historical and the systematic, the psychological and the logical, the genesis and the validity.) What the examples cited prove is relevant only for the genesis of physical theories. The idea of general gravitation is indeed absolutely new in comparison with Kepler's laws; and the idea of molecular motion in comparison with the gas laws. Thus these theories were not obtained by simple induction from experience. Nevertheless, the *validity* of theories can only be founded inductively. If we ask ourselves when theories are accepted and when they are rejected, it turns out that this depends exclusively on their inductive power, i.e., their ability to encompass certain factual material. Thus, for example, it is the typical characteristics of a fruitful theory that it also incorporates new, original facts that originally had not been considered. From the standpoint of a finished theory there is no factual difference between those facts which gave rise to its genesis and those which subsequently confirmed it. Basically both kinds of facts are in themselves of equal importance for judgment of the capability of a theory. Now, if one starts from the range of *all* facts which are explained by a theory – for a well-confirmed theory these facts are not merely possible, imagined states of affairs but facts which are really observed – then, in this case, the theory can really be established by inductive generalization.

In the generally confirmed theory of gravitation there exists a multitude of known cases in which the attraction of masses can be observed so that the inductive conclusion does not have to rely exclusively on the motion of planets as its only basis. – Similarly, in modern research there are a great number of direct indications for the existence of molecules.

Accordingly, theories, even when they are not discovered by induction in the course of research, nevertheless have to be considered as inductions as far as their validity is concerned. – The establishment of a new theory is achieved in most cases by especially far-reaching minds who, in a very small number of empirical facts, find enough supporting points for a daring construction. But only when experimental investigation yields a multitude of other confirmations, i.e., when the construction is supported by a number of further props, does our trust in its strength grow.

What we might call a factually justified trust in a theory, in other words, is its objective probability (in contrast with subjective probability, which depends on the individual and personal attitude of the investigator). Whether and how the concept of objective probability, which after all is at the focal point of the problem of verification, can be determined more closely should be our next topic of investigation.

6. THE PROBLEM OF VERIFICATION. THEORY AS APPROXIMATION. SIMPLICITY AND PROBABILITY

Various objections have been raised against the possibility of verifying physical theories and especially against a verification by a single decisive experiment (*experimentum crucis*). At first, it is pointed out that theories always can have only *approximate* validity because their experimental confirmation can only go as far as measurements are possible and as far as their precision can reach. In other words, the axioms of a theory, in their strict, precise mathematical formulation, go beyond the actual state of affairs, which is to be justified empirically, because the correlation definitions can never be given absolutely sharply as a consequence of the limited precision of measurement. Earlier, we described this relationship between theory and experience as idealization. But as the theories in their development are ever better adapted to states of affairs which are observed with constantly increasing precision, it might be more correct to talk about gradual *approximations*.

The path of experience is always the same. It gradually leads from knowledge which, at the outset, is crude and sketchy to greater and greater refinement and detail. Yet it is a fundamental trait of our knowledge of reality that all these stages can only be viewed as temporary, and that they only serve to lead us further to higher, more promising standpoints. It is precisely this fact of constant approximation which must give us the impression that the possibility of verifying theories is reasonable.

But nevertheless, for the same reason, we have to admit that verifications can never be ultimately decisive. The leeway which is always left by the inadequate establishment of the facts can be used to construct several theories which embrace the facts just as precisely. But the differences between these theories are quite uninteresting when they merely consist in quantitative differences between the number values of the constants which occur in the laws. However, it is something else when the difference goes deeper, when it becomes relevant to the entire conception of the formation of concepts. Starting from very different ideas, we erect different constructions which now compete with each other in their efforts to cover the same material of facts. We think of this case when we question the possibility of a crucial experiment.

Here, most frequently the following arguments are set forth: Theories are extremely abstract and at the same time such ramifying, complicated, one could almost say organic, structures that their capability cannot be tested by one single case. Whether a certain factual situation agrees with a theory or not can itself mostly be decided only on the basis of theoretical interpretation. The experiment which is performed for the sake of testing a theory itself requires theoretical assumptions in order to be understood. For example, when an experiment in electrical theory is performed, we already presuppose mechanical and optical theories because we work with mechanical apparatus and observe if need be with optical instruments. Therefore, the outcome of the experiment can never be decisive for one single theory; but rather it bears upon an entire group of theories. – We do not finally give preference to one particular theory because it is now decided unequivocally by experience but because it grasps the experiences with greater simplicity than other theories.

In many respects, these claims are correct but often greatly exaggerated. In principle, it certainly is impossible to decide very strictly about an individual theoretical statement. But in practice, things are by far not so bad. Namely, it is the essence of a useful experiment that it only uses techniques whose effectiveness within the required degree of precision is sufficiently well known, so that we can rely upon the theories presupposed in the experiment, at least within the range of observation needed. These theories have already been sufficiently ascertained by earlier experiences, so that only in exceptional borderline cases must they be subjected to revision again. The development of theories proceeds along the path of constant expansion, as they become *gradually secured*. Just as in everyday life, when testing some process we tend to trust our ordinary means of investigation (sensory perception, influence of body movement, etc.) when they are used with sufficient care, so similarly, at least within certain ranges of precision, we shall unhesitatingly base the examination of new and still uncertain theories upon already verified physical theories. Then, when the gradual establishment has proceeded far enough so that only one definite individual theory appears to be questionable, a single experiment can be decisive under certain circumstances.

In most cases, the *experimentum crucis* resembles indirect proof in mathematics. It carries one theory *ad absurdum* by the contradiction that exists between one of its conclusions and experience: thereby, the probability of another theory increases. For example, Foucault's experiment on the speed of light in water was decisive for rejection of Newton's emission theory. For one of the conclusions of this theory stated that the speed of light had to be greater in water than in air. But this claim was proven wrong by Foucault's experiment. Michelson's experiment by which he disproved the hypothesis of a stationary ether is another well-known example of an *experimentum crucis.*

But the situation is different for theories which in their construction are far from the facts, i.e., in which the unverified hypothetical component is still large. Here, indeed there is greater leeway for arbitrariness; the construction can be executed in quite different ways. These are the theories for which the viewpoint of simplicity as a principle of choice is normally emphasized.

If the application of the principle of simplicity is to be understood correctly, we must clearly distinguish among its different meanings. We can best succeed in this if we state in what respects two competing theories can be distinguished from one another. Here we find three possibilities.

1. The theories differ merely in their conceptual form, i.e., they only differ from one another linguistically. In that case, they are entirely translatable from one to the other; their difference can never be established by their statements about concrete experiences, for it is not factual but purely formal. Thus, the two theories can differ merely in their *formal simplicity*.

2. The difference is of a *factual* nature, i.e., the states of affairs which are assumed in the two theories are different. Then, at least in principle, it must be possible to discover empirical facts which correspond to one theory and contradict the other. But two theories of this kind can at times be compared with respect to their simplicity, because at least we can decide in passing whether one theory contains a greater number of untested hypotheses than the other, i.e., whether the states of affairs assumed by one theory are more complicated, more varied, than those assumed by the other. We call this a difference in *material* simplicity.

3. The two theories are not only different in formal or factual respects but also their difference lies in assumptions which, in principle, escape empirical examination. Therefore, these assumptions are metaphysical. In general, nowadays there is a growing and healthy tendency to ban such assumptions from physics altogether. Thus, if we want to give preference to one of two such theories, it will be the one which contains fewer – or at best no – metaphysical components. Remembering Occam's guiding principle we could speak of Occam-simplicity.

As far as this last case is concerned we are confronted with such metaphysical constructions in all those theories which assume absolute or essential natures: for example, Newton's doctrine of absolute space and absolute time, but indeed all substance theories, as well as for example of a stationary ether whose last defender was H. A. Lorentz. Here such qualities were explicitly attributed to the ether that make it fundamentally impossible to observe its state of motion: every imaginable effect is destroyed by an identical negative effect. It is clear that in such a theory the concept of the ether is reduced to a meaningless word. The theory of relativity finally rid itself radically of this ballast which for so long had been dragged along unnecessarily.

Before turning to material simplicity, we still have to discuss formal simplicity. There is merely a formal difference between two theories when they differ from each other only in their manner of mathematical representation. For example, if in modern physics we prefer the vector-analytic treatment of mechanics and electro-dynamics to the earlier mode of representation (by coordinate equations), then this merely shows a transition from a more complicated to a formula-like shorter and more lucid kind of formulation. The content is exactly the same in both cases. The example of the Ptolemaic and the Copernican world-systems is equally instructive; the two theories differ merely in formal respects when we interpret them purely kinematically. Descriptions in the heliocentric and geocentric systems of coordinates for positions and orbits of the planets can unequivocally be translated from one to the other. The advantage of the Copernican system lies in the simpler geometric form of the planetary orbits when they are based upon this system.

Now we are better equipped to discuss the question as to how one can distinguish between theories which are factually different. We stated before that factual differences, even if they have not yet shown themselves empirically, in the current state of research, in principle, at any rate, must become empirically provable somehow and at some time. For theories which are factually very different, it is therefore quite improbable that in the long run we should not find facts which make a decision possible. The history of physics fully confirms this; in most cases, this sort of conflict is quickly resolved by new observations.

The application of the criterion of *material simplicity* becomes relevant only where, even before a conflict between two factually different theories is under resolution by experiments, we wish to estimate the advantage that one theory might have over the other. It is clear that physics is *always* confronted with this situation because, due to the eternal imprecision and piecemeal character of empirical knowledge, in principle *several* theories can always be imagined which equally cover the experiential material that is already known but which, beyond this point, go their separate ways in respect to their contents. Yet, in practice, it is mostly true that one theory possesses such a great advantage over the others that the latter are not considered at all. Nevertheless, investigating the criterion of simplicity is essential for philosophical considerations and for those cases where the advantages of different theories are to some extent comparable.

The connection which exists between simplicity and the probability of theories must not be misunderstood in such a way that the basically simpler theory is in every case the more probable as well. Quite to the contrary, we have to realize that in the course of the development of theories, their adaptation to facts is achieved by an ever increasing refinement and by complication of the mathematical form of the theory. Yet, simplicity of the conceptual form of the entire structure of theories as such is not what we have in mind when we speak about material simplicity. Rather, this simplicity lies in the *relationship* between the amount of conceptual means and the range of experiences covered. This relationship to probability, as we discussed it earlier in respect to functions, applies only to this relative simplicity which is measured against the entire wealth of available experiences [Part I, Ch. 6].

We already indicated there that the relationship between theory and experimental law is somewhat similar to that between the experimental law and the individual fact of experience. And just as the probability of an experimental law increases with the number of measurement results for which it can be used successfully, so one can assume likewise that the same will be true for the probability of a theory which apparently will be the greater, the more experimental laws it explains. The simpler the assumptions of a theory covering a greater number of different experimental laws, the more probable it is that it will apply to a series of further experimental laws. But on the other hand, if a theory has to be modified at every step when new facts become known, if we have to support it by a series of auxiliary assumptions so that it may continue to do justice to these observations – then we must consider its probability to be low. This case resembles a defendant who is justly accused, and who, by inventing innumerable excuses which are suitable for each individual accusation, finally traps himself in such a tangle of unbelievable statements that the inaccuracy of his testimony is discovered even though he manages to avoid open contradictions. The desperate efforts to save the theory of a stationary ether from the results of Michelson's and others' striking experiments by using auxiliary hypotheses provide a good example for this case; the efforts toward the establishment of a mechanical theory of electricity offer another case in point.

Naturally, every individual theory has only a limited range of validity, so that there will always be cases for which the theory suffers from 'apparent' exceptions. The fact that these exceptions, indeed, are merely apparent can only be proven when the relevant deviation itself can be attributed to a natural law. For example, although the laws of free fall do not exactly apply to free falling in air, this does not constitute a mistake in the laws of falling because the deviations which are caused by air resistance themselves obey another group of carefully tested, particular laws, with an effect that is well known. Thus, when 'exceptions' are explained not by *ad hoc* invented hypotheses but by other empirical material, it is well justified to consider the theory as to some extent probable.

After all this, meaning of the concept of probability now appears to be the same as for the experimental laws. Even the deeper founding of these relationships on the fundamental character of our nature, namely that it normally does not tend to confuse us by stubbornly holding on to chance occurrences, will prove to be correct here. For it is hardly reasonable to assume that by an accidental intertwining chain of circumstances we are led to a simple theory which, due to a strange consistency of these chance occurrences, is at first frequently confirmed but must be rejected as useless in additional areas of experience.

At any rate, in spite of these agreements, we have to be extremely careful in the application of probability considerations to theories. For the probability of experimental laws (as for example in their interpolation), an exact numerical determination was not possible, and we could confirm only a certain monotone dependence on the degree of simplicity; for theories, the situation is essentially even more difficult. Here it is not even possible to give an unambiguous definition for probability in the sense of relative frequency. To do that we would have to be able to count those facts which are independent of one another, and which confirm a theory, in order to compare it with the equally counted mass of facts which do not agree with the theory. Such numerical confirmations could be possible only in a highly arbitrary way, especially since independence is not precisely defined either. Even an entirely secondary and incidental relationship between simplicity and probability is confirmed in theories only when we take all relevant circumstances thoroughly into consideration. Thus, we must not unhesitatingly draw conclusions from partly very exact agreements because these agreements are frequently chosen artificially; for example they originate by applying certain mathematical methods. We saw, so to speak, an area which at first is only that which we might see due to the theories of another area (which had been studied earlier). – Often the ways of reasoning of theories are so novel that it is hard to find any reference points from which to estimate their probability.

Although we have to make all these restrictions, the good sense of that which so many great physicists understand by 'simplicity' or the 'uniformity of nature' can be expressed satisfactorily in the form of a statement about the formation of theories, as follows: That theory which covers an area of facts by using the relatively simplest measures (means), namely which introduces a minimum of arbitrary assumptions, i.e., assumptions which are not justified by this area of facts and furthermore which has already proven itself in the manifold expansion of this area of facts, in most cases will be capable of continuing to cover new additional experiences [in the future].

Now we are in a better position to discuss the thought process of conventionalism. It says that in choosing theories we shall always choose the simpler; but thereby it refers to the *formally* simpler one. Thus, it intends to consider only those theories as valid whose *axioms* have a simple mathematical form. It is clear that by this procedure the problems are only pushed to another area, namely toward the correlation definitions. If experience does not turn out to be as simple as expected, in a formally simple system of axioms, then complications will arise when formal theoretical concepts are applied to reality. Then the definitions of correlation have to be expanded very considerably, and supplemented by rules of application which contain natural laws so that the connection between theory and experience can be preserved. – From a totally abstract, theoretical standpoint, we seem to be able to choose: either simple axioms or simple correlation definitions. But *in praxi*, physics chooses the second way because in theories it tries to find the most adequate possible expression for real relationships, something that could not be achieved by using the first alternative because the true form of reality would be hidden in the correlation definitions, and thus could not prevail as such in the theory. At close scrutiny, the first possibility turns out to be mere window-dressing because some elements of the correlation definitions have been arbitrarily included which we would normally express in axioms if we had adopted a more open and objective attitude. But the circumstance which is really decisive for the decision to prefer simple correlation definitions to simple axioms is the fact that the correlation definitions are mainly intended to allow the *direct* application of the theory to reality, for example for the purpose of predictions.

This situation is very clear in the problem of *physical geometry*. Euclidean geometry, in comparison to non-Euclidean geometries, certainly has the advantage of greater simplicity. But if the theory of general relativity is correct in stating that in our world it is as impossible with rigid bodies as it is with light beams to succeed in exactly exhibiting the relations of Euclidean geometry, and that each of the structures always has the positional relationships of a specific non-Euclidean geometry, then the only natural way to express this non-Euclidean geometry is in *axioms*. For, if we wanted to preserve Euclidean geometry *per conventionem*, then we would be obliged to compensate for the deviations (which our empirical measurements must demonstrate) by complicated rules for application which would have to be given *ad hoc* from case to case and entered into the correlation definitions. And if we tried to establish a general law for these deviations, then nevertheless we would have to apply non-Euclidean geometry in some form or other – thereby merely hiding the factual condition without otherwise achieving anything. – Another way out which normally is offered by the defenders of conventionalist freedom of choice is to explain the deviations from the Euclidean in the positional situations by physical forces. But this as well leads at most to a renaming, because these forces – they have very appropriately been called *metric* forces – are invented for this purpose only, and they differ from all others, namely from genuine physical forces, by the fact that they influence all matter in the *same* way. Whether we now say that the bodies are deformed in one way or another by forces which change from place to place, or whether we say that the bodies behave in a non-Euclidean way, the result is the same: the difference between the two statements is purely formal. But as the first one has the disadvantage of using the concept of force in another way than is normally the case everywhere else, it was only logical for Einstein, in his general theory of relativity, to formulate non-Euclidean geometry in *axioms*.

Summarizing our results about the relationship between theory and experience, we can say: Theory consists in a logical construction of general assumptions from which we can deduce the experimental laws. Although theories are rarely obtained by simple inductions, their validity is nevertheless inductive. Accordingly, the probability of a theory increases with the extent of its confirmations. Therefore, absolutely final decisions about the correctness of theories do not exist. Theories can never be considered as more than mere approximations. Every theory contains arbitrary elements but we try to reduce their number to a minimum. A freer arbitrariness only exists, first in the conceptual-mathematical form of theory, but here the restrictive viewpoint of formal simplicity holds sway; and second, in theories, which in the contents of their statements reach far beyond experience gained so far – in this case, a contest between two or more theories is possible. Here, material simplicity becomes effective as the principle of choice which, in general, is decisive for the probability of a theory.

7. THE VALUE OF THEORIES. PHENOMENOLOGICAL AND ATOMISTIC PHYSICS. THEORY AND REALITY

By pointing out the viewpoint of simplicity, we have tried to show that the arbitrariness of physical theories, although it exists in principle, is rather weak and little used in practice.

Einstein emphasized this point very clearly when he said: "... the evolution [of physics] has shown that at any given moment, out of all conceivable constructions, a single one has always proved itself absolutely superior to all the rest. Nobody who has really gone deeply into the matter will deny that in practice the world of phenomena uniquely determines the theoretical system, in spite of the fact that there is no logical bridge between phenomena and their theoretical principles; even more, the system of concepts which is unambiguously correlated to the world of experience can be reduced to a few fundamental laws from which the entire system can be logically developed. Here, the scientist sees his expectations surpassed by every new important progressive step when, under the pressure of experience, those fundamental laws become more and more simplified. With amazement, he sees how apparent chaos becomes integrated into a sublime order which cannot be attributed to the workings of his own mind but to the nature of the world of experience." (Speech for Max Planck's 60th birthday)

The increasing unification of physics in the sense of a progressive approximation to truth has often been doubted. Especially to those removed from physics it appears as if the constant innovations in the realm of theories signify an eternal state of revolution and anarchy. But that is a serious misunderstanding of the actual state of affairs because the replacement of an older theory by a newer one does not mean that earlier insights are completely abandoned. In the development of theories, we have to pay attention to a number of circumstances in order to be able to completely understand their meaning: first of all, the relationship to facts. Although thanks to new experimental possibilities and discoveries new theories come to life, these theories nevertheless, in respect to the old facts, have to lead at least approximately to the same results as the earlier theories. That causes a certain continuity in the expansion of the theoretical horizon. – But it is interesting that often the slightest empirical differences suffice to justify a very powerful theoretical innovation. But here as well, it is true that the earlier theory for the field for which it was fruitful will continue to have its justification and must be replaced by the new theory only when the greatest precision or validity for an expanded area of experience are demanded. – Furthermore, the progress of a new theory (even when this progress is not formal; in that case, one should not really speak of a new theory) can rest at least partly on the fact that superfluous, for example metaphysical, elements of the earlier theory are being eliminated. But in that case, these elements were irrelevant for the empirical contents of the earlier theory as well, so that there as well, from a factual standpoint, the difference is mostly smaller than was apparent at first sight.

Although it certainly should be emphasized that the change in basic ideas of physical theories at times is extremely profound, we must point out, on the other hand, that the development of theories, at least in one respect, shows a definite unambiguous direction: The adaptation of theories to experience constantly increases in extent, and in precision. This is the marvellous fact of convergence in our knowledge of nature. In our theories, we elevate ourselves to dizzying abstract heights and simultaneously with increasing perfection command a view of a constantly growing part of the basic structure of reality.

Especially this achievement may well constitute the *value* of theories. – But nevertheless we have answered the question about the value of physical theories in different ways. Many thinkers see it for example merely in the fact that, with the help of theories we give a simple and economical description of the world so that thereby we can comfortably grasp a great number of facts and we can find our way around the connections more easily. From this standpoint, theories are a kind of large-scale memory-aids which serve exclusively to enable us to orient ourselves in the world for all practical purposes. In a more general expression, the principle of economy of thought (Mach) agrees with it, according to which we have to organize our knowledge about nature in such a way that, with a minimum of mental effort we can manage a maximum of facts. As far as the result is concerned, this principle is probably identical with the principle of simplicity although without doubt there is a different emphasis. The emphasis on practical interest evokes the impression of dealing merely with a mental principle of convenience. But such is certainly not involved in the formation of theories because the scientists do not shy away from any mental effort, however great it may be, when they see their goal as the unique theoretical integration of facts.

The rejection of theories as *explanations* (Kirchhoff, Mach, Duhem, among others) is also connected with a more practical-economic interpretation of the value of theories. Theories would just have to provide simple and clearly arranged descriptions: every effort to see something more in theories must fail. This interpretation turns against a metaphysical concept of explanation which in any case has nothing in common with the one we have used. If we expect to gain more than empirical insights from the explanation – then of course physical theories are not explanations. But if by explanation we understand the reduction of a state of affairs to more general connections, then without doubt we can agree with what Kirchhoff and Mach understood by simple and economic description.

Nevertheless factually a difference still remains insofar as these scientists, who nowhere want to lose the firm ground of experience, radically rejected all theories which, with a less restricted construction, rose above this ground. The real stumbling block was the atomic theory. In the statements of this theory, these scientists [Mach *et al.*] saw explanations of a metaphysical nature. To talk about the existence of the atoms without being able to offer experiences by which their existence might be proven seemed to be pure fantasy (we must keep in mind that at the time these objections were raised, the direct indications for the reality of atoms as we have them today [1929] had not yet been found).

The opposition to the theory of atoms was then carried over to the more general connection of the philosophical struggle between *idealism* and *realism*. The difference between these two conceptions lies in this: idealism (or, to use a less ambiguous expression, the philosophy of immanence) sees the only reality in life experiences (perceptions, imaginations, feelings, etc.) of creatures endowed with consciousness, while realism considers the (transcendental) external world also to be something real. For the idealist, the external world is nothing but his conception [*Vorstellung*], and for him the concepts with which the realist believes he represents real objects of the external world are nothing but short, handy symbols for certain connections in the stream of his perceptions. So, for the idealist a physical object, a stone for example, is at first merely a word which designates a rather consistent complex of interconnected simultaneous sensory qualities such as gray, angular, hard, heavy, etc.

Therefore an idealist cannot consider a physical theory to be able to give knowledge of an objective reality of the external world which exists independently of his consciousness (and this is the opinion held by the realist), but rather he sees in it merely a tool for the economic description and ordering of his experiences, one by which his ability to form expectations concerning future experiences is improved.

But obviously it is not possible to discuss the problem of reality in more detail here. Nevertheless I would like to make a few remarks suggesting the basic outlines of a satisfactory solution which, in contemporary philosophy, is coming to the fore more and more strongly. In the dispute between idealism and realism the ambiguity of the concept of reality obviously plays a role. On the one hand, we understand by reality or actuality, a quality of existence which cannot be further described, a quality of existence as we find it in our immediate experience and as it unavoidably imposes itself upon us, as in the conviction of our own existence. This is the concept of reality held by idealism. On the other hand, realism interprets reality not only as this quality of experience but it talks about real things and events, especially when it wants to differentiate them from merely imagined or dreamed ones. Thus it refers to a kind of objectivity which, to a certain degree, is independent of the perceiving consciousness. – A more clear-cut characterization of these objects which the realist recognizes as real, probably takes place when we point out that they are in an especially clear and lawful connection with each other, one which always presents itself in the same way to the individual knowing subject as well as to many such subjects. In this sense, we can call those objects 'real' which are subject to an intersubjectively valid lawfulness (or – if we allow ourselves to use a mathematical metaphor – which are invariant in respect to all intersubjective transformations).

It would be totally fruitless to fight one concept of reality with another. The solution can only be found when both parties understand that there are two different concepts of reality and that, with a certain degree of justification, each of them may be applied in its own way. (I would like to remark briefly that from a strictly logical point of view, claims or denials of reality as they are normally given by these two standpoints are completely senseless.) Further, the realist will have to recognize that in the final analysis all his knowledge about the external world is derived from his perceptions, and that consequently he can retranslate all statements dealing with objects of the external world back into statements about his perceptions. And vice-versa, the idealist will notice the important fact that at a certain state in the formation of knowledge from experiences such a clear order is reached that he, too, will doubtless turn his special interest to this stage (which is the external world of the realist). After these insights have been won by both sides then the problem of reality can be nothing more than an idle dispute over words because by then everything that can be formulated comprehensibly has been said. Therefore it is advisable to drop all general claims or denials about reality altogether, and to use the neutral language of positivism which, by using the expression 'reality of the external world', does not understand anything else but that stage in the construction of objects of knowledge at which the especially clear connections which we mentioned before can be found (comp. Carnap, *Scheinprobleme in der Philosophie* [1928]).

If we deal with the concepts of reality in this cautious manner, we are free to explain the meaning of physical knowledge in two ways: first, as a conceptual adaptation to our experiences, but just as well as knowledge of the external world. All those characteristics which are used in everyday life as well as in the sciences when testing the reality of objects, simply mean when seen from this new standpoint that they permit us to decide whether certain objects belong to that organized stage of knowledge which we discussed earlier in more detail. For example, when a chemist must verify the actual presence of a substance by chemical analysis, this, in the final analysis, will be nothing more than an investigation as to whether under certain conditions certain processes of perception appear which, in their special connections, exhibit that intersubjective invariant lawfulness which we have set forth as the criterion of reality. In principle, all verifications of reality are based on the same procedure. We can state in a general way that the reality of an object of the external world is considered to be confirmed when certain statements about the process of perception have been shown to be true.

But as all our judgments of experience go beyond the field of that which can be perceived as immediately present, by directing themselves, either to things past by memory or to things future by expectation, nobody can protest when objects are called real which are not being perceived right now but which are merely *perceivable in principle*. For example, when Leverrier drew the conclusion from the disturbances in the orbit of Uranus as to the existence of another planet, he had not yet seen it but of course he considered it to be perceivable, something that, indeed, was confirmed soon thereafter. Thus, Leverrier was able to make precise statements about place, velocity and mass of a celestial body nobody had ever seen before, and that merely by relying on laws which he considered to be generally valid; similarly, the founders of the kinetic gas theory succeeded in calculating for the first time the size, weight and velocity of molecules also by trusting in general laws, and by shrewdly applying that which experience in gas phenomena had taught them to reach more profound conclusions. – If we wanted now to put forth those frequently used objections against the existence of atoms, namely that we always perceive only that which the atomic theorist calls the effects of atoms but that we shall never be able to see the atoms themselves, then we could reply that we can make the same claims about the planet Neptune, and just as well about our own table. For these, too, we perceive merely due to their effects. (From the physical standpoint our perceptions can be integrated into the structure of the world as well; in that case they are brain-processes and, as such, links in physical causal chains.) For good measure, modern experimental technique has succeeded in making the effects of individual atoms perceivable in manifold ways, so that nowadays we are as firmly convinced of their existence and of information about their size and weight as we are of the existence of the sun and the planets and of information about their size and weight.

The controversy about the reality of atoms has been completely settled thanks to progress in modern physics, and even the staunchest opponents of the atomic theory have been won over; yet at the bottom of this dispute (for which the slogan of the contrast between the phenomenological and the atomistic attitude has been coined) there is a problem that is of a more general importance, and one which goes beyond the question about the existence of atoms. If we consider how the problem of the existence of atoms presented itself at a time when only the first few clues were available, we find that then the atomic theory could not be seen as more than a very daring assumption. At that time, the standpoint of the anti-atomists could only be explained by the fact that their objections were aimed against a theory to which one really could not yet attribute great probability. We normally express great differences in probability, i.e., differences as to the degree of justification of a theory, by differentiating between hypotheses and theories (in their more limited sense). Well, these names are not decisive, and we shall never be able to use them in a very well-defined way in this field. But the underlying question which is really fundamental for the entire dispute between phenomenological and atomistic physics is how far one should get involved in theories which contain a large, unverified component. However, this question is purely practical in nature and its answer therefore cannot come from a theoretical grounding. It must be left to the scientific tact and the instinct of the individual investigator whether he wants to deal with constructions of theories which go far beyond the field of confirmed facts. But by asking ourselves this question, although we have already recognized it as non-theoretical, the state of affairs that it deals with will probably become clearer: There is no sharp border between phenomenological and atomistic theories. When phenomenologically-oriented physicists thought that in their method which directed itself to pure facts they did not use any hypothetical elements, they were subject to illusions. As we have seen earlier, there are no 'pure facts' – except in momentary total experiences of a consciousness. Everything else is already generalization, interpretation, theory. Measured against the facts of consciousness, our everyday statements are already quite hypothetical. And measured against that which, in everyday life, is unhesitatingly presented as 'fact', the 'confirmed facts of physics' again are mere assumptions – and thus the scale of uncertainty continues, up to those theories which are just beginning to germinate in physics and whose viability still has to be tested.

One thing seems to be certain in the development of physical theories: Although theories often begin as daring constructions, their unverified elements are narrowed down from two sides. On the one side, by the discovery of confirming experiments, on the other side, by the improvement of the theory itself, in the sense of eliminating arbitrary and unconfirmed elements. Thus, the final goal that a theory tries to reach is, on the one hand, the closest possible contact with the facts of experience, and on the other hand, the completeness of a logical system which wherever possible permits us to include within it the lawfulness of the entire world.

Therefore, the value of a physical theory does not exclusively lie in its efficiency with respect to mental economy, but above all, in the increase in knowledge which it enables us to achieve, and by which it enables us to gain insight into the miraculous workings of nature.

8. CONFIRMABILITY AND CONFIRMATION

Some Comments on the Empiricist Criterion of Meaning and Related Issues

[1951a]

The discussions of the last twenty years concerning the empiricist criterion of factual meaningfulness have increasingly emphasized the need for a more liberal formulation. Positivists, old and new, adamant in their repudiation of metaphysics, have tended to overshoot their goal. The motivation was of course understandable and, on the whole, admirable. The decision to eliminate from scientific and philosophical discourse the kind of problems which, because of their very conception cannot possibly be solved in a responsible manner, had an eminently salutary and cathartic effect. There is certainly no point in worrying one's head about questions which are supposed to concern matters of fact and yet cannot conceivably be answered (at least partially or indirectly) by appeal to the data of observation. We are all too familiar with the dialectic devices designed to protect the claims of transcendent knowledge against refutation. Such devices have been utilized in the arguments not only of outright theology and metaphysics but frequently enough also in the frontal areas of the expanding scientific enterprise itself. Speculations about absolute space and time, substance, the ether, causal necessity, entelechies, telefinalities, groupminds, etc., are apt to become devoid of whatever (no matter how vaguely) specifiable meaning they may have had to begin with. They are rendered proof against disproof by a simple but often rather concealed or unwitting decision to make them immune against the outcome of any conceivable test.

If the empiricist criterion of meaningfulness merely eliminates such in-principle-unanswerable questions, there can be no quarrel with it. At least there can be no objection from anyone holding a scientific outlook in philosophy. It is of course regrettable to have aroused ire and indignation on the part of the tender-minded and other-worldly thinkers. But, contemporary empiricism, pragmatism, positivism and naturalism in their various forms are all continuous with the ideas of the Enlightenment of the eighteenth century and cannot possibly help offending those engaged in the search for absolutes or in the quest for certainty. Whatever significance unanswerable questions may appear to have is easily explained in terms of their purely formal structures or their expressive and evocative (pictorial, emotional, motivative) functions.

The meaning of words and sentences depends upon the kind of use to which we put them. An isolated spoken or printed sentence may mean anything at all – or nothing – depending on the rules according to which these words are made to function. Confronted with a given assertion, the first task in ascertaining its meaning consists in finding out how the speaker or writer uses his words in connection with other words of his vocabulary and how (some of) the words of his language are related to the data of experience. To illustrate first by a perfectly obvious example: An old facetious German 'rule' for weather prediction, when stripped of its trappings, reads: 'The weather will either change or remain the same.' It scarcely needs to be pointed out that this assertion is a substitution instance of the law of the excluded middle and thus a glaring tautology. If there had been a serious intention to pronounce a law of nature (a genuinely synthetic general proposition), one can immediately see that this intention was not fulfilled by the given formulation – always taking for granted that the terms occurring in it are taken in their usual meanings. If some of the 'laws' of physics were to be understood in the manner in which Poincaré and other conventionalists construe them, they would indeed reduce to purely analytic propositions. Similar dangers of confusion are well known in the attempts of some older psychologies to formulate a law of human motivation in terms of the pleasure principle, and in the formulation of some of the laws of economics and sociology. It is a matter of purely terminological decision whether we choose to characterize analytic propositions as factually meaningless or as the null-case (lower limit) of factual content. At any rate it is indispensable first to find out, by the use of the sort of procedure that might be called 'socratic', 'dialectical', 'casuistic' or 'diagnostic', just what, if anything, the assertion achieves in asserting. Formal analysis in the sense of a reconstruction within an ideal language merely consolidates and systematizes the results first disclosed by the socratic procedure of informal explication.

Another example will bring out the essentials of the diagnostic method as regards transcendent metaphysical assertions. Newton's conception of absolute space was subjected to a searching scrutiny already by Leibniz. To be sure, the famous correspondence on this issue between Leibniz and Clarke (Newton's disciple) is replete with theological and metaphysical arguments on both sides. Yet it is obvious that in some phases of the dispute Leibniz proceeded deliberately, and strikingly like a pragmatist or positivist. He asked in effect what conceivable difference in observable consequences there could be between the assertion and the denial of the existence of absolute space. (It should be remembered that the issue concerned exclusively the space of kinematics. In regard to the dynamic aspects of space Newton not only made a meaningful assertion but was – at the time – even justified in his assertion of a privileged coordinate system, or, as he *should* have put it, a *class* of such systems.) Clarke had no other way out than to remove the concept of absolute space beyond the possibility of any sort of test whatever. We find the same sort of situation again at the beginning of our century when Lorentz, Fitzgerald and others protected the ether hypothesis (a latter-day incarnation of Newton's absolute space) by special safeguards from refutation. These theorists did not – but Einstein did – realize that thereby electrodynamics had come to include a set of sentences which were not only superfluous but strictly speaking devoid of factual meaning. (Some of the earlier versions of the ether hypothesis were of course meaningful, but were proved false by the joint evidence of a number of fairly decisive experiments.) The diagnostic procedure reveals in this instance as well as in many others that confirmability, i.e., at least indirect and/or incomplete testability-in-principles is *logically* excluded by some of the assumptions of the system of which the assertion under scrutiny is a part. In other words, factual meaninglessness can be charged against an assertion only if the premises of the hypothetico-deductive system stand in contradiction with the very idea of a test which would either confirm or disconfirm the assertion in question.

Conceived along these lines the empiricist meaning criterion draws a perfectly sharp distinction between sense and non-sense. I believe that Professor Hempel's scruples [1950a] do not apply to the criterion if it is understood in this way. For example, I cannot take very seriously the qualms about the meaning of statements involving the use of irrational numbers. Since the assumption of the ubiquitous limits of exact measurement is inconsistent with the ascertainment of precise numerical values (rational or irrational) for any physical magnitude, there is indeed 'no difference that makes a difference' and hence no difference in *factual* meaning. The use of the system of real numbers in physics is therefore to be regarded as a convention whose sole justification lies in the expedience of the *formal* aspects of mathematical representation and computation, rather than in any empirically discernible fact.

As regards the introduction of theoretical constructs, we must first decide whether they are explicitly definable in terms of less complex empirical concepts. If so (as in the case of the Laplace or Poisson operators, Hamiltonian functions, tensors, etc.) they are in principle eliminable, i.e., they can be replaced by their definientia and the difference the introduction of those constructs makes is again purely formal and *not* factual. Theoretical constructs which cannot be explicitly defined on the basis of 'observables' of course present us with the very core of our problem. If the diagnostic or logico-analytic procedure reveals that addition or omission of these constructs cannot possibly make a difference in the deducible empirical laws (and if they are not explicitly definable in the sense just indicated) then they are metaphysical chimeras and make in principle unconfirmable any assertion of which they are essential constituents. But it must be admitted that the diagnostic method does not always succeed in disclosing a definitive logical contradiction between the confirmability of a given assertion and the rest of the system of which the assertion is a part. In such more indefinite cases it is advisable for empiricists to adopt an attitude of tolerance and open-mindedness. Science at its growing edge continually issues promissory notes not only in hypotheses formulated in terms of fully defined concepts, but also in the vague and tentative introduction of new concepts which are at first only tenuously connected with other already legitimized concepts of science. For example, when Schrödinger introduced his ψ-function in 1926 into quantum theory, there was some indecision regarding the physical meaning of this new construct. It had all the earmarks of a fruitful idea but, as is well known, the first interpretation given by Schrödinger himself (in terms of electric density) had soon to be replaced by Born's statistical interpretation. If some of the data of psychical research (concerning mental telepathy or psychokinesis) were to be taken as authentic (rather than as illusory owing to experimental, statistical error, or outright fraud), imaginative theorists might well attempt to introduce new concepts and laws in order to account for those peculiar phenomena. The impact of the pragmatist-positivist-operationist thinking of the last fifty years makes it rather unlikely that scientists will gravely mislead themselves by the introduction of absolutely unconfirmable assertions.

The sort of doubt that may justifiably be raised in connection with the introduction of new concepts and hypotheses will then more usually be concerned with their indispensability – or at least with their fruitfulness. Having made sure that the requirement of meaningfulness is satisfied[1], the only other relevant conditions which have to be fulfilled are those of formal and factual parsimony or simplicity. Since formal simplicity concerns differences between theoretical systems which are logically equivalent, i.e., intertranslatable, it has no bearing on the factual meaning of the systems. Factual or inductive simplicity however is of paramount importance in the appraisal of the factual meaning-content and the explanatory power of theories.

The principle of factual simplicity is very likely what Newton intended with his first *Regula Philosophandi*. It furnishes a criterion for the acceptability of *existential hypotheses*. This principle must of course not be confused with the criterion of meaningfulness. Meaningfulness is presupposed when the question of inductive simplicity is raised. Furthermore it is worth noting that considerations of inductive simplicity arise only in the context of *methodology*, i.e., in the critical analysis of the growing, unfinished science. The criterion of factual simplicity can provide no more than very tentative directives. It is to be understood as a principle of inductive logic in that it (generally, i.e., *ceteris paribus*) ascribes a higher probability to simpler hypotheses. To illustrate: Leverrier and Adams had in principle the choice among an indefinite number of more or less complex existential hypotheses by means of which (together with the presupposed laws of Newton's mechanics) to account for the perturbations of Uranus. They could have hypothesized a numerous group of heavenly bodies to be responsible; and with suitable assumptions about the individual orbits of these bodies, could have derived the same conclusions as from the assumption of a solitary disturber (Neptune). After the telescopic discovery of Neptune and prolonged observations of its orbit, the probability of the simpler original hypothesis increased to a practical certainty. The case of Neptune illustrates however only one type of existential hypothesis – the particularized or singular form. More important, but also more problematic in regard to their philosophical interpretation is another, the generalized or universal type of existential hypotheses. The introduction of a new basic magnitude or variable into a thus augmented theoretical system may from a purely formal point of view be regarded as a genuine enrichment of the language of science. The introduction of the concepts of the electric and magnetic fields (Faraday, Maxwell, etc.) is a good illustration of this case. Over and above the well-established concepts and assumptions of classical mechanics, optics and thermodynamics these new concepts were soon recognized as indispensable for the explanation of phenomena whose confirmation base was, however, completely contained in those older disciplines. It is obvious that the introduction of the concepts of electromagnetism can by no stretch of the analytical imagination be construed as a matter of explicit definition on the basis of the older concepts (serving as definientia). The introduction of generalized existential hypotheses ('besides the mechanical masses, forces, etc., there are also electrical and magnetic forces') must be interpreted in terms of *laws* which connect the new magnitudes with the old ones in a strictly *synthetic* manner. I shall not here discuss the controversial question in which sense, if in any, considerations of inductive probability apply to the introduction of generalized existential hypotheses. In my article 'Existential Hypotheses' [Feigl, 1950b], I argued that questions of inductive probability can be significantly raised only if the total frame of concepts is presupposed and that it makes no sense to ask questions about the probability of the frame. But that argument was made from the point of view of a semantical reconstruction of the language of science. It is conceivable that in the context of a methodological analysis some sort of interpretation of the probability of generalized existential hpotheses could be specified.

Quite generally it seems to me of the utmost importance not to confuse analyses (such as of meaning, confirmability, confirmation and degree of confirmation) in the context of the semantical reconstruction of an idealized language of science with analyses in the context of the methodology of the procedures of a developing science. An idealized language of factual science is so conceived that questions concerning the precise meaning of its terms, or concerning synonymy, deducibility, analyticity, etc., can be answered with the same finality (and the same limitations) as the analogous questions regarding purely formal systems of logic or mathematics. The ideal-language-reconstruction, as it were, freezes a given stage in the development of a scientific discipline. This 'freezing' involves of course some more or less arbitrary decisions, since there are always alternatives of rational reconstruction. Just how faithfully a given reconstruction portrays the structure of the discipline will thus be a perennially debatable question. The doubts recently expressed by C. G. Hempel, W. V. O. Quine and M. G. White concerning the possibility of a sharp distinction between meaningful and meaningless (as well as of analytic and synthetic) sentences seem to me to be justified only in the context of methodology. Science in its living, evolving progress of course continually shifts and redefines the meaning of its terms. In its growth toward the open horizon of a (never attainable) ideal completeness, coherence and adequacy, innovation and revision are of the order of the day. A concept may at first be only very tentatively introduced and very incompletely specified. Confronted with such a concept we may decide altogether to refrain from an idealized reconstruction. The semanticist could justifiably say that there is no concept there to be reconstructed. A concept, after all, is specified only if it is sufficiently fixed by a set of rules. On the other hand the semantical reconstruction may be undertaken if it is understood in the sense of an explication which renders the portrayal of a concept much more precise than the rather ambiguous and amorphous original. In some cases (as in the reconstruction of theoretical systems like classical mechanics, relativity theory, etc.) the discipline in question has assumed a sufficiently definite form to allow – if not for one unique reconstruction – then at least for a limited set of alternative reconstructions. A given formula, e.g., in classical mechanics, may thus be interpreted as an analytic proposition in one reconstruction and as synthetic (a genuine law) in another.

But wherever a scientific theory (or, *a fortiori*, ordinary discourse, or vague speculation) has not crystallized into such definite forms, logical reconstruction will always seem to do violence to its subject. It is in the context of methodological (in contradistinction to ideally reconstructive) analyses that it makes sense to speak of *degrees* – *not* of meaning or meaningfulness, to be sure – but of *specifications* of meaning.

Ever since Pierre Duhem's classical critique of the possibility of *experimenta crucis* his point of view has been re-emphasized by various writers, most recently by Hempel and Quine. Since statements in isolation from others are regarded as indefinite in meaning, it is maintained that only whole systems of statements can be subjected to test. It would take more space than is available here to disentangle what is right and what is wrong with this view. Permit me to remark, somewhat *ad hominem*, that logicians (and I am sure, among them Hempel and Quine) would look askance at any scientific theory which presented itself with the claim: 'Take me or leave me, but you can't pick out any parts.' If I recall correctly that is precisely what Freud once said about psychoanalytic theory and its 'monolithic' character. But Freud himself has repeatedly modified parts of his theory – presumably on the basis of clinical evidence. Science would be in a sorry condition if its theories could not be stated in terms of logically independent postulates. It is precisely for the sake of systematic examination through empirical testing that we must unravel the knowledge claims of a theory into a maximal number of independently confirmable postulates. For example, only after disentangling the various components in the principles of special relativity can we say which experiments confirm which laws. The experiments of Michelson and Morley (Trouton and Noble) confirm one component, the observations on double stars by de Sitter another; and for the confirmation of auxiliary hypotheses the measurements of aberration and the experiment of Fizeau are equally indispensable. – A view that maintains that the whole body of a scientific theory (if not of all science) confronts experience and that modifications may be required in *any* part of the system if it does not 'fit' – such a view obscures dangerously what is of the greatest importance for the progress of science: the successive testing and securing of parts of science – at least in the sense of an approximation. Naturally, no part can be considered as established with finality – but this insight which impresses the pure logician should not blind him to the recognition of the method of successive confirmation. It is curious that it should be the logicians who undermine first their own enterprise by a denial of the sharp distinction between the empirical and the logical (because here they are struck with the vagueness of ordinary discourse and of scientific methodology) and then proceed to blur the methodology of science beyond recognition (because from a purely formal point of view a theory is in principle adjustable in any one of its parts). I suggest that if our logical analyses and reconstructions of the scientific enterprise are thus far unable to do justice to its most important and indispensable criteria (of confirmability and of confirmation) that we must look for better analyses and reconstructions instead of giving a distorted picture of scientific method. Perhaps it will help in this connection to remember the distinction between ideal reconstruction and methodological analysis – to point this out was the major purpose of the preceding informal remarks.[2]

NOTES

[1] I.e. having shown the concept or hypothesis in question is not 'isolated' by (implicit or explicit) devices which make it *logically* impossible experimentally to test the difference between their admission and omission.

[2] Since I have elsewhere [Feigl, 1945a, 1950a, 1950b, 1950c, 1950d], dealt at length with other aspects of the problem of meaning I shall here only refer the reader to my critique of the restrictive ('negativistic') implications of phenomenalistic positivism and operationism. In this connection it is indispensable to distinguish between 'meaning' as the positivists usually understand this term, i.e. the evidential base or the 'method of verification', and 'meaning' as the factual referent or designatum of sentences. The confirmability criterion actually requires such a more realistic interpretation in terms of pure semantics and pragmatics [W. Sellars, 1948a and 1948b]. An important corroboration of this position from the point of view of scientific methodology may be found in the illuminating analysis by L. W. Beck [1950]. Among the discussions of the meaning criterion by non-positivists that have nevertheless arrived at roughly the same conclusions, I mention those of Weyl [1949, p. 117ff] and Margenau [1950, Chapter 5]. A clear formulation of the realistic interpretation of the field concept is given by Weyl [1949, p. 114]. The topic of inductive simplicity is discussed by Reichenbach [1938, Section 42]. A promising beginning in the analysis of partial specifications of meaning may be found in the article [1946] by A. Kaplan. A stimulating discussion of the relation of synthetic and analytic truth in physical theory is given in A. Pap's book [1946]. The view according to which the distinction between analytic and synthetic propositions rests on dogma or prejudice has been set forth persuasively, but to my mind by no means conclusively by M. G. White [1950] and W. V. O. Quine [1951]. However, my recent publications indicate that I agree with Quine's rejection of the positivistic-phenomenalistic doctrine of reductionism. Along similar lines there is at least a partial agreement also with Nagel [1950], Hempel [1950a], and Reichenbach [1938]. The distinction between explicitly definable concepts and hypothetical constructs, as it concerns especially psychological theories, is forcefully pointed out by MacCorquodale and Meehl [1948].

9. THE LOGICAL CHARACTER OF THE PRINCIPLE OF INDUCTION[1]

[1934a]

The purpose of this paper is to make clear (1) that the widely recognized formulations of the principle of induction do not express the most fundamental rule of induction; (2) that the current view concerning the probability of induction must be revised in terms of a frequency theory of probability; (3) that on this basis the problem of induction in its traditional form is a pseudo-problem; and (4) that the principle of induction must be interpreted as a *pragmatic* or *operational* maxim.

I

Let us begin with a brief summary of those views concerning the problem of induction which seem to have received the most general approval among contemporary logicians and philosophers.

(1) Induction is essentially different from deductive inference. It can never attain certainty. All attempts to transform inductive into deductive inference fail because they necessitate the introduction of *inductive* premises. Hume has shown that induction can be proved certain neither on logical grounds nor on the basis of its own success.

(2) Induction is the indispensable foundation of all factual science, although it is admitted that the more advanced factual sciences do not actually proceed by inductive generalization. Their method consists rather in the construction of hypothetico-deductive systems. The strength of such systems lies in the high degree of internal connectedness by which the various parts of a system reinforce one another. But logically, if not genetically, a theory is inductive. This is clear from the fact that any verification establishes the truth only of singular or particular propositions but not of general hypotheses.

(3) The principle of induction expresses the increase of the *probability* of inductions in dependence upon the accumulation of factual evidence. Such evidence consists in the elimination of irrelevant circumstances, as well as in the positive confirmation of a specific connection, uniformity, or regularity.

(4) This principle of induction is not a consequence of the purely logical axioms of the calculus of probabilities. It can be demonstrated only on the basis of *assumptions concerning the general constitution of nature*. Thus,

Jevons took nature to be something like an urn to which we can apply Bayes' Theorem. Peirce, similarly, assumed that our observations represent 'fair samples' of a thoroughly statistical world. Zilsel, Broad, Keynes and Nicod introduce more refined formulations of the 'Principle of the Uniformity of Nature' such as the 'Principle of Limited Depth and Variety', or at least the antecedent probability of such assumptions [E. Zilsel, 1916: C. D. Broad, 1927–28; J. M. Keynes, 1921; J. Nicod, 1930].

(5) The quantitative value of probabilities and their convergence toward certainty can be derived only from the presupposition of rather arbitrary and artificial conditions whose fulfilment is by no means warranted in any case of scientifically significant induction. The idea of determining the numerical value of the probability of scientific theories seems preposterous.

(6) Since these general hypotheses underlying induction are interpreted as significant assumptions concerning the structure of reality, they must be logically synthetic and, therefore, themselves *inductive*. This is the fundamental difficulty. *What can be meant by the probability of these presuppositions of the probability of all particular inductions*? It is held that even these presuppositions can be rendered increasingly probable by the verification of their consequences. Keynes and Nicod believe that they have proved this argument to be free from circularity. To many thinkers the whole issue appears highly problematic.

II

The unsatisfactory state of the problem of induction seems to me to be due to the preoccupation with the problem of the *validity* of induction. The more fundamental question concerning the *meaning* of the principle of induction is rather neglected and it is this which needs a careful, logical analysis.

The chief difficulty, undoubtedly, lies in the interpretation of the concept of probability. To Hume, the probability of induction was a subjective or psychological matter. It was a degree of belief or an intensity of expectation, based on habit. In contrast with this reduction of probability to something irrational and in opposition to the classical *subjective* interpretation of mathematical probability, two types of *objective* interpretations have been advanced:

(1) *Probability as a Logical Relation* – the theory of Leibniz, Bolzano and W. E. Johnson, most fully expounded in Keynes' *Treatise*, and accepted by C. D. Broad, Nicod and others.

(2) *Probability as the Limit of a Statistical Frequency* – the theory of Venn and Peirce, rejected by Keynes, but recently restated, defended and

mathematically systematized by the Berlin mathematician, R. v. Mises.

According to the logical interpretation, which still seems generally favored, probability is the relation of partial or inconclusive implication between one proposition and another. But this account is for Keynes merely a characterization, not a definition, of the fundamentally unanalyzable and indefinable probability relation. Here, I believe, Keynes is fundamentally in error. I do not mean to deny that the probability relation can be chosen as a primitive notion in an axiomatization of the probability calculus. That can of course be done, and has in fact been done by Keynes, and more recently by Reichenbach [1932]. But probability is also applied to empirical facts and in this case we need rules in order to determine the value of the applied probability. These rules, if they can be stated at all, are then the *definition of the empirical or applied probability concept*. I have not time here to prove in detail that in any significant application of the probability concept its essential meaning is statistical. However, I shall discuss a few of the relevant points.

If the principle of indifference operates not on the basis of equal ignorance, as did its predecessor, the 'principle of insufficient reason', but on positive grounds, it must inevitably make use of statistical assumptions. Often these assumptions are tacit and in many cases their statistical character is not recognized, but they are the true source of every fruitful probability argument. The 'indifference' or 'irrelevancy', which is the crucial concept in the principle, means generally *causal irrelevance*. Causal irrelevance, however, is identical with random distribution, and random distribution is a fact which can be established only by statistical investigation. The essential and final test for the correctness of any estimate of probabilities is always the comparison with the statistical frequencies. Confronted with an '*a priori*' probability, one can take only one reasonable attitude, and that is to ask the direct question: What bearing does it have on observable facts? It is understandable that Peirce, who introduced this pragmatic question as a general criterion of meaning, was at the same time perhaps the most convincing advocate of the frequency theory.

III

If probability is to be a significant guide for our expectations and predictions, inductive probability must be interpreted in terms of the frequency theory. Once it is granted that induction is not an infallible procedure, all our care must be directed toward attaining success at least with a maximal frequency. Mill's famous question – Why are the experimental methods (as stated in his own canons) much more efficient than induction by pure enumeration? – means precisely: Why are the experimental methods successful more frequently than simple induction in the discovery of a law? The fact that they are more successful has stimulated the desire for a more profound explanation in terms of assumptions about the structure of the world. The assumption that there are at the bottom of nature strictly deterministic laws, on the one hand, and complete independencies on the other, seemed to account for the superior reliability of experimental methods. For if we follow the simpler method of pure enumeration, we can always be deceived by strong statistical correlations which we may mistake for laws. Only the experimental methods are capable of splitting up such correlations into their strictly causal and strictly random components.

This hypothesis of the 'All or None' character of nature has been one of the most fundamental and fruitful guiding principles in almost every field of science. But the recent development of Quantum Physics has proved that it is not an *a priori* or necessary truth. According to Quantum Physics some of the elementary laws of nature are statistical correlations which will probably never be reduced to a deterministic scheme. The change in attitude is fundamental, and even if determinism should be reestablished as a result of surprising new discoveries, the lesson taught by modern physics would remain of great importance to the theory of induction. Any assumption or 'Inductive Hypothesis' (in the sense of Broad and Keynes) about the general constitution of nature is subject to possible correction in the light of new experimental facts, and can therefore be regarded only as a tentative frame-work for more special research. Any such assumption is simply one of an infinity of possibilities, and unless it is accompanied by still more general and precarious suppositions, it can not be assigned a finite probability. On this point the theories of Keynes, Broad and Nicod are seriously in error. Even on the basis of the logical interpretation of probability, the assumption of the finite antecedent probability of an 'Inductive Hypothesis' is untenable. These able thinkers are mistaken when they assert that a singular fact can confer a finite probability upon a general assumption. This is possible only by the exclusion of alternative assumptions. Therefore, the whole issue is prejudged. It can never be demonstrated that the principle of induction has the faintest probability except by a *petitio principii*.

But even if we accept the 'Principle of Limited Variety' as a necessary condition of induction, it is easy to see that it is by no means sufficient. Unless we are allowed to infer from the probabilities of the chance coincidences of causally independent characters (or events) something concerning their corresponding frequencies, the principle of induction can have no significance. It is precisely the assumption of the stability of statistical frequencies which is necessary here. But of this assumption we can never be sure. The occurrence of a long chain of extremely improbable coincidences can always mislead our inductions. And there is no way to make sure that we are not living in just such an unfavorable world epoch. If we actually believed that we were so situated we would terminate all investigations and wait until the world passed into a more propitious stage. But the peculiar fact is that we are optimists and refuse to abandon the belief that we can obtain 'fair samples' of the world.

The probability of induction is therefore established on the basis of generalizations for which there is no probability at all. These generalizations extrapolate statistical frequencies, but only more special hypotheses can acquire probability with reference to such frequencies. The probability of a natural law is determined, roughly speaking, by the success-frequency of the inductive method by which it was discovered. The principle of induction, formulated in terms of the frequency theory, states simply that those regularities which have held so far without exception will be found to hold most frequently in the future.[2] According to this analysis, the probability of induction is always secondary and hypothetical, and can never be a genuine attribute of pure generalization.

But if, as we have seen, this most general presupposition of all induction cannot be shown to be appreciably probable, is there any other justification for accepting it? The usual reply – and here the influence of Kant is noticeable – is that such assumptions are necessary conditions for the possibility of knowledge in general. It is true, of course, that knowledge of nature would be impossible if there were not a certain amount of order and simplicity. But what are we to infer from this? That in our scientific investigations we must always begin with the postulate or demand of order and simplicity? But obviously it does not follow from the fact that we demand something that we get what we demand!

Our critique must seem very destructive, and it is destructive as regards illusory solutions of the problem. After the failure of all these attempts to achieve anything like an objective vindication of induction are we finally driven back to Hume's scepticism?

There are thinkers, however, who deny that Hume's analysis of causality and induction has any sceptical consequences. R. E. Hobart [1930] has most convincingly shown that Hume's arguments appear sceptical only to those who desire to prove what cannot possibly be proved. Moreover, he has shown that almost everything that we call reasonable, rational or justifiable in our active life is absolutely dependent upon belief as the ultimate basis of all our significant knowledge. This ultimate belief, according to Hobart, is present as an immediate fact in every cognitive situation. Although I fully agree with him, yet I feel that for the purposes of a logical analysis of knowledge his formulations are not adequate. For, as he admits, to speak of belief is to speak in terms of psychology. Yet psychology itself is possible only through the belief in induction. Every explanation of belief in the context of psychological or biological theories presupposes induction, because it is by induction that we establish explanations. It is quite legitimate to study the phenomenon of belief from the scientific point of view, but in a systematic logical account of the structure of knowledge the principle of induction is prior to the recognition of its embodiments in psychological or biological processes.

IV

What then is the nature of this principle? Its peculiarly elusive character is startling. If it is a meaningful assumption about the world, then it is no longer the most general principle of induction, it is itself inductive. And if it is stated in such terms that it can never be verified or proven false, then it does not say anything at all. How can we escape this dilemma?

A glance at the logic of deduction will provide us with an instructive analogy. In any axiomatic, deductive system the starting point of our deductions is a set of primitive propositions or postulates whose truth is either 'evident' or assumed. From these we derive other propositions. But in order to do this we must have methods or rules of deduction. Important examples are the Rule of Substitution and the Rule of Inference. The one allows us to substitute logically equivalent terms for each other, the other allows us to drop true premises and assert the conclusion. These rules are not commands, but anyone who wants to perform deductions must employ them. (Professor Sheffer of Harvard calls these rules "prescripts" in contradistinction to the postulates which are "descriptive" either of facts or of logical structures.)

Analogously, the principle of induction is not a bit of knowledge, it is neither analytic nor synthetic, neither *a priori* nor *a posteriori, it is not a proposition at all. It is, rather, the principle of a procedure, a regulative maxim, an operational rule.*

According to the viewpoint of logical analysis, all empirical knowledge is a construction erected upon immediate experience. What this immediately given really is can be disputed, but that there must be some such 'ground floor' of knowledge is necessary if any empirical proposition is to have a meaning. If it is the possibility of verification which establishes meaning, then verification itself must consist in the comparison of elementary or atomic propositions with the given. These elementary propositions are the raw material of knowledge. Moreover, as in the case of deductive systems, inductive science too has its prescriptive rules, and the principle of induction is undoubtedly the most significant among them. Its nature, just as the nature of the rules of deduction, can be determined only through the recognition of the function that it fulfills with regard to the goal of science.

Now the ultimate goal of science is not the achievement of a loosely connected miscellany of descriptions, but the establishment of a systematic structure of laws as a basis for explanation and prediction. The prescriptive rule, which is a direct consequence of this objective, is then the real principle of induction. It reads: '*Seek to achieve a maximum of order by logical operations upon elementary propositions. Generalize this order (whatever its form be: causal, statistical or other), with a minimum of arbitrariness, that is, according to the principle of simplicity.*' The condition of simplicity is essential, because it restricts the ambiguity of the procedure. But, since simplicity is measureable, if at all, only with great difficulty, there will usually be several ways of generalizing. This explains the case of competing scientific theories. Only when new experimental evidence is supplied, can it be determined that the one or the other theory is more complicated in that it employs more arbitrary hypotheses.

If foreknowledge is to be distinguished from arbitrary or capricious guessing, if it is to be different from dream and inspiration, no other definition can be given of the procedure of science. However, the principle does not carry in itself the guaranty of its own success. In this it is radically different from the rules of deductive inference. Here the analogy breaks down. Hume's scepticism is irrefutable if it simply emphasizes this difference. But with regard to operational rules doubt has no meaning. As long as there is knowledge in the sense in which we have hitherto understood knowledge, the principle of induction will be its inescapable guiding maxim. This is in itself an analytic proposition, the sheerest tautology, because it merely makes explicit the definition of knowledge. The attempt to know, to grasp an order, to adjust ourselves to the world in which we are embedded, is just as genuine as, indeed, is identical with, the attempt to live. Confronted with a totally different universe, we would nonetheless try again and again to generalize from the known to the unknown. Only if extended and strenuous efforts led invariably to complete failure, would we abandon the hope of finding order. And even that would be an induction.

NOTE

Questions and criticisms of those who have read the manuscript of this paper made it clear to me that it is too condensed to carry conviction. It would have to be expanded into a little book (which I hope to present sometime) in order to fully substantiate both the critical and the constructive tenets here set forth rather dogmatically. My consolation for the meantime then must be: *Sapienti sat*.

I wish to state also that essentially the same solution of Hume's problem as suggested here has been expounded most recently by Prof. H. Reichenbach in *Erkenntnis*, Vol. 3, pp. 421–425. It is particularly gratifying to me that Prof. Reichenbach, after an odyssey of attempts to found induction on probability has finally recognized our (Viennese) criticisms and is joining us now in the pragmatic view of inductive generalization.

NOTES

[1] This paper was read at a meeting of the Western Division of the American Philosophical Association at Ann Arbor, March, 1932. In what is presented as the constructive part, I am indebted to the criticisms and suggestions of Professor P. W. Bridgman and Professor C. I. Lewis of Harvard University, as well as to my friends of the Vienna Circle, especially Professor R. Carnap and Professor M. Schlick.

[2] It should be noted that the frequency theory is still in process of completion and reconciliation with the logical theory. There are difficulties, but they do not appear insurmountable.

[The preceding article was discussed in a subsequent issue of the same journal in an exchange between H. H. Dubs and Professor Feigl. – Ed.]

THE PRINCIPLE OF INDUCTION

Dear Sir:

Since methodology is fundamental to all science and logic, may I be permitted to make a remark upon the methodology of Professor Feigl's very interesting paper, 'The Logical Character of the Principle of Induction'? This remark is not intended in any sense as derogatory to what is an excellent paper; it is merely an attempt to assist in solving the important issue he has raised.

He founds his paper on the statement that the unsatisfactory state of the problem of induction is due to philosophers' preoccupation with the problem of the validity of induction, whereas the more fundamental question concerning the meaning of the principle of induction, to which he addresses himself, has been rather neglected (p. 154).

But are these two problems really distinct? For the problem of the meaning of induction (in which the chief difficulty, according to Prof. Feigl, lies in the interpretation of the problem of probability) is, stated more exactly, the problem of the *correct* interpretation of these concepts. There are many interpretations of induction and of probability; this paper is itself an attempt, not to define previously undefined concepts, but to discover their *correct* interpretation. Then he must logically obtain, previous to securing any 'correct' interpretation, some adequate method of determining correctness. Since, now, as he says, "Induction is the indispensable foundation of all factual science" (p. 153) some theory of induction becomes the presupposition of obtaining any 'correct' account of induction. Then previous logicians, who concentrated upon the problem of the validity of induction, were quite right in their procedure. Instead, therefore, of the problems of the validity and meaning of induction or probability constituting two problems, they are but two sides of the same problem. The validity and the correct meaning of an ultimate concept such as induction must be determined at the same time by the same procedure.

In conclusion, may I suggest that the solution of the problem of induction may be found in a direction indicated by Prof. Feigl, though not worked out by him? He admits that "The more advanced factual sciences do not actually proceed by inductive generalization. Their method consists rather in the construction of hypothetico-deductive systems." (p. 153) It is then possible that what has been called 'induction' is not the true foundation of the more advanced and therefore more adequately established sciences. 'Induction' may be a procedure adopted by the less advanced sciences because of their own inadequacy. The fact then that induction "can be proved certain neither on logical grounds nor on the basis of its own success" (p. 153) does not prevent finding an adequate logical foundation for the factual sciences in the hypothetico-deductive method. It is this consequence of generally accepted logical views that I have worked out in my theory of 'rational induction,' to which the distinction between what has been called 'induction' (which I style 'empirical induction') and the hypothetico-deductive method (which I style 'rational induction') is fundamental. It may then be true that scientists have really been using a logically correct method; logicians have merely failed to recognize its precise nature and the grounds of its correctness.

Marshall College,
Huntington, W. Va.
HOMER H. DUBS

Dear Sir:

In replying to Professor Dubs' gratifying remarks I should like to elucidate a few important points of my pragmatic view of induction.

I quite agree with Prof. Dubs' contention that the problems of the meaning and the validity of induction "are but two sides of the same question." My purpose in differentiating the two "sides" was only to point out that almost all previous formulations of the principle of induction gave it such a meaning that its validity could not be considered anything but *inductive*. – Also Prof. Dubs' own solution of the problem [Dubs, 1930, pp. 450–457] in which he substitutes for the traditional postulate of the uniformity of nature his "principle of the eternity of truth" suffers from this defect. For as he admits (*ibid.*, p. 453) "unfortunately for philosophy and science it must be confessed that at present the permanence of truth is merely an empirical statement. We are merely able to say that what truth has been discovered in the past is still true; we cannot say *why* it has remained true, nor be absolutely sure that it will continue to remain so, since no merely empirical law can attain to certainty. Consequently we have to confess that our armor of defense against skepticism is not complete, for this empirical element is mixed in with all Knowledge." This passage shows clearly that Prof. Dubs' claim of infallibility for the method of "rational induction" (i.e., the hypothetico-deductive method) is not seriously upheld even by himself. His attempts to demonstrate the certainty of rational induction (*ibid.*; chapter IX, X) and to refute Hume's skepticism (p. 454 f) rest on rather questionable dialectical arguments and on the promise of the future discovery of a metaphysical proof of the "eternity of truth." Now, supposing such a proof could be given, it would have to be based on premises which, by logical necessity, would have to contain *universal* propositions. But, even disregarding my antimetaphysical convictions, I find it plainly inconceivable that any metaphysical method could validate most general factual propositions without recourse to empirical induction.

It was just this rather obvious logical situation which made me determined to reject any of the formulations of the induction principle which express a factual meaning, i.e., make a statement concerning the characteristics of the world. In my opinion there can be no guaranty for the validity of such and other generalizations, be they simple enumerative inductions or hypotheses of the more advanced scientific type. At any stage of scientific progress (as we know it) there will be outstanding premises, from which the more specific statements can be derived, with – indeed – (*deductive*) certainty; but those premises in themselves are *assumptions*, ever ready for revision, valid only 'until further notice.'

The principle underlying all induction can therefore have only *pragmatic* significance. Its meaning is not factual but 'motivational' (directive); it simply tells us 'to go ahead' with our generalizations from past experience and with the establishing of hypotheses-systems. There is no proof that these procedures, even if conducted as carefully as they may, will lead to the success aimed at, i.e., the truth of predictions. But the principle is nonetheless more than an arbitrary prescription and it is certainly not a categorical imperative. To use classical terms, its nature is best expressed as the *hypothetical imperative: If you intend to predict correctly use the method of simplest generalization on the basis of as broad an experiential (observational, experimental, statistical, etc.) background as you can secure.* This imperative, however, does not and cannot promise the desired success. Its only claim is: If you desire to proceed according to a *method*, i.e., if you are not interested in capriciously guessing, gambling or mystically intuiting the future (more generally the yet unknown, but knowable) then you have *no choice*. Any other 'method' like expecting systematic deviations from past regularities or not employing all available experience) would be 'madness' because it would not in any typical sense be distinguished from an indefinite number (a continuum, virtually) of equally arbitrary 'methods.' (The old principle of sufficient reason construed as an operational maxim.) The *justification* of the inductive procedure lies precisely in this uniqueness (extremum character) of the only rational method capable (but not guaranteed) of success. This then is the small step we can go beyond Hume's view of the matter. The step may seem too small to those who still cherish hopes for a better vindication of induction. But those who acknowledge the very simple fact that we don't know what we really don't *know* will be satisfied with having at least a rational way of taking a chance of knowing what we don't know.

I hope that these remarks will suffice to make clear why I thought it so important to formulate the principle of induction in such a way that its meaning would express the basic regulative maxim of factual knowledge. If, in the analysis that led me to my result, I have made use of induction, such as in taking it for granted that the essential features of human knowledge will remain the same in the future as they have been in the past, I would reject the charge of vicious circularity. I was not forced to use some theory of induction as a *presupposition* of my account of induction. And this for the simple reason that my aim as well as my method were analytic and not inductive. I did not try to establish the certainty or probability of inductive generalization – this I had shown to be an illusory problem – but my goal was merely to analyze factual knowledge and to disclose one of its most fundamental rules of procedure.

H. FEIGL

10. WHAT HUME MIGHT HAVE SAID TO KANT

(*And a Few Questions about Induction and Meaning*)

[1964a]

The aim of this brief essay is to present some conjectures as to how Hume might have criticized certain central epistemological doctrines of Kant.[1] By a natural extension of this theme I shall also raise a few questions concerning Karl Popper's views on induction and his critique of the empiricist meaning criterion. I do this in the spirit of my very great admiration for Popper's work and his friendship, which I cherish sincerely. If I have misunderstood his basic outlook, I hope that he will set me right in future discussions.

There can be little doubt that one of the main purposes of Kant's theory of knowledge was to overcome Hume's skepticism, and to establish the apodictic validity of a number of basic principles of factual knowledge. It has been said that though Kant admitted to have been 'awakened from his dogmatic slumbers' by Hume, he nevertheless went quickly to sleep again by blandly accepting the synthetic *a priori*. It is true that Kant hardly questions the existence of such knowledge, and that his primary aim was not to demonstrate its existence, but rather to *account* for it. Hume, had he lived to read the first *Critique* and the *Prolegomena*, would hence have had to assess Kant's reasoning in the transcendental deduction.

Following current custom we may distinguish the 'anthropological' from the 'presuppositional' strands in Kant's epistemology. Kant's own explicit programmatic remarks in the preface to the *Critique of Pure Reason* indicate that he was very clear about the distinction. *Quid facti* and *quid juris* are the terms Kant himself used in this connection. But, as is evident also in many other great thinkers, it is not easy to avoid the psychologism which results from mistaking questions regarding the validation of knowledge-claims for questions regarding their origin. In the 'anthropological' strands of the *Critique* Kant attempted to explain the existence of synthetic *a priori* knowledge of space, time, substance, and causality on the basis of the structure of the human mind. There we are told that the forms of pure reason, i.e., the forms of intuition and the categories of the understanding, are impressed upon the raw materials of sensory experience; and that knowledge, being the product of the combination of these two factors or components, could not conceivably fail to exhibit the *a priori* forms which entered into it in the first place and inevitably. If, for example, we interpret with Kant the principle of causality as the assertion of the strict lawfulness of nature, then Kant's transcendental deduction in the anthropological version boils down to the well-known claim that it is the human mind which imposes this order upon experience.

How would Hume have reacted to the anthropological version? I think it is fairly obvious that he might well have confronted Kant with the following dilemma. If the order of nature is to be a product of the organization of the human mind, then Kant must have assumed that this organization is fixed and unchanging. And if this assumption itself is to be justified, it can be done *either by induction or by deduction*. In the first alternative, our confidence in the order of nature cannot rest on *a priori* grounds, although it remains synthetic in that it consists in an assertion about matters of fact. Hume's skepticism would apply with its full force to the extrapolative assumption that the human mind will essentially retain the same forms for all time to come. Such an assumption is then clearly synthetic *a posteriori*, and could not possibly warrant apodictically true knowledge of the uniformity of nature. If Kant were to avail himself of the second alternative, i.e., to justify the constancy of reason by deduction, he (according to both Kant and Hume) could do this only by incorporating such constancy in the very *concept of reason*. But then we are merely explicating what we have put by definition into this concept, and we arrive at a statement which in Kant's own terms would be *analytic*. It is obvious that the synthetic principle of causality could not be derived from an analytic statement regarding the constancy of human reason.

Hume would certainly not have allowed Kant to escape between the horns of this dilemma by declaring the constancy assumption itself to be synthetic *a priori*, for in that case Kant would be faced with an infinite regress in transcendental deductions. Hence, Hume might well rest his case against the anthropological version by contending that in the light of the criticisms just presented, Kant could arrive at either synthetic *a posteriori* or analytic *a priori* formulations of the principle of causality, but never at a synthetic *a priori* formulation. Paraphrasing the well-known saying of Einstein's, Hume could maintain that, inasmuch as the principle of causality makes an assertion concerning the world, it is in principle always open to doubt; and inasmuch as it is not open to doubt (viz., as an analytic proposition), it provides no information about matters of fact.

What about the 'presuppositional' version of Kant's *Critique*? Here we are told that the validity of the principle of causality is a necessary condition for the very possibility of knowledge concerning matters of fact. It is easy to understand how tempting this interpretation has been to Kant and many of his followers. Knowledge at its best, as in the great theories of the physical sciences, is knowledge in terms of *laws*. Scientific explanation, explicitly or implicitly, always utilizes lawlike premises. Science would be reduced to mere descriptions and narratives if there were no regularities of nature. Science as we understand it, at least since the Renaissance, is essentially a body of knowledge claims which formulate the regularities of nature in lawlike statements, be they deterministic or statistical, 'low-level empirical' or 'high-level theoretical.'

It may well be granted that this is the current common understanding of the significance of the term 'scientific knowledge.' It certainly was also Kant's understanding. For him Newtonian physics was the paradigm of factual knowledge. Now, all that the presuppositional version of Kant's epistemology warrants is the claim: 'If there are no regularities there can be no scientific knowledge.' However, from what we have said above, this must be recognized as an *analytic* statement, i.e., as a simple consequence of the meaning of the term 'scientific knowledge.' We have no assurance that scientific knowledge in this sense will always be possible in all domains of experience. The optimistic confidence of scientists that such knowledge will always be attainable is defensible only on inductive grounds, and hence, if formulated in a proposition, would be synthetic *a posteriori* – making it again open to Hume's skeptical doubts. Hume's reply to Kant's presuppositional version would once more confront him with an ineluctable dilemma.

(I realize that I have – for brevity's sake – presented only the bare bones of the issue. But I don't think that it is necessary for my purposes to enter into the finer details of exegesis of either Hume or Kant. In what follows I shall, in the same spirit, concentrate on fundamentals only, and discuss some questions concerning Popper's critique of inductivism, and related issues. This is pertinent since Popper's point of view appears as a *tertium quid* between Hume's and Kant's epistemologies. I expect he would agree with the criticism advanced by Hume *redivivus* – as I construe them.)

In his brilliant and powerful analysis of scientific knowledge Popper has emphasized again and again that there can be no justification of inductive inference. By explicating the logic of science in terms of the hypothetico-deductive model, he arrived at the position that laws and theories can be refuted (and that, indeed, if they were not conceivably refutable, they would not be scientific), but that they can never be verified. Of course, Popper allows for corroboration in the sense that we may say that a theory which has withstood very severe tests is, at least until further notice, acceptable as part of the justified corpus of scientific knowledge-claims.

It is precisely on this point that I wish to ask Popper whether he does not in effect fall back on the much maligned inductivism. Here is a simple illustrative example: The law of the conservation of energy is one of the best and most severely tested principles of physics. Although there have been occasional suspicions (as by Bohr, Kramers, and Slater in 1924) that the law might not hold strictly on the micro-level of atomic physics, such suspicions have been quite effectively removed by further experimental testing. I should think that the law deserves to be considered extremely well corroborated. To any adherent of inductive logic this would provide excellent grounds for the assumption that the energy law will not be refuted in the future. If 'corroboration' does not entail this sort of justified expectation, I don't know what reasons we could give to would-be inventors of perpetuum mobiles (of the 'first kind') to abandon their efforts. If Popper feels entitled to give such (negative) advice to inventors, he is trading on the *inductive* implications which the word 'corroboration' has in common language. (These inductive implications are the same as those connected with such words as 'confirmation', 'evidential support', 'substantiation', etc.) In the light of his critique of inductivism, is Popper really justified in utilizing these customary connotations of 'corroboration'? All he is entitled to say on the grounds of his own analysis is that the energy principle is thus far unrefuted though it has been subjected to very incisive and multifarious tests.

My point can also be made by an opposite type of example. Some experiments are generally considered to be extremely decisive refutations. For instance, the Michelson-Morley experiment and some of its more recent replications are considered as highly conclusive refutations of the stationary ether hypothesis. (Actually, the situation is more complex, but I am sure there is no disagreement between Popper and me on the significance of other hypotheses, experiments, and observations in this context. For example, I agree with Popper that even observation statements are revisable. Still, there would be no 'corpus of science' if we did not rely on some observation statements in a given context and 'until further notice.') Of course, according to the simple model of *modus tollens* reasoning, a universal statement can be definitively refuted, once and for all. In its purely logical form this is unassailable. However, consider the situation in the empirical sciences. Popper's policy of the critical, rational approach must (and does) leave open the possibility that the Michelson-Morley type of experiment might give positive results beginning tomorrow and forever after. It is only by *induction* that we can assume that a well-refuted theory will *stay* refuted. After all, it is logically conceivable that such a 'knocked-out' theory might begin to 'stand up' at any time; and from there on out for all *future* concerns this would be just as good as a theory which had never been refuted.

Perhaps I cannot have what I am asking for: an inductive logic which provides an *objective* justification for the confidence which goes with corroboration as well as with refutation. Popper and I surely agree that Hume's psychological account – even if modernized in the light of recent psychological theory – does not give us a justification. The element of 'animal faith' remains. Moreover, any psychological account of belief, expectation – in short, of induction – in relying on psychological laws makes use of inductive extrapolation. Hume knew this very well, and the Kantians who took him to task on this point misunderstood or distorted Hume's intentions. The various attempts in the direction of an inductive logic by Zilsel, Keynes, Nicod, Jeffreys, Reichenbach, Williams, Kneale, Braithwaite, Carnap, and others are, in their different ways, all subject to serious philosophical scruples. The sort of vindication of inductive inference that Reichenbach, Kneale, Salmon, and I[2] have been proposing is a very weak one, it must be admitted. Perhaps nothing stronger can be attained than such a pragmatic justification of the adoption of inductive procedures. It rests on the analytic truth that if any method of extrapolation succeeds, then inductive extrapolation will, too. This yields the 'degenerate' justification: If there are regularities of nature, not too complex and not too deeply hidden, they are discoverable by the hypothetico-deductive method. Hence, it is reasonable to employ this method and tentatively, i.e., 'until further notice', to rely extrapolatively on its results, be they corroborations or refutations. Despite its 'weakness', such a vindication appears to me philosophically more enlightening than the short-cut of the Wittgensteinians (Strawson, Edwards, Hanson, Toulmin, and others), who essentially fall back on the common usage of such words or phrases as 'good reasons', 'strong evidence', etc. The problem is precisely to show what entitles us to use these honorific descriptions.

My final questions concern the issue of criterion of demarcation versus criterion of meaning. The logical empiricists, very much in the spirit of Hume (as well as of Peirce), and more radically than Kant, have opposed a certain type of metaphysics by declaring its questions as well as its answers devoid of factual meaning. Popper has been at pains to separate himself from this radicalism. He proposed a criterion of demarcation, drawing a line between questions that pertain to empirical enquiry, and those that fall outside this domain. He considers the proposal of a meaning criterion fruitless and even harmful. His arguments deserve our fullest attention. Popper is surely right in repudiating the narrow meaning criterion of the early logical positivists. With painful clarity he pointed out that *that* criterion not only eliminated metaphysics but, alas, science as well. This was not intended by the positivists, of course. As I see it, the shift to logical empiricism involved a welcome and indispensable liberalization of the criterion. I wonder if Popper fully appreciates why the logical empiricists, while agreeable to the demarcation criterion, wish to go further and insist on a criterion of factual meaningfulness. If we make it impossible 'in principle' to test a given hypothesis, i.e., if the manner in which an hypothesis is construed logically excludes refutation or corroboration, we have made it impossible to give empirical reasons for either the rejection or the acceptance of such an hypothesis. This may be achieved by making it immune to tests either by its outright formulation, or by hedging it about with special additional *ad hoc* assumptions. Such was clearly the case, for example, in the final stage of the ether hypothesis of H. A. Lorentz. While the earlier formulations of the ether hypothesis were clearly testable, the negative results of such tests motivated Lorentz to safeguard the hypothesis absolutely and unconditionally by his peculiar assumptions. It became 'in principle', that is, logically, impossible to provide evidence for or against any assertion concerning the velocity of a body with respect to the stationary ether. Paraphrasing Peirce and James, there is not – and there could not be – any difference that makes a difference between the assertion and the denial of a given proposition concerning velocities relative to the ether so conceived. To be sure, the differences in question pertain to observational evidence; and since I am in full agreement with Popper's realistic epistemology, I would insist on distinguishing sharply the truth conditions of statements from their confirming (or corroborating) evidence. Nevertheless, if we isolate statements radically from whatever evidence might speak for or against them, do we still know what the statement is about? I believe that the appearance of meaning in this case is engendered by the pictorial appeals of the statement – but its factual meaning seems to me to have been 'emptied out' just as decisively as in the case of tautologies or contradictions.

I admit that many statements which the logical empiricists diagnosed as factually meaningless are open to more charitable interpretations. Often it is the vagueness so characteristic of theological and metaphysical assertions which makes it possible to interpret them as having at least a modicum of factual meaning, or as presenting some sort of 'promissory note' of a specification of meaning to be 'made good' later.

In its least offensive form, the meaning criterion, very much like Popper's criterion of demarcation, may be construed as a *proposal*. (Carnap settled for this many years ago.) Personally, I think a somewhat stronger case can be made. If we distinguish the cognitive functions of language from the noncognitive (pictorial, emotional, motivational) functions, the acceptance of the meaning criterion amounts to preventing confusions of the latter with the former. This still seems to me an eminently clarifying procedure, and thus a contribution of logical empiricism which has enduring value. I admit that Popper's demarcation criterion is adequate for the distinction of scientific from nonscientific enterprises. But the illusions and the confusions of transcendent metaphysics can be more effectively exposed by attention to the functions of language and the meanings of 'meaning.' In this way, I think we can steer clear of Hume's reductive fallacies, i.e., his skepticism and his phenomenalism, and nevertheless retain what is valuable in his empiricism and his critique of metaphysics. Popper's position seems in this respect closer to Kant than to Hume. With Kant he considers problems of transcendent metaphysics meaningful but undecidable. Logical empiricists admit that many questions will forever remain unanswered. But they insist that questions which can be shown to be unanswerable because of immunization to tests of any kind are not merely transempirical but are devoid of precisely the sort of meaning which certain metaphysicians impute to them.

NOTES

[1] I have attempted a similarly anachronistic exercise in my article 'Matter Still Largely Material' [1962a], in which a 'Locke *redivivus*' criticizes Berkeley.

[2] See my article 'On the Vindication of Induction' [1961c], as well as the references listed at the end of that article.

11. OPERATIONISM AND SCIENTIFIC METHOD

[1945a]

INTRODUCTION

[About a year ago Professor E. G. Boring wrote to suggest a symposium on operationism and appended a few questions which he said he would like to have answered. The Editor welcomed the idea and invited a number of persons interested in operationism to take part in the symposium. They were sent Professor Boring's questions and asked to add questions of their own to the list. Professors E. G. Boring, P. W. Bridgman, Herbert Feigl, Harold Israel, Carroll C. Pratt and B. F. Skinner accepted the invitation and submitted additional questions. The complete list was then sent to each participant with instructions to write on any or all of the questions. When the galley proof was ready, all of it was sent to each person with the request that he write a short rebuttal. All but one participant consented. These replies, which are more than rebuttals, follow the main articles under the title 'Rejoinders and Second Thoughts.'

These are the questions. They are referred to in the text by their numbers.

1. (a) What is the purpose of operational definitions? When are they called for?

Since it is obviously impossible to explicate an operational definition for every construct-term used in scientific discussion, there must be some principle which determines when operational definitions are useful.

(b) Logically, operational definitions could form an infinite regress, since the construct-terms used in describing an operation are themselves in need of definition.

How is this regress limited in scientific practice?

2. When the same construct is defined by two independent operations, should it be said that there are really two constructs? For instance, it has been said that tape-measured distance and triangulated distance are really two kinds of distance and should perhaps have different names.

Against this view it can be argued that there are operations for showing the equivalence of operations, *e.g.*, for demonstrating the identity of taped and surveyed short distances.

3. (a) Arc hypothetical operations which are physically impossible with

present available techniques, of scientific use? Is the other side of the moon what you would see if you went there?

It is arguable that an unperformable operation has value in stating the conditions by which a construct could be validated. Such a statement shows that the construct is not at the moment valid.

(b) Is there a use for hypothetical operations that would define constructs which are actually at the moment nonexistent?

Red and green are supposed to be derived from yellow in the course of evolution. The discriminatory operations which would establish the existence of two new colors, derived similarly from blue, could be stated, although they could not be performed at the present stage of evolutionary development. The operations which would define a new invisible planet are similar.

(c) Is there a use for hypothetical operations which could never be performed?

The definition of infinity depends on operations which can never be completed.

4. Is *experience* a proper construct for operational definition?

It has been held that experience is ultimate, subject to immediate intuition but not to operational definition.

5. Are there scientifically good and bad operations, and how are operations evaluated if they differ in value?

Objectivists hold that the data of experience can always be operationally defined if the data become public, because the operations of publication define the datum. It is, however, argued further that the operations of verbal report are 'poorer' than the operations of discriminatory choice (C. R.; jumping stand) because the verbal response itself involves terms that are less rigorously defined.

6. Is operationism more than a renewed and refined emphasis upon the experimental method (as understood already by Galileo, if not even by Archimedes) – i.e., a formulation of modern scientific empiricism and pragmatism (especially of the Peirce-Dewey variety), mainly of criteria of factual meaningfulness and empirical validity?

7. Must operationists in psychology relegate theorizing of all sorts to the limbo of metaphysics? Bridgman in physics is perfectly aware of the value of theories as long as they are in keeping with his operational requirements. The Gestaltists, particularly Köhler and Koffka, have repeatedly attacked positivism (an identical twin of operationism), reproaching it for its (alleged) opposition to theoretical construction. C. C. Pratt [1939, pp. 147–154] on the basis of his operationism maintains that all theoretical explanation is

circular or tautological. Köhler [1940, pp. 107–125] holds a strictly opposite view. Which position is the most adequate for psychological research?

8. Some radical operationists assert that the meaning of a quantitative concept lies exclusively in the set of measuring operations which determine the application of the concept. (E.g.: 'Intelligence is what the intelligence test tests.') But how can we then know what it is that we are after in constructing tests; and what possible meaning is there in talking about improving or revising tests and measurements if there are no criteria outside the chosen test methods?

9. Are *all* scientifically legitimate definitions operational in character? This is (at least in part) a terminological question, but certainly one that it would pay to settle (not only) among psychologists.

10. What is a definition, operational or otherwise? It is important to know whether one is presupposing a logical apparatus for dealing with the language of science or intending through a psychological analysis to justify such an apparatus.

11. For the purpose of operational definition, what class or classes of events may be used properly as defining-operations? Specifically, can a phenomenon be identified or its properties be defined in terms of the events (operations) which are effective to produce, or occur as results of, the phenomenon?

THE EDITOR
The Psychological Review]

[PART I]

Question No. 6. 'Operationism' is a new name for certain fairly generally recognized aspects of scientific method. It is indeed no more than a refined and modernized emphasis upon the requirements that scientific concepts must meet if they are to be meaningful and fruitful. Descriptions, laws, hypotheses, and theories may be critically examined on the basis of such criteria applied to the concepts they contain. In the perspective of the history of science and the history of philosophy, operationism represents a recent formulation of some of the essential features of the experimental method and of empiricism generally, accentuated in the direction of pragmatism and instrumentalism (Peirce, James, and Dewey). Bridgman's formulations of the criteria of empirical meaning, though probably quite original with him, have much in common especially with C. S. Peirce's in 'How to Make Our Ideas

Clear.'[1] There are two questions with which we are (or at least *should* be) concerned in any cognitive enterprise: '*What do we mean by the words or symbols we use?*' and '*How do we know that what we assert in these terms is true* (*or confirmed to some degree*)*?*' The characteristically pragmatic turn becomes manifest when these two questions are scrutinized in the light of two further questions, 'What do we *do* in order to find out whether a term is legitimately applicable?' and 'What are the *fruits by which we shall know* whether the introduction of a term is scientifically useful?' The first of these two latter questions stresses the element of active intervention in measurement and experiment. The second intends to remind us that concepts are scientifically worthwhile only if they help in the task of prediction (and, possibly, practical control). The essence of the operationists' emphasis may thus be seen in the formulation of criteria of scientific meaningfulness and fruitfulness for concepts and of criteria of validity for factual statements. While these criteria were implicitly respected in much of the scientific work of recent centuries (and even occasionally to some degree explicitly formulated), the trend of our age toward a fuller awareness of and critical reflection upon the underlying presuppositions and guiding principles of our thinking and doing has, among other results, promoted a more penetrating analysis of scientific method. The particular interest in operationism may be understood as arising from the need (1) of purifying scientific method by the elimination of pre-scientific and non-scientific (e.g., metaphysical) elements, and (2) of understanding more clearly the meaning of the highly complex concepts employed in the more abstract and constructive levels of modern scientific theories. Such critical awareness is especially valuable when one needs to examine science where it is in the making. The numerous approaches in recent psychological theory are most certainly a case in point.

Questions 10, 1, and 9. Clarification or analysis of meaning is pursued by *definition*. Definition, i.e., the specification or delimitation of the meaning of a term or symbol, may be considered a statement of the *rule concerning the use* of a term or symbol. The need for definition practically arises only when (*a*) we are not sure what a given term or symbol means, if it means anything at all (obscurity); (*b*) where there is a plurality of meanings (ambiguity) – to be removed by multiple definition; (*c*) when the term or symbol in question is used so vaguely that a definition giving it greater precision is demanded; (*d*) when the term or symbol, though defined clearly, unambiguously, and precisely in some respects, is to be given a place in a wider context and thereby enriched in meaning. This wider context may be a system of symbols only (such as we find in a pure calculus) or it may include symbols with empirical reference – in this latter case the definition may be 'coordinating', i.e., relating a symbol of an abstract system to terms which through previous definition already possess empirical reference; (*e*) finally, definition is needed whenever a situation in research calls for the coining of a new term as an abbreviatory convenience for more complex aggregations of terms either already in use or logico-mathematically so aggregated for the first time. This latter condition arises particularly in the context of the discovery of new elements or relationships in the subject matter of research.

Definition as here conceived is *nominal* (i.e., definition of *terms* or *symbols*). So-called *real* definitions (of *things, properties*, etc.) reveal themselves either as empirical descriptions with all terms understood (by previous nominal definition) or as characterizations of things for the sake of identification. Since, in this latter alternative, identification occurs through labelling of things on the basis of observational test it is tantamount to *nominal* definition of the label. Just how precisely, completely or directly operational definitions enable us to identify objects is a matter of great methodological importance, but also surely a matter of degree.

To demand definition of *every* term used in a piece of scientific discourse would not only be unduly pedantic (besides being incapable of practical fulfilment and thus utopian) but also quite unnecessary. The adult and sane use of common language is on the whole sufficiently definite to permit intersubjective communication and intelligibility as regards terms representing things and their observable properties. Doubt as to meaningfulness or as to precise meaning arises usually only regarding higher-order constructs and/or terms of fairly clear subjective meaning but lacking sufficient determination for successful intersubjective testability. Doubts of this second sort are rare in physics but plentiful in psychology. Doubts of the first sort may arise in any science which organizes its subject-matter by means of constructs above the level of observable-property predicates.

Aside from these practical considerations, it is obvious also from a logical or epistemological point of view that there can never be an occasion for an unlimited regress in definition. A series of definitional steps may be long, but it will terminate with definientia which are linked to something outside the realm of terms and symbols – namely, items of direct observation. As to whether this last step in the definition of any empirical concept is to be considered itself a kind of definition is a mere question of terminology. It is rather fashionable nowadays to speak of '*ostensive definitions*.' Psychologists in particular should not have much trouble with this: our use of language is after all a product of learning and we learn the intra-linguistic relations of words to one another as well as the extra-linguistic application of *some* words to items of experience. What from the viewpoint of logical analysis appears as a rule for the use of symbols represents itself as a habit (or rule-regulated pattern) of symbolic (i.e., verbal) behavior to the psychologist. An ostensive definition, then, may be considered either as a designation rule formulated in a semantical metalanguage or as a piece of practical drill in the learning of the 'right use' of words. Quite generally, an exact logical analysis of the meaning of scientific terms requires the use of the apparatus of syntax and semantics. The corresponding psychological analysis can be carried out in the object language of (preferably a behavioristic) psychology, where words in use are described as physical events, e.g., of emission of sounds, and are thus included in a general study of the behavior of human organisms. These two ways of studying definitions, and the uses of language generally, supplement each other very well, are entirely compatible, and fulfill, each in its way, important functions. The logical analysis examines given uses of language in the light of critical standards, such as consistency, non-circularity, sufficiency, etc. The psychological study is essentially a description and causal analysis of verbal behavior. Each can be made to reflect the other on any level of investigation. The psychologist in his study of definitional behavior uses, at least implicitly, definitions himself – and thus provides subject-matter for logical appraisal. And the logician in appraising definitions (or in any other syntactical or semantical pursuit) is, after all, *behaving* in specifiable ways and thus furnishes material for the psychologist.

If all definitions amount to rules delimiting the use of terms of symbols, aren't *all* definitions operational in character? This is indeed a purely terminological question. But terminology is a significant weapon in the strategy of scientific enlightenment. It is my personal conviction that the battle-cry of operationism can have its intended beneficial effects only if the meaning of 'operational' is confined in its application to the definition of *empirical* concepts. The problems of concept formation in pure logic and pure mathematics are of a very distinct character, and rash transference, by analogy with the empirical sciences, of conditions for restrictive meaning is, to say the least, of highly dubious value. When we demand operational definability for the terms of the factual sciences, we may (and must!) indeed include purely calculational operations – but these operations, called variously 'mental', 'paper-and-pencil', or better 'logico-mathematical', should be applied to root-terms which have *empirical* reference. Without this restriction the term 'operational' would become synonymous with, say, 'functional', implying that any definition that is worth its salt must specify the functions that the term to be defined is to fulfill. Such 'functional' definitions could be given for the terms of pure mathematics on the one side but, on the other hand, also for terms used in the most transcendent speculations of theology and metaphysics. As Bridgman has repeatedly stated, the original impetus toward his operational analysis in physics came from a consideration of Einstein's procedure in the theory of relativity. Einstein had realized that certain terms, like 'absolute length', 'absolute duration', 'absolute simultaneity', in Newtonian physics were devoid of empirical meaning because no observational or experimental procedures were or could be specified for their application. The operational criterion here serves to distinguish physics from metaphysics. It will seem obvious to many psychologists that, for example, Freud's 'death instinct', though possibly of some emotive or literary value, is devoid of factual, scientific meaning. To put it briefly, if crudely, operational analysis is to enable us to decide whether a given term, in the way it is used, has a 'cash value', i.e., factual reference. If it does have factual reference, operational analysis is to show us precisely *what* that factual reference is, in terms, ultimately, of the data of direct observation.

Question 3. The velocity of a uniformly moving vehicle may be determined by means of a yardstick and a stopwatch. The I. Q. of a person may be determined by the Binet-Stanford testing procedure. In both cases a simple arithmetical division (s/t; mental age/chronological age) yields the final result. In the case of more complex concepts or higher-order constructs, the mathematical operations by means of which these constructs are defined are correspondingly more complicated. The factual reference of scientific concepts in physics or psychology, however, depends not so much upon the purely mathematical operations but rather upon the observational and manipulatory (mensurational, experimental) operations. These establish the link between the empirical (or descriptive) terms of our scientific language and the data of experience. Again, it is a merely terminological question as to whether simple acts of perceptual discrimination and identification by themselves should be labeled 'operations.' Such acts are certainly involved in the terminal ostensive steps of any definition of factual terms. Most of the cases in which definitions are called for, present situations of greater complexity. Thus the question concerning the *possibility* of operations practically always amounts to asking whether certain measuring or testing procedures referred to in the definition of a term can be carried out. But 'possible' and 'can be' are notoriously ambiguous words. We must distinguish between (*a*) logical ('in principle'), (*b*) empirical (natural), and (*c*) practical (technical) possibility. Since the limits of practical possibility are relative and, as a matter of fact, receding with every advance in the techniques of observation and experiment, no scientist restricts factual meaningfulness to testability within the bounds of the technical facilities of the moment. Bigger and better microscopes, telescopes, electroencephalographic instruments, etc., are obvious reminders. Of course, there is always an element of risk involved if definitions of terms hinge upon operations that are technically not feasible at the moment. It is like issuing or accepting a promissory note. But, if the operations are compatible with well-established natural laws, merely practical impossibility of the testing procedure does not deprive the term in question of its scientific meaningfulness. All we may have to wonder about is the fruitfulness of concepts whose definition necessitates reference to operations too far removed from technical feasibility at the time. Concepts of atomic structure were very far beyond practical testability only forty years ago. Concepts regarding the nature of cerebral memory traces are in the promissory-note stage today. Scientific research as an ongoing process involves a continuous scale of degrees of technical testability. Considerations of a very practical and inductive sort only enable the scientist to draw a line between operationally satisfactorily and unsatisfactorily defined terms. Much more serious is the limitation of empirical (or natural) testability. A testing procedure that is incompatible with well-established natural laws will never define a scientifically acceptable concept. For example, the very well-confirmed laws of quantum mechanics exclude operations which would enable simultaneous determination of the speed and the location of electrons. Therefore it is now generally agreed that some of the customary concepts of classical mechanics as applied to electrons are not false but scientifically meaningless. Absolutely or downright meaningless (i.e., devoid of factual reference) are terms whose definition would involve contradictions with any testing procedure. Vitalistic conceptions of 'entelechies', for example, belong in this last category (of by definition unconfirmable ideas).

Question 7. Operationism has occasionally hypertrophied into a radically anti-theoretical attitude. Even since Galileo replaced the question 'Why?' by the question 'How?' and since Newton pronounced his (much misunderstood) '*hypotheses non fingo*' positivistic scientists have been inclined to restrict their endeavors to pure description and correlation. *Explanation* is considered a metaphysical misfit. Mach and Ostwald rejected the atomic theory in physics and chemistry. Some outstanding psychologists of our day still reject as non-operational the psychoanalytic theory of the unconscious; others consider neuro-physiological hypotheses when used as explanations for behavior as so much metaphysical verbiage. If the student of the history and the methodology of science be permitted to mediate in this quarrel, he would say that, as so often, the truth lies somewhere in the region of the golden mean. The issue is closely connected with the one discussed in the preceding section. The positivists are objecting to the use of terms (and theories employing such terms) which are only very tenuously and indirectly connected with the evidence technically accessible at the moment. They are, temperamentally, perhaps because of a general fear of the intangible, extremely wary and reluctant in accepting promissory notes. But their lack in confidence is more than made up by the enthusiasm of the theorists. And, just as in other matters involving guesswork, it is hard to tell whose attitude will prove more fruitful in a given case. In physics the atomists won out over the positivists. The ether-theorists lost. In psychology it is perhaps still too early to tell, but it looks as if substantial components of psycho-analysis as well as of neuro-physiological behavior-theory were to stay with us and to be developed more extensively. The case of psychoanalysis is particularly favorable if some of its outright mythological and metaphysical features are dismissed and the remainder translated into behavioristic terms. Such systems as those of Hull, Tolman and Skinner are generally quite in keeping with the narrower operational criterion of meaning. Relatively few references are made to operations of a purely hypothetical sort. The essential point worth noting is that even in these highly positivistic approaches *explanation* does have a legitimate place. On the basis of only indirectly confirmed theoretical assumptions (not only of experimental laws) more specific descriptions of phenomena are logically deduced. The difference between these approaches and the theories condemned as speculative seems to this impartial observer one of degree. W. Köhler, in his *Dynamics in Psychology*, may be said to issue a great number of promissory notes. But who can tell how soon they may be backed by independent verifications coming from neuro-physiology? His critique of C. C. Pratt's conception of scientific explanation and the function of theories seems perfectly adequate. All one can say in favor of the conservative side is that the knowledge of empirical laws concerning behavior in terms of macro-concepts is still far from being complete and that a long and fruitful period of investigations on this level of analysis will yield significant results – which will in any case be required in order to test neuro-physiological 'micro'-theories.

Questions 8, 2, and 11. If the linear expansion of a mercury column in a glass tube of even width furnishes the basis of an operational definition of 'temperature', then the question whether mercury expands in linear proportion to temperature (so defined) must be answered with 'yes' as a matter of logical necessity or tautology. Analogous considerations apply to similar questions regarding psychological concepts (such as I.Q., habit strength, excitatory potential, etc.) operationally defined by standard testing procedures. Out-and-out operationists have often resorted to such easy and dogmatic quips as: 'Intelligence is what intelligence tests test'; 'temperature is what thermometers measure.' Statements of this sort may be intended to intimidate inquirers of a somewhat mystical or metaphysical bent. But they do not even begin to give an adequate account of the meaning of concept formation in the sciences. Quantitative concepts of the type just mentioned are usually the product of long labors of adjustment by repeated redefinition. Thermometers or intelligence tests did not arise in an historical vacuum. They were devised in a context of problems that arose out of a background of previous qualitative and semi-quantitative knowledge. Problems of description and prediction led to a search for suitably precise, objective, and fruitful concepts. The concepts thus designed, in their logico-mathematical (functional) relations with other concepts, were to represent *empirical laws*, i.e., relationships between the various measurable (or at least testable) variables (or factors). In the light of these considerations it makes perfectly good sense to ask whether a mercury thermometer measures temperatures adequately. There are empirical laws, such as the First Law of Thermodynamics, which relates temperature to mechanical energy. If we wish to give this law its most universal and simple form, we are forced to consider thermometer readings of *any* sort merely as an approximation which is to be corrected or replaced by the Kelvin scale. Similar but usually more complex considerations apply to psychological magnitudes.

Empirical laws enable us to define the same concept by different operational routes. The fact that length may be defined by the yard-stick as well as by the triangulation techniques is a consequence of the empirical laws of (applied) geometry according to which hard and solid bodies as well as light rays exhibit (in terrestrial experiments with a high degree of accuracy) Euclidean relationships. Nevertheless, Bridgman's warning to the effect that different operational routes define *different* concepts is not unjustified: The convergence of operational results is to be taken for granted only until further notice, i.e., until evidence to the contrary emerges. But until then, as a policy of typically inductive procedure and with all the provisos just made, it is one of the most helpful devices of scientific method to *identify* the concepts corresponding to results of convergent operational routes. This becomes especially clear when we consider the role of scientific theories.

A *theory* may be regarded as a set of assumptions from which empirical laws are derivable by logico-mathematical deduction. From physical theory we can deduce the electrolytic, magnetic, and thermal effects of electric currents. From a theory of learning we can deduce regularities of the emission or omission of responses. Through the unifying procedure of theoretical explanation we 'understand' what on the level of empirical law is a mere brute fact of functional dependency or correlation. Such 'understanding' is, of course, bought at the price of assumptions which are under the jurisdiction of confirming or disconfirming evidence. Deduction merely explicates what is implicit in the premisses. The theoretical constructs therefore contain in compressed form the empirical concepts which can be extracted by explicit definition.

Theories and the constructs which constitute them may significantly, even if not too sharply, be grouped in two classes. In physics it is customary to distinguish phenomenological from atomistic theories. Neither characterization is terminologically fortunate. We shall here call them quite neutrally 'theories of the first kind' and 'theories of the second kind' and distinguish them by illustrations. Chemistry with its concepts of elements, compounds, compounding weights, affinities, etc., is a theory of the first kind. So is classical thermodynamics with its concepts of energy, thermodynamic potential, and entropy. Atomic theory and statistical thermodynamics are theories of the second kind. In psychology such theories as those of Hull, Tolman, Skinner, and Lewin are of the first kind. Neuro-physiological theories such as those advanced by Sherrington, Adrian, and Köhler are of the second kind. The constructs of the theories of the first kind are homogeneous with the operationally defined terms in the empirical laws of the given fields. The constructs of the theories of the second kind are in this respect heterogeneous: Atomic structure, which explains chemical properties and reactions, is a construct mainly derived from spectroscopy. Concepts of nerve-currents and their patterns are distinctly physiological. If these physiological concepts are ever to furnish an explanation of behavior, they will have to be 'identified' with constructs of pure behavior theory. As a matter of program, but not of actual achievement, this is anticipated in Hull's system. What remains to be shown is the actual convergence of operational routes for both types of constructs. On the physiological side this is still in the promissory stage and may remain there generally still for a long time. This means that the operations for the definitions of constructs in theories of behavior of the second kind are largely hypothetical (i.e., at present technically impossible). It will not do to consider them sufficiently defined by the molar behavior route anymore than it would do to consider atomic structure sufficiently defined by operations available to chemists at Dalton's time. Yet, as pointed out before, these are matters of degree. The prototype of an operational definition such as we can advance for directly measurable magnitudes should not mislead us into banishing all concepts which do not come up to this high level of methodological aspiration. Even in physics we have to define many concepts (not only highly indirectly but) sometimes only very partially. Before any theory of x-rays was developed, x-rays were simply 'what you got when cathode rays impinged upon metal surfaces', and 'that which produced photographic images of a certain kind.' Only as we advance in discovery and technique are such very sketchy definitions supplemented by fuller qualitative, quantitative, and far-flung relational characteristics. Operationism wisely understood and applied must take account and render account of the level of precision, completeness, and fruitfulness reached at the given stage of concept formation.

By way of an appendix to this section I should like to analyze by means of a simple schema the 'economic' function of concept construction. If we as yet have not reached the level of theory, we often prepare a place for theoretical constructs through the introduction of operationally defined empirical terms. The term 'electric current' had a sufficiently definite meaning already before the Maxwell electromagnetic theory or the electron theories had arisen. There were numerous different conditions which were said to produce an electric current, and numerous types of effects which were said to be caused by the current. The current itself seemed to many an intangible mystery. Similarly, there are many ways of acquiring a habit. And there are many ways in which this habit manifests itself. But what *is* the habit? These foolish questions disappear if one adopts the operationist outlook. The constructs ('electric current'; 'habit') are introduced in order to save statements. If there are m causal conditions and n possible effects we would need mn statements in order to formulate all possible observable relations. If, however, we introduce our auxiliary concepts, the number of statements required shrinks to $m + n$. For large numbers m and n the conceptual economy is accordingly quite considerable. Since the established scientific procedures keep all constructs in principle open for additional definitional routes, m and n are never limited in a dogmatic manner. This feature in an enlightened operationism prevents the typical ultra-positivistic (I would call it 'negativistic') fallacy according to which things (particularly those inferred entities designated by constructs) 'are simply what they are known as' or 'are nothing but fictions introduced in order to speak more conveniently about certain sets of data.' Things are, rather, what they are known *and knowable* as; and the sets of data are in principle capable of unlimited extensions in various (sometimes even surprising) directions. Once the empirical constructs (i.e., these auxiliary concepts) are linked to one another, either through empirical discovery or by deduction from theoretical assumptions of further functional relations between them, a whole network of variously connected concepts arises and the scientific discipline in question has attained a high degree of maturity.

Operational definitions then, or the concepts which they define, may be classified into various sorts: Purely qualitative; semi-quantitative ('comparative' or 'topological' – such as the hardness scale; or introspective concepts of 'more' or 'less'); fully quantitative or metrical (e.g., temperature, loudness); causal-genetic concepts (e.g., habit); theoretical constructs (e.g., electric fields, excitatory potential). It depends entirely upon the level of research attained which of these forms or which of their possible combinations may be applied. Concept formation in psychology, and particularly in psychiatry, is on the whole still in a relatively unfinished stage. The situation is here similar to the one in medicine where the definition of diseases is at first primarily in terms of symptoms, later in terms of more reliable causal-genetic data and, perhaps, finally in terms of a micro-account of the physiological type.

Questions 4 and 5. Since science is, as one might say, by definition, a social enterprise, it must insist upon operations which are repeatable not only by one observer but in principle performable by any properly equipped observer. A statement is scientifically meaningful only if it is intersubjectively testable. Subjective peculiarities, subjective differences can, of course, become the subject matter of scientific study – *if* – as in the psychology of individual differences – we can verify them in a perfectly objective manner. Private, immediate experience as such is only the raw material, not the real subject-matter of science. One's own immediate experience, the actual-lived-through stream of data, may therefore be conceived as the epistemological basis of all concept formation and theoretical construction in the empirical sciences. In that sense it is not a construct but that small foothold in reality that any observer must have in order to get at all started in his business of exploring the world of things and organisms surrounding him. If one wishes to convert this narrow realm of directly given experience into subject-matter for description one is limited to a purely phenomenological approach of the introspective type. The scientific value of such a study is very insignificant as long as it remains unrelated to a study of extra-dermal and intra-dermal stimuli (physical and physiological processes) or to behavior responses. If, however, one does so relate the phenomenally given to these processes inferred (or constructed conceptually), it can itself be conceptualized as overt or potential behavior (or by way of promissory hypothetical anticipation as cerebral processes). That is precisely what happens in the 'psychology of the other one', where one constructs on the basis of one's own data the other one's 'experience.'[2] And here again we have the choice of various conceptual systems or languages for description: mentalistic, behavioral, or – ultimately – physiological. In this sense, then, we may say that 'the other one's experiences' (or better: the concepts describing it) are constructs. In this sense, also, and in this sense only, concepts referring to one's own experience are constructs. They designate processes which at least to a very small extent fall within the scope of direct phenomenal acquaintance.

Much as the epistemologist or phenomenologist may wish to clarify and formulate the meaning of 'experience' along these lines, there is the alternative of a strictly physicalistic or behavioristic approach right from the start. From the point of view of the methodology of science this is preferable, since it eliminates with one stroke the pitfalls of the traditional metaphysical pseudo-problems of solipsism, the mind-body puzzle, etc. If only intersubjective operations are admitted, one's own experience is in every respect on a par with that of the other one. Introspection itself is then described as a response to previous responses ('early retrospection'). The only asymmetry that remains is that an individual person is able to predict some (but by no means all) of his own behavior better than another observer could. But that is a matter of degree. Many of the intra-dermal and particularly the cerebral conditions are not as yet technically as directly accessible to outside observers as they are (however vaguely and diffusely) to the individual subject himself.

CONCLUSION

Operationism is not a system of philosophy. It is not a technique for the formation of concepts or theories. It will not by itself produce scientific results. These are brought about by the labor and ingenuity of the researchers. Operationism is, rather, a set of regulative or critical standards. In the light of these critical standards the meaningfulness and fruitfulness of scientific concepts may be appraised. It seems that the outstanding requirements which operationism has quite justifiably stressed may be formulated as follows: Concepts which are to be of value to the factual sciences must be definable by operations which are (1) logically consistent; (2) sufficiently definite (if possible, quantitatively precise); (3) empirically rooted, i.e., by procedural and, finally, ostensive links with the observable; (4) naturally and, preferably, technically possible; (5) intersubjective and repeatable; (6) aimed at the creation of concepts which will function in laws or theories of greater predictiveness.

The degree to which these ideals are approximated varies from one science to another. But it would seem that all of these criteria are applicable not only to a well-developed and systematized science such as physics but also to a science still largely in the making such as psychology.

NOTES

1 The epochmaking article by C. S. Peirce, 'How to Make Our Ideas Clear' appeared first

in [1878]. For the relations of operationism to modern positivism, see my paper 'Logical Empiricism' [1943a]. The logic of empirical constructs is discussed in some technical detail by R. Carnap in 'Testability and Meaning' [1936a]. For a sketch of the same, more elementary, and applied to psychological concepts, see S. Koch 'The Logical Character of the Motivation Concept' [1941, pp. 15–38; 127–154]. The operational criteria of measurement are very clearly formulated in G. Bergmann and K. W. Spence, 'The Logic of Psychophysical Measurement' [1944]. A very penetrating analysis of the epistemological problems of psychology has been given by H. Reichenbach in *Experience and Prediction* [1938, see especially Sections 26–28].

2 Operations which would enable one to 'inspect' the other one's private experience are by definition, i.e. logically, impossible. What would happen if nervous systems were connected is a matter of conjecture but certainly the logical impossibility of *having* the other one's experience will thereby not be removed.

[PART II]

Rejoinder by Professor Feigl

[1945b]

Owing to limitations of space, the theme of my rebuttal will be restricted to the concept of scientific explanation. Instead of presenting a point by point polemical discussion of some of the views of my fellow-symposiasts I shall try to clear up the issue mainly by a concise statement of my own analysis of the explanatory procedure. The reader may then compare this account with that of Bridgman and especially the one of Pratt. [In the original symposium.]

Some positivistic and operationistic definitions of 'scientific explanation' have all too narrowly stressed: (1) that there is no fundamental difference between *description* and *explanation*; (2) that all scientific explanation is *circular* or *tautological*; (3) that in explanation we *reduce the unfamiliar to the familiar*. I would urge that there is something basically wrong, or in any case, something very misleading, in all three contentions. A modern logical empiricism may retain the valuable anti-metaphysical tendency in the older point of view while at the same time giving a methodologically more adequate reconstruction of the explanatory process as actually employed in the various sciences. It is agreed that scientific explanation differs sharply from the pseudo-explanations of the animistic, theological or metaphysical types in that the explanatory premises of legitimate science must be capable of test, and must not be superfluous (i.e., not redundant in the light of the principle of parsimony). The significance of the premises and verbalisms of pseudo-explanations is usually purely emotive, i.e., pictorial and emotional. It is also agreed that all legitimate explanation is never absolute but *relative*, in the following two regards: (1) any given explanation proceeds from premises which, although possibly capable of further explanation, are *assumed* or *taken for granted* in the given case. It is only at the price, and in the light, of such assumptions that we can account for the explicanda; (2) the explanatory premises, as regards their validity, are relative to the confirming evidence, and therefore subject to revision.

My terminological suggestion thus amounts to the definition of 'explanation' as the inductive-deductive or (on higher levels) hypothetico-deductive derivation of the more specific (ultimately descriptive) propositions from more general assumptions (laws, hypotheses, theoretical postulates) in conjunction with other descriptive propositions (and often together with definitions). 'Explanation' is thus taken primarily as a *procedure of inference* (just like the closely related 'prediction'), with the only admissible alternative of the more substantival use that calls the required *set of premises* in those deductions 'the explanation' of the facts to be explained (as formulated in the conclusions).

The 'necessity' which is bestowed upon the facts by their explanation is the *logical necessity* of the *implication* underlying the inference from assumptions to conclusions. Neither the premises nor the conclusions in explanatory inferences of the empirical sciences are logically necessary in and by themselves. Only in a purely mathematical proof, such as we find in arithmetic or algebra, premises as well as conclusions may in themselves be logically necessary (analytic).

It is very helpful to restrict the meaning of 'description' to singular statements representing fully specific facts, events or situations. Such descriptions may appear as the conclusions of explanatory, i.e., deductive inferences. Some of the premises of these inferences must then be scientific laws or theoretical assumptions. Since laws and theoretical assumptions are (or at least contain) generalized statements (i.e., unlimited universal propositions), they are not here classified as descriptions. They are the premises of explanatory or predictive (deductive) inferences and thus are themselves essentially of inductive validity. Sometimes these laws or assumptions may be more familiar than the conclusions. The whole trend of mechanistic explanation manifests this tendency toward familiarization. But since, even in physics, this mechanistic trend found its very definite limitations, and since, particularly in the scientific achievements of the last eighty years, the trend has often been reversed, we may say that very frequently the well-known, long familiar facts have been explained by principles only much more recently discovered and lacking the tang of familiarity. (Among dozens of examples bearing out this point I will only mention the electromagnetic explanations of the familiar properties of light; the quantum-mechanical explanations of chemical processes; or the neuro-physiological explanations of sensory or reflex phenomena). That which matters is thus not the familiarity but the *generality* of the explanatory premises.

Since generality is a matter of degree, or rather of level, it is useful to distinguish *levels of explanation*. The empirical (i.e., experimental, or else, statistical) laws which function as premises in the deductive derivation of strictly descriptive conclusions may in turn become the conclusions of a super-ordinated deductive derivation from higher theoretical assumptions. In principle this process could repeat indefinitely but in practice it is usually found to stop at a second or third level. There is neither a danger of nor a need for an infinite regress. The top level at any given stage of theoretical research (in the ideal case) simply covers all relevant and available descriptive data; and there is no need for climbing higher on the tower of constructs if all the data one cares to see are within sight.

It seems convenient to represent the levels of explanation along the lines of the following scheme (read from bottom up!):

Theories 2nd order	Still more penetrating interpretation (still higher constructs)
Theories 1st order	Sets of assumptions using higher-order constructs (results of abstraction and inference). (Deeper interpretation of the facts as rendered on the Empirical Law-level)
Empirical Laws	Functional relationships between relatively directly observable (or measurable) magnitudes
Description	Simple account of individual facts or events (data) as more or less immediately observable

In actual scientific practice the distinctions, as well as the number of levels are neither quite as sharp or fixed as suggested here. The scheme is offered merely as a suggestion toward a first orientation. The question 'why' (in the sense of a demand for explanation) is answered by deduction either from empirical laws or from theories. Deduction from empirical laws may be styled 'low-grade' explanation. It merely puts the fact to be explained into a class of facts characterized by the same empirical law. Thus the explanation for the fact, e.g., that there is a mirror image of a bridge in a river, is achieved by subsuming this fact under the law of reflection in geometrical optics. This law is simply the common denominator of all the various phenomena in which light-reflection is the essential feature. A 'higher-grade' explanation we find in the Maxwell-electromagnetic wave theory, which serves as a basis for deduction for a variety of optical phenomena: reflection as well as refraction, diffraction, interference, dispersion, polarization, etc., etc. It is on this theoretical level (the 'row of genius' as I like to call it) that we gain a 'real insight into the nature of things' (as metaphysicians call it). What we give on this level are interpretations concerning the *structure* of light, of matter, of electricity, etc. The constructs of this theoretical level usually concern the micro-structure of the observed macro-phenomena, i.e., they involve existential assumptions (atom, electron, photon-hypotheses) or constructs of the abstract mathematical order (energy, entropy, tensors, probability functions, etc.). No wonder that the 'Aha-experience' is much stronger for these deductions from theories than for the much simpler deductions from empirical laws.

Once the theoretical concepts are properly introduced, they can be used also for purposes of description on the lowest level – e.g., Einstein can describe the physical state of a given volume of space in terms of 14 highly theoretically defined magnitudes. Similarly Tolman or Hull can *describe* the behavior of an organism in terms of the intervening variables of their respective systems.

The question regarding *circularity* may be resolved by defining what is usually called an *ad hoc* explanation. Now, an *ad hoc* explanation is deceptive because it has only the external form of a 'real explanation.' It is *ad hoc* in that it explains *only* the fact which it was to explain (i.e., for the sake of which it was introduced). It may be either purely verbal, e.g., 'Birds build nests because they have nest-building instincts.' Or it may be unscientific in that it assumes entities which do not manifest themselves in any other way (explanation of Gravitation by Lesage: particle-radiation, etc.) or it may be down-right metaphysical, if the explanatory hypotheses are in principle incapable of test (such as the assumptions of entelechies, vital forces in vitalistic and animistic biology and psychology). Of course, everything depends on how the explanatory phrases are interpreted; the use of the word 'instinct' can be quite legitimate (and more than purely verbal) if, e.g., it is meant in the sense of an empirical regularity in the behavior of a species. *Then* it is a 'low-grade' explanation, possibly preparing the way for a 'higher-grade' theoretical explanation (say on the basis of a physiological theory of heredity, maturation, etc.) Similarly, explanations of rapidity of learning on the basis of 'intelligence' are not purely verbal (or *ad hoc*) but *low-grade* explanations on the basis of empirical laws. That is, as long as the 'intelligence-quotient' refers to various types of capacities, various types of learning-activities, it enables us to relate the ones to the others *via* the common factor, 'I.Q.'

In some cases the reproach of 'circularity' is made against 'low-grade explanation' *if* it pretends to be 'high-grade.' But it seems there is no absolutely sharp line between the two – because sets of empirical laws sometimes function very much like theoretical assumptions of the higher construct type.

More fundamentally and logically speaking the contention of circularity or tautologicality in scientific explanation is right in one interpretation but definitely wrong in another: It is right if it stresses the analytic (i.e., strictly logical, sometimes called 'tautological') character of the deductive inference leading from premises to conclusion in any explanatory argument. In a more precisely definable sense it can be said that the conclusion is 'contained' in the conjunction of the premises. The charge of circularity or of *petitio principii* is justified only if either the conclusion appears literally as one of the premises or if the truth of one of the premises is proved by appeal to the conclusion. The customary procedure of the hypothetico-deductive method in the empirical sciences is perfectly capable of avoiding both sources of circularity. The (psychological) novelty sometimes amounting to surprise (Eureka!), in the more advanced and worthwhile instances of scientific explanation shows that the conclusion was not one of the premises. And the truth of the explanatory assumptions is always only suggested (i.e., confirmed to some degree) but never fully proved by evidence which is distinct from the facts to be explained. Newton's law of gravitation together with his laws of mechanics were already highly confirmed by the facts of planetary motion, by the orbits of satellites, comets, and many other items of evidence, when Leverrier and Adams used those laws as explanatory premises, together with the existential hypothesis regarding the orbit of another up to then not observed planet, in order to explain the irregularities of Uranus' motion. True, by a 'tautological' (better: deductive) transformation the conclusion (concerning Uranus' path) was derived from premises (laws of mechanics, law of gravitation, etc.) but the major premise says infinitely more than the conclusion and it is therefore not possible to deduce the premises from the conclusion. Pratt overlooks the inductive leap, the leap from 'this' to 'all' in explanatory generalizations. By declining to differentiate sharply between explanation and description Pratt views generalizations as descriptions. 'Description' thereby loses its ordinarily precise meaning and the distinction between facts on the one hand, and law or theory on the other, is in danger of being blurred or even obliterated.

Moreover, the hypothesis of the existence of a further planet (after its telescopic discovery called 'Neptune') was suggested by the analogy with the facts regarding the then known planets. The inductive probability of this existential hypothesis therefore (at the time of Leverrier and Adams) did not rest exclusively on the *ad hoc* or circular procedure described by Pratt but had an independent foundation, no matter how weak or strong, in the already established body of astronomical knowledge. Similarly in psychology: However vague and uncertain Freud's original hypotheses regarding the repressed or unconscious parts of the mind may have been, they were *not* circular in the sense of *ad hoc*. He was guided by analogies of the conscious and preconscious and was able to unify through his hypothesis a great number of previously unrelated facts, such as certain types of forgetting, slips and lapses, dreams, hysterical and neurotic symptoms, etc. – A methodologically similar situation prevails also in psycho-physiology.

To summarize: A scientific explanation is free from objectionable circularity or *ad hoc* character if it helps connecting hitherto unconnected specific facts ('low-grade' explanation) or laws ('high-grade' explanation). While the deductive part of the hypothetico-deductive procedure may be said to be 'tautological' or analytic (in the sense in which the classical syllogism is valid only if a denial of the conclusion strictly implies a denial of at least one of the premises); but it is not *circular* (in the sense that the conclusion be logically equivalent with one of the premises or that the conclusion itself be the sole basis of the inductive probability of the hypothetical premise).

NOTE

Very briefly replying to the critical questions of Harold Israel, particularly those raised in his joint article with Goldstein (*Psychol. Rev.*, 1944, **51**, No. 3), I should like to state that I see no such cleavage between the operationism of the physicists and that of psychologists as is there emphasized. It is true that Bridgman's exemplifications usually involve the mensurational aspects of physical constructs. But aside from these aspects there are others: purely qualitative (classificatory), causal-genetic or dispositional aspects of concept-formation, often preceding the quantitative-functional or the theoretical-constructive. Such physical concepts (to choose at random) as: magnetism, radio-activity, photo-electric sensitivity, phosphorescence, etc., before they attain metrical status or theoretical interpretation, are introduced (if you will: 'defined') on the level of empirical laws by conditional definitions (involving reference to test situations). This is very much as it is in psychology with such concepts as personality traits, habits, drives, latent dispositions, etc. The fact that the introduction of these concepts requires reference to experimental or testing situations has been misinterpreted by pragmatists and some operationists in construing it as a 'production' of the pertinent properties or reactions. But there is no *basic* difference between the purely observational or mensurational procedures of a science, say like astronomy, and the experimental sciences where, according to the pragmatist interpretation, the intervention or interference of the scientist's actions is essential and indispensable. What really matters in either case is that we observe what happens under what conditions – be these conditions 'naturally given' or 'artifically set up.'

BIBLIOGRAPHY

The confusions, particularly in J. S. Mill's *Logic* [1930], regarding the circular character of deductive inference were well criticized already by J. M. Keynes in his *Studies and Exercises in Formal Logic* [1904, Section 381, p. 474ff], and again, very pointedly, by R. M. Eaton in *General Logic* [1931, pp. 140ff]. A very lucid discussion of the concept of scientific explanation with special reference to the social sciences may be found in C. G. Hempel's article 'The Function of General Laws in History' [1942, p. 35ff]. For a very elaborate discussion of explanation in physics, see N. R. Campbell, *Physics: The Elements* [1921] and also G. Bergmann, 'Outline of an Empiricist Philosophy of Physics' [1943a, pp. 248–258, 335–342]. Further excellent discussions of operationism and explanation in psychology are in the articles by G. Bergmann and K. W. Spence, 'Operationism and Theory in Psychology' [1941] and by K. W. Spence, 'The Nature of Theory Construction in Contemporary Psychology' [1944]. Nor should the, by now classical, article by K. S. Lashley, 'The Behavioristic Interpretation of Consciousness' [1923] be forgotten.

12. EXISTENTIAL HYPOTHESES

Realistic Versus Phenomenalistic Interpretations

[1950b]

The intention of the present essay is to urge a reconsideration of the Realism-Phenomenalism-Issue, mainly and primarily in regard to the interpretation of scientific hypotheses; secondarily also relating to the basic problems of epistemology.

In order to initiate discussion I shall first very briefly set forth the major contentions and questions. (These will be elaborated in what follows further on.)

(1) There is a need for a more adequate account of the meaning of existential hypotheses* than the currently widely held phenomenalistic-nominalistic (positivistic, operationalistic) interpretation. More specifically: Can the analysis of existential hypotheses and theoretical concepts in terms of postulate systems, coordinating and operational definitions be upheld in the light of the actual procedures of science and their underlying semantical presuppositions?

(2) Can we avoid both the reductive fallacies of phenomenalism and the redundancies and confusions of metaphysical realisms?

(3) Does not the notion of the probability of existential hypotheses presuppose a 'realistic' frame which cannot itself be meaningfully justified by considerations of probability?

(4) What is the proper logical form for the definition (introduction) of hypothetical constructs which cannot be introduced by explicit definition in terms of observables?

(5) If there is a parallel (or homology) between the issues of scientific methodology and those of the more basic theory of knowledge, what can be learned for the latter from the suggested revisions of the former?

The view singled out for critical discussion is, very briefly, the one suggested in Bertrand Russell's (by now classical) programmatic pronouncement of his "supreme maxim of scientific philosophizing"; *Wherever possible, logical constructions are to be substituted for inferred entities* [Russell, 1929a, p. 155]. According to the conclusions of the non-controversial parts of *Principia Mathematica* the concepts of arithmetic and analysis are definable in terms of purely logical concepts; and the truths of those mathematical disciplines are translatable into purely logical statements. Russell himself [1929a, 1929b], and later, more emphatically, Carnap [1928b] argued an analogous doctrine for the field of empirical knowledge: The concepts of the factual sciences, certainly the concepts of physics, so they maintained, are definable by a series of steps terminating with concepts that refer to directly given qualities or relations; and the statements of the empirical sciences were thus considered as translatable into statements concerning the data of direct experience. Obviously, a great deal depends upon just what is being referred to as 'the data of direct experience.' At the moment it will suffice to remember that these 'data' may be located *either* on the level of the phenomenally given; (and this involves the notorious difficulties of a phenomenal language with its particulars, predicates and relations characteristic of the visual, tactual, kinaesthetic, auditory, etc. 'spaces'); *or*: on the level of ordinary commonsense objects, described in what Carnap [1936a] has called the "physicalistic thing-language (in which at least fragmentary finite coordinate systems form the frame, and predicates of the naive-realistic kind, like 'blue', 'cold', etc. are applied to spatio-temporal regions)." It is this latter alternative that is more directly relevant to a study of scientific method. The former alternative comes into consideration if we pursue the reconstruction of knowledge to its phenomenal rock-bottom (if there be such!).

A generation ago the bone of contention between realists and positivists (or idealists) was the 'independent existence' of the objects of science. This issue has since been reformulated in 'the new way of words': Those who hold the translatability-thesis may now be called 'phenomenalists'; those who oppose it are 'realists' in some new sense of this ambiguous word.

The glib and easy dismissal of the issue as a pseudo-problem will no longer do. No doubt there are ways of putting the issue that makes it into a question devoid of any specifiable significance. But the advance of modern syntactical and semantical techniques enables us not only to restate the problem in a new and sharpened fashion; it also offers some hope that the issue may now be more responsibly and more satisfactorily adjudicated.

It is time to take stock and examine the value of Russell's maxim in the light of recent developments in the logic of science.

A PRELIMINARY ILLUSTRATION OF THE ISSUE AND THE RAPPROCHEMENT OF THE TWO INTERPRETATIONS

The purpose of the discussion in this section is to introduce our main issue by means of some simple illustrative material from elementary physical theory and to make plausible how far a sufficiently critical phenomenalism (operationism) can accommodate an equally critical realism.

The customary paradigm for illustration of our issue is, of course, the atomic theory. We shall, however, choose electromagnetics for our example – not only because the evidential basis is here less complex and multifarious, but also because it may be well to discuss *re*alism vs. phenomenalism in a domain where the notion of *thinghood* is irrelevant and the attendant dangers of confusion (engendered by misplaced picturizations) are more easily avoidable.

The test-basis of electromagnetics (similarly of the older ether theories; also of certain parts of atomic, kinetic theories) is the macro-mechanical level of (intersubjectively) observable phenomena.

There is one class of phenomena (processes) for which the laws of classical statics and (Newtonian) dynamics are approximately sufficient – usually called 'conservative' processes, i.e., processes in which *mechanical* energy remains constant. In these processes no other forms of energy are transformed into or out of mechanical energy. The laws of classical mechanics are clearly insufficient for the prediction and explanation of non-conservative processes in which thermal (classical level!), chemical, electrical, magnetic, or optic phenomena take place.

To simplify and focus the issue let us consider only the most elementary facts of electro-statics. We shall disregard any test bases, such as the optical (e.g. the observation of sparks) except the mechanical, i.e., forces measured by means of accelerations of bodies of known masses, or deformations of bodies of known coefficients of elasticity. A strictly phenomenalistic description of the so-called electro-static phenomena on the idiographic level would run something like this: A glass-bar after being rubbed with leather is brought in contact with an electroscope, at which moment the gold leaves assume a divergent position. Contingent upon the independently measurable moisture saturation of the surrounding air this divergence will vanish more or less rapidly. Other experiments, most of them of similar simplicity, described in every elementary physics text, reveal further well-known phenomena. All these phenomena, under well-controlled (constant) attendant conditions, exhibit a rather strict repeatability. Laws can be formulated which tell us (and enable us to predict successfully) what happens upon what more or less complex conditions. If, for the time being, we avoid the introduction of hypothetical constructs, i.e., if we restrict ourselves to the use of concepts exclusively of the mechanical (aided by the everyday chemical)[2] concepts then we can formulate a, to be sure, fairly complex and unwieldy, but nevertheless predictively sufficient set of laws which represents the regularities of the phenomena concerned. Many of these laws will have 'historical' ('mnemic') character, in the sense that causal relations between what occurs at a time t_2 ($t_2 > t_1$) is made dependent upon what happened at t_1. E.g. if and only if the glass bar was rubbed at t_1 do we get goldleaf-divergence at t_2. It may be possible to eliminate geographical characterizations (such as might otherwise be required on a very primitive level of magnetics, as in statements concerning the provenience of the lodestone) by the introduction of dispositional concepts. These concepts are simply our modern way of formulating the much maligned 'powers', 'faculties', 'capacities', or 'occult qualities' of ancient and medieval thought. We may leave it undecided here through just what form of definition dispositional concepts are most adequately to be introduced. (Carnap's "reduction sentences" or some sort of nomological conditional have been suggested.) In a similar manner even the historical character of the laws may be eliminable. Momentary dispositions, defined by test-condition → test-result conditionals, might thus assimilate the laws in question to the differential equations of the customary type. In certain cases, however, this may make our laws dangerously trivial, and in the extreme limit tautological. (A glass bar which repels or attracts a goldleaf will have the well-known effect on a goldleaf electroscope.)

The phenomenalist can then express all the regularities, which, from the level of hypothetical construction we have so thoroughly become accustomed to explain as – and therefore to call – 'electrostatic', by means of the mechanical *concepts* (aided by the commonsense chemical *concepts*). While it is admitted, then, that the *laws* of classical mechanics are insufficient for the prediction of these phenomena, these laws can be supplemented by a set of further laws, and using the very same concepts which figure in classical mechanics together with the common life chemical concepts. (The latter appear in the practical applications of classical mechanics just as well, when, e.g., the elasticity coefficients of various materials are to be determined.)

The 'cash value' of all science, after all, consists in the reliable correlations (functional dependencies) on the level of the directly observable. No matter by what extraneous fascination theoretical constructions may appeal to us, the decision as to their adequacy depends exclusively upon the agreement or disagreement of the derived theorems with that cash value – the empirical (or experimental) laws. – This is not to deny that in the actual progress of science not only considerations of formal simplicity but also of aesthetic appeal, fashion of the times, personal bias and other factors may play an important role.

We turn now to the arguments in favor of theoretical (hypothetical) construction. Beginning with arguments traditionally accepted by phenomenalists (positivists, radical empiricists, operationists) we shall proceed gradually to arguments which, though compatible with a broader empiricist outlook, may encounter some resistance in the phenomenalist camp. To put my cards on the table, I am going to try to make realism a little more tempting and palatable than it has hitherto been with phenomenalists. We list the arguments as follows:

(1) Hypothetical construction brings about a considerable formal simplification. That is, it enhances, in the genuine sense of Mach's principle of economy, the perspicacity of science. Along this line it is easily seen that the introduction of a new concept enables us to reduce the number of laws. This may be shown by a simplified scheme: if there are m different causal conditions and n corresponding possible effects, we need mn statements in order to state all possible testable relations (laws). If, however, we introduce an auxiliary construct, the number of laws can be reduced to $m + n$.[3] Simplifications of this type are to be distinguished from: a) simplifications arising out of a more felicitous choice of a coordinate system: as, for example, in the preference of the heliocentric over against the geocentric kinematics of the planetary system; – b) simplifications arising from the adoption of more powerful logico-mathematical techniques, such as vector and tensor analysis; Hamiltonian equations; factor-analysis; etc., etc.

The bundling together of the facts is achieved already on a lower level by means of the empirical laws, especially when in metrical form. For example, instead of stating for countless angles of incidence the corresponding angle of refraction (for two given optical media) we introduce the law of refraction (with its characteristic constants for the respective media). Closely connected with the simplification obtained through the introduction of auxiliary constructs is the deductive fertility of theoretical (postulate) systems. This represents the formal aspect of the factual unification discussed in the next point.

(2) Another quite generally accepted and emphasized justification for hypothetical construction lies in their explanatory power and heuristic value. Hypotheses are fruitful in suggesting further avenues of research. This is, as it stands, a pragmatic statement. It concerns the temporal progress of scientific investigation. But the logical core may be identified with the aspect of unification or convergence achieved through theoretical synthesis. A large manifold of originally separate and apparently quite heterogeneous facts (or, on the next level, laws) are all brought together. They become deducible from a common basis of theoretical premises. Even in the very limited field of electrostatics this is clearly the case. We are so used to thinking and describing the multifarious facts in terms of successful theoretical constructs that we should

find it generally quite difficult to enunciate these facts (or the empirical laws of a low level) without availing ourselves of the language of constructs. Once the construction has been introduced and proved fruitful, it becomes so much an essential element of our language in the given field that we no longer remember or appreciate the unifying effect produced by its original introduction. A good example is the Newtonian synthesis which has been absorbed to a surprising extent even in the ordinary life of the educated. Similarly accepted in our industrial age are the explanatory concepts of electricity and electromagnetism. They tie together countless qualitatively diverse phenomena of the modern household, of the industrial production and use of power, of radio, television, radar, etc.

(3) Closely connected with the foregoing point and really only a psychologically more persuasive corollary to it lies in the derivability (predictability) of *more directly* confirmable consequences of constructive hypotheses. This is especially obvious in the theories of micro-structure. Dalton's atomic theory, first introduced to account for the chemical laws, and the molecular and kinetic theory of matter first introduced to account for the gas laws, found the most surprising and much more direct confirmations later on in modern experimental atomistics. Such experiments, for example, as those of Rutherford, Millikan, Stern and Born, C. T. R. Wilson, and countless others, in which we observe the effects of single molecules, atoms, and subatomic particles, testify convincingly to the fruitfulness of the corresponding theoretical assumptions. Quite analogously recent research in genetics bears out the 'truth' of Mendel's original hypothesis of units of heredity. – Scientific realists (Boltzmann, Planck, Reichenbach, Bavink, etc.) have confidently argued that in view of these independent findings micro-structural hypotheses can no longer be considered as 'mere convenient models or purely conceptual constructs.'

(4) A further extremely important justification for the introduction of hypothetical constructs lies in the spatio-temporal continuity they afford; and in the nomological (causal) coherence thus achieved. In simple explanations, in electrostatics for example again, it is assumed that electric charges are distributed over the surface of conductors. The distribution functions themselves are formulated in the form of laws. Furthermore, there are at least crude and semi-quantitative laws concerning the generative conditions of electric charges as well as concerning the observable effects of such charges. The hypothetical momentary states of charge distributions on conductors become thus part and parcel of our theoretical construction. Hence the discontinuous and historical character (action at a spatial and/or temporal

distance) of the phenomenalistically restricted account vanishes and is replaced by a spatio-temporally continuous (contiguous) and nomologically coherent formulation on the level of hypothetical construction.

(5) One of the most remarkable features of the more fully developed stages of the theoretical construction may be loosely characterized as a sort of *Copernican revolution*. It consists in the derivation, with corrections coming from the theoretical scheme, of the peculiarities of the very basis of confirmation. In the case of electrostatics this is so obvious that it may easily be overlooked. The divergence of the goldleaves of the electroscope which epistemically serves as an indicator of the presence of electric charges is immediately deducible from the theoretical assumptions of electrostatics, i.e., primarily from the Coulomb law of attraction and repulsion. The corrections mentioned might come from considerations of the dielectric constants of various media or the conductivity of moist air, etc. This corresponds to what on the much deeper epistemological level a proper theory of perception accomplishes: the derivation of the immediately perceived data from physical assumptions about a state of affairs (the stimulus situation), physical laws (such as concerning the propagation of light, sound, odoriferous substances, etc.) and psycho-physical or psycho-physiological laws relating the stimuli to the reaction elicited in organisms of varying structures and conditions. The corrections in this case concern not only the perceptual illusions, but quite generally they amend, augment and make more precise the inaccurate, incomplete and vague knowledge that would otherwise accrue from the use of the raw sense data. Such causal theories of perception, just as in the causal theories of indication in the case of physical instruments, must of course themselves be confirmable on the perceptual (or indication) basis. As long as a science has not attained a very high level of explanation the process of indication may not be deducible from theoretical premises. This corresponds to the epistemological situation in pre-scientific times when extremely little was known about the mechanisms of sense perception. Or, on the scientific level again, this corresponds to the theory of acids and bases in chemistry at a time when the process involved in the well-known litmus paper test was not itself logically derivable. (I don't know whether this has been achieved as yet.) – The behavior of such thermometric substances as alcohol or mercury was not theoretically deducible until the kinetic (molecular) theory of heat put these indicator processes on a par with countless other thermodynamic processes *as interpreted* on the micro-level. Again the correction 'from above' comes about here through a realization that the concept of temperature required on the level of thermodynamics is only approximated but never fully satisfied by

the indications of thermometric substances of any kind. In a fairly analogous sense it might be said that the behavior of the planets as formulated in the Kepler laws serves as an indicator of the gravitational force of the sun. But to these same Kepler laws a correction accrues 'from above', i.e., from the very law of gravitation (in the theory of perturbations) which they served to establish in the first place.

The Copernican turn then consists in relating the observer to the observed, the indicator to the indicated – not epistemically – but so to speak cosmologically. What epistemically must be looked at as the confirmation bases of the hypothetical construction, will in the fullfledged theory be given a place within the cosmos of which the theory treats.

(6) By way of corollary to the foregoing we may say that there is quite generally a certain correspondence between the basic (logical) frame principles of knowledge and some of the broad features of the cosmos as represented in the results of knowledge. This is to be expected in a sound epistemology. If knowledge (as behavior) is not to remain an utter mystery or miracle, it is clear that the knowing organism itself must find a place in the world it knows. Whatever object can be reached by empirical knowledge must, no matter how indirectly, be related (yes, causally related) with the processes in the knowing organism. This corresponds clearly to the principle of confirmability in the pure pragmatics of science. An existent conceived as completely isolated, i.e., cut off from any possible causal influence upon the knowing organism, remains of course (by definition, as it were) unknowable. In the epistemological frame of reference this is reflected in the decision to rule out as factually-meaningless any assertion which by its very construction is in principle incapable of confirmation or disconfirmation.

(7) On the epistemological level, perhaps the most significant feature of the Copernican turn lies in the elimination (through proper allocation) of the experientially unique roles of the 'I,' 'Here' and 'Now', as well as of any other of the alleged predicaments of this sort. What in the methodologically solipsistic reconstruction appears as the base of all cognition corresponds to a relatively late product, very limited in scale, of the process of cosmic evolution. The epistemic uniqueness of the base corresponds only to objective specificity and focal character of a spatio-temporal region in the cosmological account.

The foregoing discussion deliberately attempted to show how a sufficiently open-minded and critical phenomenalism (or operationism) can, by degrees, be assimilated to a critical (or empirical) scientific realism. In order to determine what discrepancies, if any, remain when this impressive rapprochement has been conceded, we must sharpen the issue by means of a closer logical analysis.

SOME REQUISITE DEFINITIONS AND DISTINCTIONS

In our characterization of the hypothetical super-structure of science we shall find it convenient to distinguish between *hypotheses in the form of laws* and *hypotheses in the form of singular statements*. The customary and paradigmatic conception of scientific laws is of course that of a functional relationship between two or more independently accessible factors (variables, parameters). This functional relationship is either deduced from more abstract theoretical assumptions (or models); or it is established by observation, measurement, curve-fitting – i.e., by (inductive) interpolation and/or extrapolation, statistically corrected and formulated in the formally simplest of the various most approximate mathematical functions. However this characterization of the typical *empirical law* is apt to deflect attention from the following philosophically important distinctions:

(1) *Hypotheses in the form of laws.*[4]

A) Relating directly observable properties (magnitudes) to one another.

B) Relating directly observable to indirectly testable properties (magnitudes).

C) Relating indirectly testable properties (magnitudes) to one another.

Examples:

A) The anatomical structure of crows as related to the black pigmentation of their feathers. (Quite generally: anatomical laws of co-existence of structures or traits.) The rapid compression of a gas as related to the increase in (felt or measured) temperature. Archimedes' law of the lever. Galileo's law of free fall. Laws of light reflection and refraction. Laws of unconditioned and conditioned responses. Learning curves. Mendel's laws of heredity.

B) The degree of (felt or measured) temperature as related to the mean kinetic energy of molecular motion. Wilson cloud chamber tracks as related to the trajectories of subatomic particles. The deflection of a magnetic needle as related to the intensity of the magnetic field. Pathological symptoms as related to the presence of filterpassing (microscopically invisible) viruses. Inheritable phenotypical traits as related to the microstructure of genes. Freudian lapses or neurotic symptoms as related to unconscious motives or conflicts.

C) The relation of thermal and electric conductivities (Wiedemann-Franz law). The relation of dielectric constants and refraction indices (Maxwell). The relation between electric and magnetic field vectors according to Maxwell's equations. The energy-entropy relations of classical thermodynamics. The probability laws of statistical thermodynamics and quantum mechanics. The curvature of space (and the gravitational field potentials) as related to the matter-energy-tensor according to Einstein's equations.

(Examples from psychology are of questionable validity. But there are adumbrations of the theoretical level in such systems as those of Freud, Lewin, Hull and Tolman. More pertinent illustrations may be found in such approaches as Rashevsky's mathematical biophysics.)

(2) *Hypotheses in the form of singular statements (or conjunctions thereof)*

The customary explication of this type of hypotheses defines them as specific descriptive statements which have (at least up to the given moment) been verified only incompletely and/or indirectly. A first example may quickly provide an illustration typical of the simplest kind of conjecture we have in mind here. A prospector finds traces of gold in some rocks and thereupon makes the hypothesis that there is much more of it further inside the mountain. For obvious reasons we may briefly call this type of hypotheses *existential hypotheses*. (Logicians may be reminded that there are scarcely any important cases of hypotheses in the empirical sciences that make sole and essential use of unlimited existential quantifiers. But since every descriptive singular statement entails an unlimited existential one, even the pedant may not resent too severely the terminological liberty taken here.)

The important distinction to be drawn here is between existential hypotheses (type A) that assert *directly testable* (though not actually directly tested) states of affairs and those (type B) that assert only *indirectly testable* states of affairs. Illustrations for each type are as follows:

A) There are some matches in this match box. There is oil underneath Houston, Texas. There is a brain in Lord Russell's head. There is organic life on Mars. There is a further planet beyond the orbit of Pluto.

B) This rock contains uranium oxide. There is calcium vapor on the surface of the sun. This light beam is plane-polarized. This room is traversed by radio waves. John's lungs are infested with filter-passing viruses. John's brain contains memory-traces. There are more than 10^{19} molecules in this cubic-centimeter of air. Electrons are concentrated on the surface of this copper bar. Protons are moving rapidly through this cloud chamber.

The examples mentioned in the first group are considered directly testable because the presence of the asserted state of affairs is ascertainable or refutable by means of ordinary sense perception as soon as an opportunity for unhampered observation of the specified spatio-(temporal) area is afforded. We understand the given statements in terms of the predicates that refer to directly perceptible properties. This may be acceptable at least as long as we are not too exactingly scientific in our demands for proper identification and as long as we don't bother ourselves epistemologically about the reliability of our perceptual or mnemic performances. We shall assume, for example, that we know oil when we see, touch or smell it; or that we would recognize something as a living organism when we watched it for a while. We also imply by 'directly testable' that the obstacles that may at the moment prevent the (direct) testing of the given hypothesis are of a merely practical or technical character. In other words, we know that the relevant test is possible, precisely because it is the kind of test that has been performed on previous occasions (drilling a well, opening a skull) or that could be performed if the required occasion were afforded or the required means perfected (rocket ship for the trip to Mars, etc.).

It may of course be argued that the difference between the examples of type A and of type B is not one of kind but merely one degree (– often of considerable degree, this might be conceded). Is there, it might be asked, really a fundamental difference between such operations as opening a man's skull (to determine whether it contains a brain) and applying chemical analysis to a specimen of rock (to determine whether it contains uranium)? It may further be urged that, for example, the utilization of telescopes, optical (or electron-) microscopes somehow fills the gap, if there is any at all, between the directly testable and the only indirectly testable hypotheses. Similar questions may be raised about hypotheses concerning the specific heat, electric conductivity, solubility in specified media, of given specimens of materials. Experimental and mensurational procedures are indeed required for the testing of such hypotheses. And yet, the pertinent operations of confirmation in these instances do not seem fundamentally different from those involved in direct verification. – These questions foreshadow in a very rough and preliminary form the main issue of our discussion. In the following section we shall proceed to focus and illustrate it more fully.

CONSTRUCTION, INFERENCE OR POSTULATION?

A Survey of Nine Points of View

As the patient reader may have expected, our major concern is with the laws (type B) that relate directly testable with only indirectly testable predications; and correspondingly with existential hypotheses (type B) which obviously require for their confirmation an appeal to laws of type B. Quite clearly, any confirmation (short of independent direct verification, wherever this be possible) of existential hypotheses (type A or B) must make use of a confirmation rule. There is no question that the confirmation rules for existential hypotheses of type A are laws of type A which, according to our definition, may themselves be directly, though never completely, confirmed by favorable instances. For example, we may infer from footprints the earlier process of their production; from smoke, fire; from the visual appearance of a fat man, his great weight; from the presence of a backbone, the existence of a central nervous system; etc., etc. In justification of such reasoning we should unhesitatingly quote the law (deterministic or statistical) that is drawn upon as a ground of validation. The better, i.e., the more completely, the law is confirmed, the higher the degree of confirmation for the given existential hypothesis.

Now what about laws of type B? How are they to be confirmed? Common sense as well as methodological and epistemological considerations seem to demand that an item of evidence and the state of affairs for which it constitutes evidence be clearly distinguishable and independently ascertainable. Just this demand meets with difficulties in the case of laws and existential hypotheses of type B. For example, we are apt to say, glibly enough, that the deflection of a magnetic needle is evidence for the presence of a magnetic field.[5] But how do we ascertain the validity of the law (type B) that relates the behavior of needles to magnetic fields?

In what follows we shall list and discuss some of the typical (naive or sophisticated) answers to this question:

(I) Since the behavior of the needle must have a cause, we maintain that that cause (the magnetic field) exists even if this cause is not independently accessible to direct verification. The existence of the magnetic field is required by the principle of causality and is confirmed by the deflection of the needle. (We shall call this the position of Naive Physical Realism.)

(II) The observable behavior the needle displays, is in every respect *as if* there were an independently existing (but forever unknowable) reality: the magnetic field. The concept of the field is a useful fiction (Fictionalistic Agnosticism).

(III) The independent existence of the field cannot be asserted with certainty. But it can be inferred with probability from the behavior of the needle and other items of evidence. Quite generally, all inference that proceeds from observables to (directly) unobservables must be based on inductive probabilities. The proponents of the statistical conception of probability interpret these as weights based on estimates of limits of relative frequencies. The proponents of the logical conception of probability submit various definitions of the strength of the evidence in terms of a degree of confirmation. According to C. D. Broad [1914, 1923, 1927–8] existential hypotheses, as well as realism generally, may acquire a high degree of probability if they possess some 'initial' or 'antecedent' probability greater than zero. This somewhat obscure notion of 'initial' probability corresponds perhaps to Reichenbach's 'blind' or 'anticipatory' posit – provided that this notion be applicable to hypotheses of a sort more advanced than simple inductive extrapolations. Existential hypotheses possess a surplus meaning over against their evidential basis; they are not equivalent with or reducible to (by translation) any set of actual or possible confirming statements [cf. Reichenbach, 1938, esp. pp. 212–217]. We shall call this the view of 'Probabilistic Realism.'

(IV) There is no conceivable way of independently and directly testing the existence of the field. Therefore to speak of a law (type B) relating the needle's behavior to the field, is extremely misleading. The concept of the field is *defined* by the behavior of the needle and has no meaning over and above what could be stated (more cumbersomely, to be sure) about the actual behavior of the needle. 'Laws' of type B are nothing but definitions. Hypothetical constructs are thus regarded as strictly circular. Our hypotheses are so chosen that they parsimoniously (i.e., in more succinct language) summarize what could in principle be formulated as regards the *actually* observ*ed* facts (Naive Conventionalistic Positivism or Phenomenalism).

(V) The tests for the presence of a magnetic field are not limited to the behavior of needles. A magnetic field 'manifests' itself also in the effects upon electric currents and upon the trajectories of electrically charged particles; in the rotation of the plane of polarization of light beams (Faraday, Kerr); in the Zeeman effect, etc. It should also be remembered that we know something about the conditions (lodestone, magnetized iron bars, electric currents, etc.) which generate magnetic fields. Part of the meaning of assertions of the presence of a magnetic field may therefore be taken to consist in a reference to these possible causes. Moreover, even considering solely the effects upon needles, the statement asserting the existence of a magnetic field cannot be simply translated into reports about the *actually* observ*ed* behavior of a given needle. We have to take into account not only the actual behavior of a given needle, but rather the *possible* behavior of needles of various possible sorts and sizes, in all (in principle infinitely many) possible positions and orientations. The assertion of the presence of a magnetic field is thus an extremely condensed and economical ('shorthand') formulation for sets of infinitely many directly testable 'if-then' statements; conceivably even of infinitely many such sets. The 'if-then' statements tell us what happens, or what would happen, under specified circumstances, i.e., experimentally introducible conditions. Every verified instance of these singular implicative statements'[6] establishes a (partial and usually extremely incomplete) confirmation of the hypothesis asserting the existence of the field. But the total set of sets of these singular implicative statements is strictly logically equivalent with the existential hypothesis. In keeping with Peirce's pragmatic maxim the content of an hypothesis can be no more than all the verifiable consequences that are deducible from it. According to this view of the matter the existential hypothesis connects (correlates) all the directly verifiable 'manifestations' ('symptoms') into a closely knit system of empirical laws (type A). The short-circuit circularity alleged by Naive Conventionalism (IV) is thus avoided. With reference to our specific illustration it can then be maintained that the assertion of the existence of a particular magnetic field means (over and above the specific evidence that may have suggested the hypothesis) the total system of implicative relations between all sorts of conceivable test conditions and their corresponding test results. Hence the assertion (law of type B) that a needle is being deflected by a magnetic field (of given strength, direction, spatio-temporal extension) is far from tautological. It connects by synthetic statements all the various effects that would (within the specified spatio-temporal region) be observable on needles of all sorts and all the other effects (Faraday, Kerr, Zeeman, etc., etc.) that would occur in the same region under appropriately contrived conditions. And, beyond all that, it may even relate the observed and observable test results to the observable conditions which, as we say, 'generated the field' (Critical Phenomenalism, Operationism, or Positivism).

(Va) A view which may well be regarded as a variant, or perhaps rather as an amplification, of the preceding one, focusses attention upon the role of theoretical laws (our type C) in the hypothetico-deductive structures of explanation and prediction. These laws are viewed as postulates in a calculus which is so constructed that (*more geometrico!*) all the empirical laws of a given field, e.g., electro-magnetics, are deducible from it; and which is interpreted via coordinating definitions. Either certain abstract, undefined concepts or else some concepts explicitly definable in terms of those primitives, are thus set in correspondence to the empirically or operationally defined constructs that have their place in the empirical laws. This requires a distinction either between empirical and theoretical constructs; or between empirical constructs and their mathematical idealization and formalization in a pure calculus. Although this view lends itself also to realistic interpretations, it is mentioned here as an important refinement of phenomenalism. As such it coincides fully with (V) and contributes additional plausibility to the view that the entities which figure in the laws of theoretical science are *nothing but* useful formal constructs; the theories themselves being 'nothing but' mathematical models. The upshot then is still: the theoretical constructs are auxiliary devices, they are *façons de parler*, abbreviatory schemes for the description of the complex relationships between observables. This view with its emphasis upon the pure syntax of calculi, introduces a yet more distinctly nominalistic tinge into phenomenalism. (We may call it 'Formalistic Phenomenalism' or 'Syntactical Positivism'.)

(VI) Another view, not too far removed from the two preceding ones, may be characterized as follows: Since the empirical constructs of science (e.g. magnetic field strength, electric field strength, electric charge, intensity of electric currents, electromotoric force, conductivity, etc. etc.) are all linked together in a network of relationships, it depends upon the context of experimental investigations, and is in this sense somewhat arbitrary, *which* of these relationships may be regarded as genuine laws (synthetic propositions) and *which* others are then taken to be definitions (conventions, analytic propositions). Since in actual research laws of types B and C, and hypotheses of type B are never capable of test in isolation but always in the *context* of a whole system of relationships our initial query ('How are laws of type B to be validated?') is here considered as too simple-minded. In testing one hypothesis we invariably fall back on others which in this context are construed as definitions and provide the indispensable (and for the time being unquestioned) background and presupposition without which the very notion of a test of this kind is impossible.

Even as regards classical mechanics this situation has impressed itself upon the more reflective minds. The question, for example, as to whether Newton's second law of motion is to be viewed as a definition of 'force' or as to whether it is a genuine empirical law that tells us what effects (accelerations of given masses) given forces will produce, cannot be answered in isolation. If we use either Hooke's, or Archimedes' law as an operational *definition* of 'force' then of course, Newton's second 'law' may be regarded as a genuine empirical *law*. But the logical situation is quite symmetrical and hence the roles of law and definition may be systematically interchanged.

The present interpretation of this methodological insight frankly faces the fact that in testing a law of type B we must, so to speak, remain within the system; and that, in a sense, it is the system as a whole, and not any of its isolated fragments, that is being confirmed or disconfirmed by the data. Now it is one of the essential features of the system that the relations as formulated in laws and hypotheses of type B *may* be taken as relations of causal or functional dependency, provided that other relations in the same system are accordingly interpreted as *definitions*. A certain surplus meaning for existential hypotheses of type B is thereby justified. Yet, if that surplus meaning is considered to be completely reducible to the directly testable (i.e., the evidential basis) the present view reveals itself as a variety of phenomenalism (let us call it 'Contextualistic Phenomenalism').

(VII) An alternative interpretation of the methodological situation just described is found in a view that, I think, is very widely held, but only rarely explicitly stated: The 'dualistic' (i.e., realistic) assertion of the independent existence of the referents of hypothetical constructs is an essential and indispensable feature of any satisfactory explanatory system. To quote V. F. Lenzen [1939b]: "In general the dualistic theory may be viewed as a scientific hypothesis which explains and predicts perceptions. Contemporary physical theory is characterized by the dominant role played by constructive hypotheses. Assuming for the moment that everyday things are directly given in perception, knowledge of the entities of atomic physics consists in the acceptance of hypotheses from which it is possible to deduce consequences that can be tested by experiment. The energy levels of the atoms are the objects of hypotheses from which one can predict the positions of spectral lines on a photographic plate .... In all these examples there is no direct perception of the entities considered; confirmation of the hypotheses consists in the explanation of past phenomena and the prediction of future phenomena." And in a passage preceding this one we find (*ibid.*) an epistemological application of the same idea: "The dualistic theory of perception is based on the constructive hypothesis that perceptions are caused by independent things that radiate influences to the perceiving organism. Causality may be interpreted as a functional relationship between thing and percept, but even with this restriction the hypothesis is not capable of direct confirmation. It is confirmed by its success in explaining past perceptions and predicting future ones." (We may call it 'Hypothetico-Deductive Realism' or 'Explanatory Realism'.)

While of course closely related to Probabilistic Realism (III), in its basic outlook, Explanatory Realism is not committed to the questionable justification by means of inductive probability. Considerations of probability are here, as in any case, indispensable when it comes to the choice between different hypotheses. But the decision to supplement phenomenal description at all with 'transcendent' hypotheses is not in itself based upon inductive arguments. This view, however, provides only a hint, but no definite answer, as to the precise analysis of the asserted 'independence' or 'surplus meaning.'

(VIIa) The missing explication[7] has been advanced in semantical terms. The surplus meaning is understood to consist in the *factual reference* of the constructs employed in theoretical laws (of types B and C) and the existential hypotheses (of type B). This requires a clear distinction between *epistemic reduction* (i.e., the evidential basis) and the semantical relation of *designation* (i.e., reference). This distinction is most easily illustrated and rather convincingly justified by a consideration of our knowledge of the past. This well-worn issue of epistemology is notoriously one of the most stubborn obstacles in the path of all types of strict phenomenalism or radical empiricism. Statements concerning the past (e.g., as asserted in astronomical, geological, paleontological, political or cultural history, etc.) are obviously only indirectly testable. A direct test would require a literal return to the past – such as would be afforded by something like H. G. Wells' fancied 'time-machine.' A 'trip' to the past in this sense is usually considered *logically* impossible. It involves the obvious contradiction: I did and I did not live at the time, say, of the ancient Egyptians, and observed how they built the great pyramid.

Perhaps a more accurate statement of this impossibility would pronounce the contradiction as dependent upon and relative to some of the tacitly presupposed basic features (laws) of *our* world as we conceive it. It is the kind of impossibility that, for example, could be illustrated in the geometry of physical space. A point on a geodesic line may be billions of light years apart from itself.[8] This is logically impossible in Euclidean space; but it would be a necessary consequence of the postulates of certain Riemannian geometries; it is a well-known theorem in Einstein's cosmology of spherical space.

Setting speculation aside, it is a plain fact that we can confirm all retrospective statements only by means of present or future data. A statement concerning past events is thus epistemically reducible to its evidential basis in the present (or future). But clearly we take historical statements to assert something about the *past*. We may say then that we must distinguish between the radical empiricist's meaning of 'meaning' (i.e., epistemic reduction) and another, more common-sensical meaning of 'meaning' (factual reference).

Logical Positivism, before the absorption of the semantical outlook, combined its phenomenalism with a purely syntactical view of confirmation (essentially the positions characterized under (V) and (Va)). The obvious pragmatic-methodological significance of the word 'indirect' in 'indirect verification' was explicated on the one hand by reference to the purely formal-structural relations of verified to verifying sentences; and on the other hand by a pragmatic description of the symbolic behavior of human beings (e.g., scientists) in the various occasions of their adjustments to exigencies of orientation, foresight, etc. Plausibly enough one was satisfied that the notion of reference was thus taken care of. And one could always say that whatever still seemed to be missing had to do merely with the emotive (pictorial) appeal of certain words. But it was perhaps not sufficiently fully realized that the pragmatic approach (being the psycho-bio-sociology of cognitive behavior) is itself one of the empirical sciences, and therefore just as much in need of logical analysis and explicative reconstruction, as, for example, the sciences of astronomy, physics or geology. Designation and reference cannot be adequately explicated in either (descriptive) pragmatics or (pure) syntax. This can be achieved only through the construction of an appropriate *semantic* metalanguage. The very phraseology of indirect verification (confirmation) of statements requires for its explication a conceptual model in which statements as well as the states of affairs that render these statements true, can be represented. It simply makes no sense to speak, for example, of 'the present moment' except with reference to the act of speech and its moment of occurrence within a framework of other moments, i.e., of the dimension of time. The same applies to 'here' with respect to space, to 'I' with respect to other selves, and so on, for the other egocentric particulars. 'Directly tested' likewise makes sense only if there is a theoretical model in which it is contrasted with and supplemented by 'indirectly tested.'

The factual reference of not directly verifiable statements is to be construed in such a manner that it is semantically perfectly on a par with the factual reference of directly verifiable statements. The difference between the two may be dealt with in *pure pragmatics* [W. Sellars, 1947a, 1947b, 1948b]. On the deeper epistemological level it may thus be viewed as an ultimately contingent though very fundamental feature of the world in which we find ourselves: A world which contains the organisms that 'know it'; a world, that contains the data (evidential basis) which furnish the raw material for the construction of the knowledge of that world; a world that contains the (spoken, written, etc.) language which describes (or speaks 'about') that world.

It should be noted that the apparatus of pure semantics, as it has been elaborated by Tarski and Carnap, would by itself in this issue not yield any but trivial results. Whether a descriptive term of the object language has a designatum, obviously depends upon the presence of a 'translation'-equivalent in the corresponding metalanguage. If our metalanguage is rich enough to contain translations of such terms as 'the magnetic field of the earth' then this term has a designatum. – Only when we impose the requirements of pure pragmatics do we attain the desired scope of genuinely designating terms. That is to say, that in the language of empirical science all those terms (and only those terms) have factual reference which are linked to each other and to the evidential base by nomological relationships. Concepts or constructs that designate directly observable items of the world and those which do not, but are required for the coherent spatio-temporal-causal account to which science aspires[9] are thus properly related to each other by means of the metalanguage of pure pragmatics and semantics.

The view thus outlined (VIIa) may be called 'Semantic Realism.' It is a corrected form and refinement of the empirical realism held by some logical positivists or empiricists; at the same time it represents a rapprochement with the position of critical realism (epistemological dualism, essentially (VII)).

DOES SEMANTIC REALISM INVOLVE A METAPHYSICAL TRANSCENDENCE?

We shall be asked: Does not the notion of factual reference lead us back into the perplexities of traditional transcendent realism and metaphysics? We reply: Not unless we expect of the semantical analysis some *justification* of the assertion of the 'independent existence' of the designata of hypothetical constructs. Such a justification is usually understood as the validation of a knowledge claim. In this regard semantics (not being a mysterious magic any more than ordinary logic) cannot add anything to the usual procedures of empirical confirmation. 'Is there a magnetic field in this region?' – 'Of course there is, – look at all the confirming evidence!' – 'Do you mean by your existential assertion anything over and above what the totality of conceivable tests would reveal in terms of evidence?' 'Yes and no' – and here we should have to tell the whole story over again. The semantic conception of reference does not justify (demonstrate) realism. It merely *explicates* what a cautious empirical realism can legitimately mean by 'reference', 'independent existence', etc. If we handle our concepts responsibly, we can avoid metaphysical perplexities. No concrete existential hypothesis of ordinary life or of science is factually meaningful unless it is confirmable. The essential requirement of empiricism is thus safeguarded. But the very adoption of the *confirmability* criterion (in preference to the narrower *verifiability* criterion) allows as much realism as we are ever likely to warrant.

The feeling of frustration that haunts those metaphysical realists who want 'something more' is perhaps psychologically understandable, but it lacks rational justification. The remedy for this malaise lies, as should be obvious by now, in making clear to oneself that the wish that cannot be fulfilled here is bound to be frustrated precisely because it involves a self-contradiction (one cannot have one's cake, etc.): The metaphysical realist craves for a 'proof' of the existence of entities which are not directly verifiable. But if he is loath to use theological or rationalistic methods of proof, the only proof he will admit as legitimate is empirical inference. We offer him indirect verification (confirmation). This he refuses as insufficient, thereby revealing that what he really wants is direct verification. But here is the contradiction: He insists on transcendence (independent existence of the object of knowledge; surplus meaning); in other words, he first stipulates the impossibility of direct verification and then is tempted to renounce his own stipulation.

The craving for direct verification seems cognate with the wish for immediate experience notoriously manifest on the deeper levels of epistemology in the camps of subjective idealism, radical empiricism and some of the older varieties of positivism. There the issue hinges upon two different notions of 'reality'; (and, correspondingly, two different notions of 'knowledge of reality'). One is the intuitive notion of reality – as stressed by Descartes, Berkeley and Bergson. The other is the empirical and scientific notion of reality. According to the first view the criterion of reality is direct experienceability. According to the second view reality is ascribed to whatever is required (confirmed) as having a place in the spatio-temporal-causal system. Let it be realized that assertions of existence in the second sense are vacuous if not confirmable on the basis of direct experience. And if it is equally realized that the one word 'reality' is used in those two radically different senses, the whole issue loses most of its air of insolubility and thus stands revealed as a pseudo-problem engendered by confusions of meaning.

The danger of a related confusion may be seen in the perennially fashionable utterances of scientific agnosticism. 'Even if we knew all *about* electricity (matter, life, mind) we should never know what electricity, (etc.) *really* is.' Phrases of this sort (popular with great scientists, especially at the occasion of after-dinner-speeches, presidential addresses at association meetings) may be the expression of a proper and commendable humility in view of the tremendous and obviously incompletable tasks of scientific research. The phrase in this interpretation merely emphasizes that scientific progress is a matter of successive approximation. But frequently enough it is intended as a genuine 'ignorabimus.' No matter how complete our scientific knowledge, it would never acquaint us with the essence of things. This agnosticism could indeed be overcome only by such fanciful procedures as intuitive identification (perhaps real coalescence) of the knowing subject with the to-be-known-object. As long as our direct experience is limited to the data of our consciousness, we shall indeed never be able to 'know' (by acquaintance) what electricity 'really' is, because we should have to *be* an electric current in order to achieve that crowning feat of 'real knowledge.' It is truly astounding to find how widespread and deeprooted this confusion is as regards 'knowledge' and how tenacious the wish for direct intuition.

It should not be necessary here to review the related and generally recognized dangers of picture and model thinking in science. The function of visualization and of images as mental crutches, and often enough, as extremely helpful heuristic devices, is of course acknowledged. But it is of the utmost importance to distinguish these *psychological* features of *images* and their role in the process of discovery (or in didactic contexts) from the *logical* characteristics of *concepts* and their role in connection with meaning and validation.

We conclude that Semantic Realism is not subject to the charge that it involves a metaphysical transcendence. As a logical reconstruction of the language of science, and more fundamentally, of empirical language quite generally, it may be inadequate, unfruitful or unenlightening: but it is not fraught with the vacuities of traditional metaphysics.

THE ISSUE NARROWED DOWN AND CRITICALLY APPRAISED

For convenience we list the labels of the nine points of view just presented.

(I) Naive Physical Realism
(II) Fictionalistic Agnosticism
(III) Probabilistic Realism
(IV) Naive Conventionalistic Phenomenalism
(V) Critical Phenomenalism (Operationism, Positivism)
(Va) Formalistic (Syntactical) Positivism
(VI) Contextualistic Phenomenalism
(VII) Explanatory (Hypothetico-Deductive) Realism
(VIIa) Semantic (Empirical) Realism.[10]

The order in which these views were deployed in the preceding section is neither chronological nor systematic, but was suggested by considerations of expository and dialectical efficacy. From a more systematic approach it is easily seen that (I) reappears in more sophisticated forms in (III) and again in (VII) and (VIIa). Likewise (II) and (IV) may be absorbed in the more adequate formulations of (V), (Va). Finally, much of (VI) is compatible with and assimilable to (VII) and (VIIa).

Several further views were omitted in the survey. The neo-Kantian; objective relativist; neo-realist; and the (to me, at any rate) somewhat obscure and hybrid views of Whitehead (or some of their components) seem either so ambiguous in regard to our main issue or else to overlap to such an extent with components of the views here presented that separate discussion appeared superfluous. I confess that my reaction to John Dewey's position (especially that of his Logic), is similar, though I would stress that the grounds for rejecting it are somewhat different. A view recently presented by Churchman [1948, the spiral view of scientific method] seemed close enough to either (VI) or (VII), to justify absorption there. – Among recent neo-Kantian approaches the one typically represented by R. G. Collingwood [1940], though not explicitly applied to the reality problem, would plausibly formulate scientific realism as an 'absolute presupposition.' According to Collingwood's understanding of this term this would exclude assimilation of his view to Hypothetico-deductive realism; (VII) for Collingwood absolute presuppositions are not hypotheses capable of any sort of test – but they are the ultimate assumptions without which any tests of any hypotheses would be inconceivable. We may disregard here Collingwood's own psychologistic and historicist account of absolute presuppositions. That people in certain epochs of thought and research are guided in all their questionings by some basic frame of reference, may be granted. But from the point of view of *logical* reconstruction such a frame of reference may be regarded either as a set of basic hypotheses (this is the view Collingwood rejected) or else as a system of conventions concerning confirmation principles and semantic reference (a view which he unfortunately never even considered). This latter alternative would assimilate the neo-Kantian position to Semantic Realism (VIIa). We do not take the trouble here to restate the refutation of the view that regards basic presuppositions as synthetic *a priori*. This refutation has been effectively achieved by Schlick [1925, 1938a], Lewis [1929, 1946], Reichenbach [1938, 1948], Ayer [1946, 1940], Pap [1946], Kaufmann [1944], Nagel [1956]; and many others who realized that the valid insights of Leibniz and Hume (i.e., a logical empiricism) supplanted Kant's form of rationalism even before it was formulated.

A special remark is called for regarding Probabilistic Realism (III): The view, represented in the – otherwise discrepant – forms by Broad [1914, 1923] and Williams [1933–34] on the one hand, Reichenbach [1938] on the other, has been very seriously undermined by Stace [1932, 1934] and explicitly criticized by Nelson [1936], Nagel [1939], Barrett [1939], and Rynin [1947]. The crux of the problem lies in the justification of applying the concept of inductive probability to the inference from the directly verifiable to directly unverifiable assertions. Any straightforward frequency interpretation of probability could serve here only if the success frequencies of such inferences were ascertainable. This is outright impossible if independent access to the 'Illata' is barred. Utilization of the logical concept of probability (cf. Carnap [1945b]), in this matter even more obscure and problematic, seems abortive for the same reasons. Probabilism must therefore be modified in the direction of contextualism or explanatory realism. But even so, the legitimacy of applying the probability concept to the whole realistic frame, instead of merely to inferences within it, remains painfully questionable.

Our discussion may then well be restricted to the two remaining views: Syntactical Positivism and Semantic Realism (Va and VIIa). Despite the undeniable rapprochement of phenomenalism and realism that has already been achieved in these two positions, there are equally undeniable differences. Are they differences that *make* a difference? In the usual pragmatic sense of 'making a difference' there is indeed no difference. C. I. Lewis [1929, p. 194] was quite right in saying that a sufficiently critical realism can have no quarrel with a sufficiently critical idealism, except for false issues that arise out of confusions. The pragmatic test in its customary sense, however, applies only to the determination of differences in regard to observable consequences. Our present issue, be it remembered, does not concern such differences in factual consequences. These would be relevant in the decision between competing scientific hypotheses. Our issue, however, concerns the adequacy of two different logical reconstructions of factual (and especially of scientific) knowledge. The differences that make a difference in this regard and are not easily characterized. The criteria of the adequacy of logical analyses are quite generally vague and controversial. Appeals to 'intuition,' to 'pragmatic suitability', to 'clarifying potency' or to 'simplicity' crop up with alarming innocence and frequency in the justifications offered for particular modes or results of logical analyses. The school of thought that is influenced by G. E. Moore's conception of philosophical analysis insists that any reconstruction of meaning must be in close accordance with the ordinary interpretation of the language of sound and common sense.[11] It should be remarked, however, that commonsense does not furnish an unambiguous or unquestionable criterion. Yet, a measure of correspondence to commonsense, together with logical consistency and some all-around completeness and circumspection, are the standards by which we may most justifiably judge the success of philosophical analyses.

This is what I take to be the rationale of Reichenbach's "choice based on entailed decisions" [Reichenbach, 1938; Ayer, 1946; Campbell, 1921]. More precisely, the adequacy of a reconstruction of the empirical (and especially the scientific) language by means of the metalanguages of pure pragmatics and semantics is to be judged by the approximation attained in the accommodation of the conceptual structure and the confirmatory procedures of empirical (scientific) knowledge. Instead of justifying the surplus meaning of existential hypotheses and hypothetical constructs (Reichenbach's 'illata') by means of inductive probability, I suggest that we justify the conceptual frame of the realistic language by its entailed consequence; viz., by showing that only within such a frame does it make sense to assign probabilities to existential hypotheses. The criticism implied here amounts to a repudiation of Reichenbach's justification of scientific realism by means of the argument from 'projection' (as in his model of the cubical world, [C. D. Broad, 1923 and 1927–8; Reichenbach, 1938]). Reichenbach himself comes close to embracing this point of view (in [N. R Campbell, 1921], 'Positivism and Realism as a problem of language') and I submit that he thereby shows the redundancy (I should even say, the illegitimacy) of his preceding inductive argument for realism. In his discussion of the problem of the existence of other minds [Dingle, 1932], Reichenbach (almost) completely abandons the probabilistic approach. There he realizes and formulates most convincingly the need for definitional or conventional stipulation in introducing the mentalistic language in connection with the observed or observable behavior of other persons. It seems to me that such definitional stipulations are equally required in setting up the conceptual frame of the 'physical world' (the constituted universe): Once this frame is provided, the phraseology of common sense and scientific method ('direct evidence', 'indirect evidence', 'verification', 'confirmation', 'inductive probability', 'inference by analogy', etc.) can be adequately reconstructed.

The remaining issue turns upon the merits of the respective reconstructions of existential hypotheses (type B) and laws (types B and C) in Syntactical Positivism and Semantic Realism. We turn then to our final critique of the phenomenalistic aspect of Syntactical Positivism.

The crucial points here concern the adequacy of the reconstruction of *particulars* (spatio-temporal positions) and of *universals* (predicates, relations, functors). Existential hypotheses of the particularized form involve a reference to specific spatio-temporal regions. An hypothesis, to return to our example, may assert the presence of a magnetic field in a certain volume of space and during a certain interval of time. According to the views of Syntactical Positivism this assertion is formulated in the language of a theoretical or mathematical model. Its factual meaning can be explicated only through recourse to coordinating and operational definitions. The coordinating definitions link the abstract-model language with the empirical constructs; and the operational definitions connect the empirical constructs with terms that designate directly observable data. This is the prevailing view in present-day philosophy of science. N. R. Campbell [1921] explicates the relation of hypothetical constructs to the observable data by means of the notion of a 'dictionary' that enables us to make the transition from the first to the second. His ideas have been applied to the problems of hypotheses in psychology in C. D. Hardie [1939]. In a very early essay of Carnap's [1923] we find practically the same conception (cf. however, its more recent modification in [1939]; also Bergmann [1943a, 1947]; Feigl [1945a]). Reichenbach's "coordinating definitions" [1938, p. 250] are, by and large, identical with these dictionary-translation-formulae. But Reichenbach's own realistic view of hypothetical constructs would preclude agreement in this particular point.

The *reduction* made possible by the dictionary must, however, not be confused with *translation*. This all-important criticism of an earlier positivistic (phenomenalistic) view is advanced for the following reasons: (1) Since the place and/or the date of the events that furnish the *evidence* (confirming or disconfirming data) for or against a particular existential hypothesis (of type B) may differ from the place and/or date of the event (state, etc.) whose existence is being tested, the strict identity of reference needed for *logical* equivalence as a basis for genuine translation does not (generally) obtain. Preoccupation with the logical analysis of theoretical *laws* may be responsible for the neglect of this fundamental point. But the criticism is inescapable as soon as attention is directed upon the particulars, as in the historical disciplines or in astronomy. A phenomenalistic reduction of statements about particulars, i.e., to singular descriptive propositions about the unobserved or unobservable, if it is to achieve its purpose, must locate and date the events described in the antecedents and the consequents of the factual or counterfactual conditionals which form the evidential basis. But such locating and dating indispensably requires the 'realistic' frame of positions in which both the *confirmandum* as well as the *confirmans* are equally assigned their respective places. (The temperature in the interior of a solid block of metal is in principle only indirectly testable.)

(2) The equivalence in question could therefore be only of the physical (nomological) type. This is precisely what is implied in a full statement of the theoretically assumed relations between tested conditions and test results. (This point was discussed above as the 'Copernican turn'.)

(3) Even such nomological equivalence is an idealization of the actually prevailing situations. All tests are open to question on the basis of a) the usual errors arising out of ignorance of 'disturbing' factors and as a special case of these, the inaccuracies of measurement (always admitted even within the frame of classical determinism); and b) the inaccuracies arising from the more fundamental indeterminacies discovered in quantum physics. (These may, of course, be practically neglected in most problems of macro-science.) The equivalence must therefore be supplanted by probability-implications.

(4) As already indicated above, the set of possible confirming conditions is in principle infinite (strictly speaking, of the order of the continuum). Moreover, restriction to the (at a given stage of science) known testing methods would be unjustifiably narrow. The set of infinite sets of confirming conditions is obviously one that is open to additions, often in radically new directions. Only if we were *sure* of the completeness of our knowledge of natural laws, could we restrict ourselves to one definitely circumscribed set of infinite sets.

(5) Closely connected with the preceding point is the (also previously indicated) lack of a sharp dividing line between direct and indirect tests. This holds especially on the level of scientific method (though it might be questioned on the deeper level of epistemology). The history of the atomic theory furnishes the most impressive illustration of the transition from highly

indirect to much more direct types of evidence. The implication here is simply that, short of omniscience, there is no way of telling just what specific new data may come to assume the role of confirming evidence.

(6) Syntactical Positivism acknowledges, in its manner, that the introduction of new universals (predicates, relations, functors) cannot always be achieved by explicit definition. Indeed it emphasizes that the purpose of a postulate system consists in introducing (otherwise) undefined concepts by means of a network of logical relationships. The concepts of electromagnetics, far from being reducible to (in the sense of explicitly definable in terms of) mechanical concepts must be linked with them by nomological relations. Syntactical positivism tends to regard the postulate systems of science as relating only unobservables to one another; and only some of the derived (explicitly defined) terms as coordinated with observables. While admitting the historical merits of this reconstruction, it may be suggested that, in the light of all that has been said so far, a more adequate reconstruction should treat observables and unobservables on an equal footing if they are on a par within the nomological network.

Once the phenomenalistic claim of full reducibility (translatability) is abandoned, the syntactical approach may very well be combined with the semantical one. A plausible and customary avenue of reconstruction that will no longer do without thorough qualification, explicates the logical situation in terms of two languages: *The language of constructs* and *the language of data*. (I suggested this terminology as long ago as in 1934 [Feigl, 1934b], but my views regarding their relations have been revised.) The factual reference of the constructs may never coincide with their epistemic reduction. The 'dictionary' really does not afford a strict translation, but at best only one-way deducibility. The term 'dictionary' may thus be recognized as a misnomer, and the whole doctrine of coordinating definitions will have to undergo a radical revision.

CONCLUSIONS CONCERNING SCIENTIFIC METHOD

The phenomenalistic interpretation of scientific constructs has been weighed and found wanting. The empiricist principle of meaning and of validity, however, may be (and must be) retained and combined with the semantic reconstruction of a genuinely critical realism. The system of statements and concepts that constitutes our scientific knowledge is best understood as a network that connects the directly confirmable with the indirectly confirmable. In this manner we try to achieve a maximum of nomological coherence by means of a minimum of hypothetical construction. The confirmability criterion of factual meaningfulness, properly understood, already provided the realistic frame. This frame represents itself in the context of logical reconstruction as a basic convention. Its adoption cannot be argued for or against by means of inductive probability. Quite to the contrary, the customary and legitimate uses of inductive probability presuppose the realistic frame. The strongest justification for the adoption of the realistic frame is to be found precisely in that it makes intelligible what we mean by the probability of existential hypotheses. The introduction of new *basic* and *irreducible* concepts (as, for example, in electromagnetics during the last century) may be reconstructed as an expansion of the empirical language. Only after our language has thus been enriched, can we significantly assign probabilities (degrees of confirmation) to specific predictive or explanatory hypotheses. The step of expansion of language cannot itself be justified on the grounds of probability, except perhaps in the sophisticated pragmatic sense of the question: Will this expansion be methodologically fruitful? The progress of science may be viewed as the successive trials to achieve maximum predictability of the observable facts by means of adaptations, revisions, emendations or simplifications of a network that includes the unobservable (but indirectly confirmable) facts as well. 'Nothing risked, nothing gained' – this is certainly true of the hypothetico-deductive procedure of science.

Frequently enough special hypotheses may have to be abandoned if the stipulated nomological relationships prove incapable of consistent connection with already confirmed major portions of the network. This was the fate of the ether-hypothesis. The 'facts' (really *laws*) established by Fizeau, Bradley, Michelson and Morley, Trouton and Noble, de Sitter, etc., were simply incompatible with the deducible empirical consequences of the hypothesis of the stationary luminiferous and electromagnetic ether. The desperate devices of Lorentz and Fitzgerald to safeguard the ether hypothesis against this refutation were disclosed by Einstein as strictly *ad hoc* and unconfirmable. We must therefore sharply distinguish between the older ether hypothesis which was factually meaningful, but was disconfirmed; and the later ether hypothesis (in its comatose stage) which by its very conception precluded disconfirmation as well as confirmation altogether and in principle. Such constructions, no matter how ingenious or pictorially appealing, should not even be called 'hypotheses.' They are pseudo-statements of the order of the vitalistic or animistic verbalisms in biology and psychology. Entelechies, vital forces, souls or spirits are to be ruled out *not* because they are not directly and independently verifiable (they share this trait with the legitimate concepts of the magnetic field, the atom, the nucleons, etc.). Vitalistic and animistic entities are to be ruled out as so much metaphysical ballast because they are at least traditionally so conceived that they do not and could not in the least add to the explanatory power of the extant empirical laws and theories. This is merely another way of saying that these 'hypotheses' have no factual content because they are so conceived that they are in principle incapable of test. – A really open-minded empiricism can however not afford dogmatically to preclude the formation of existential hypotheses in fields in which the empirical regularities are only very incompletely established. In such cases even the outlines of the theoretical network may not be more than vaguely discernible. If, for example, in the disputed fields of extrasensory perception (or in the still more questionable fields of mediumism) the claimed empirical regularities should prove unexplainable by means of the theoretical frame of present-day-science, some emendation or even radical alterations of the network may quite conceivably be required. (May it be noted that this concession in principle involves no commitment as regards the validity of the alleged findings in the fields of 'psychical research.') The grave incompleteness of our knowledge in the bio-psychological field compels us to suspend judgment even in the field of 'normal' phenomena as regards the relation of the mental to the physical. Methodological behaviorism, as well as the double-language view of mind and body (that I have been advocating for many years, [cf. Feigl, 1934b]) depends for its validity upon certain fundamental features of the world, and cannot responsibly be justified by mere definitions or conventions. It thus depends upon a basic feature of the world as to whether the existential hypothesis as to the presence of mental states in other persons (or higher animals) is logically, or else empirically, equivalent or not equivalent (in any sense) with statements concerning certain aspects of the neurophysiological processes in those organisms. Definitions and conventions are of course indispensable for the introduction of concepts – here as elsewhere. But their fruitfulness depends (everywhere) upon certain factual features of the world over which we cannot legislate by mere conceptual stipulation.

SUGGESTIONS CONCERNING THE ANALOGOUS ISSUES OF EPISTEMOLOGY

The analogy (or homology) that we repeatedly invoked in the foregoing discussion may briefly be expressed by the formula: (Immediate experience): (Commonsense World) = (Commonsense World): (World of Theoretical Constructs). Epistemology concerns itself primarily with the first part of the proportion, scientific methodology mainly with the second. Phenomenalism in epistemology, as represented in the earlier work of Carnap [1928b, 1928a] and in the more recent developments by Ayer [1936; 1940] and Lewis [1946, 1934], maintains the reducibility, in the sense of equivalence, of statements regarding the commonsense world with sets of implicative statements, "terminating judgments" in Lewis [1946, esp. VIII], concerning immediate experience. This equivalence has been vigorously disputed by Reichenbach [1938] and even previously in effect disclaimed in the more recent work of Carnap [1936a]. Wilfrid Sellars [1947a, 1947b, 1948b] has most auspiciously outlined the pure pragmatics and semantics of the new outlook required in epistemology. It would seem that most of the arguments advanced in this essay in favor of a clarified scientific realism may be applied, *mutatis mutandis*, to the parallel issues of epistemology. The factual and counterfactual conditionals concerning the data of immediate experience are deducible from the hypothetical assumptions about the laws and the facts of the 'real' world. But not even an infinite set (or set of sets) of such conditionals is logically equivalent with the theory and history of the world. As long as we confine the factual and counterfactual conditionals within the plausible limits of ordinary human experience (as different from a divine omnipresence and omniscience) this one-way-deducibility or lack of equivalence is inevitable. And do we not pursue epistemology for *human* beings?

Taking statements concerning immediate experience as the evidential basis for the reconstruction of our knowledge of the world, it is easily seen that the (realistic) frame of space-time-causality-matter requires introduction by means of fundamental conventions. 'Postulates' may not be the best term for these conventions because in its prevailing usage the word 'postulates' connotes the premises of either purely formal or else interpreted (and therefore testable) calculi, i.e., hypothetico-deductive systems. This becomes especially poignant in regard to the introduction of the *past*, as well as in the conventions underlying the construction of the concepts of *material objects*. The probability of specific historical assertions must not be confused with the assertion that 'there is' a past at all. The inductive probabilities we have for the dating of past events presuppose the semantical frame that is the *conditio sine qua non* of such retrospective dating. No amount of present evidence can meaningfully be said to bestow a probability (or degree of confirmation) upon the frame convention that, so to speak, provides the very possibility of ordering events and confirming their occurrence. Even the fictionalistic agnostic and the radical phenomenalist (conventionalist) certainly find it easy enough to refute a *probabilistic* realism here, as well as, *mutatis mutandis*, in the quite analogous problem of material objects. Only an Explanatory Realism, recast in the form of a semantical reconstruction, seems to be able to avoid the notorious pitfalls of the traditional psychologistic, probabilistic (let alone naive) forms of realism. Before we can significantly ascribe probabilities to the assertions of existence of the unobservables we must provide a language that links the (directly) unverifiable with the directly verifiable. It is in this reconstruction, that we can say that the distinction (made in pure pragmatics) between the directly verifiable and the merely indirectly confirmable statements corresponds to the cosmological, but cognitively fundamental feature of our world, that human beings are severely limited in their direct awareness of (or immediate acquaintance with) the universe in which they are embedded, and of which they form a natural part. The notorious predicaments ('ego-centric', 'present-moment', etc.) are thus reflected both in the semantical and pragmatic account of the meaning and validation of statements, as well as in the factual account of the processes of cognition.[12]

PERSONAL POSTSCRIPT

Critical readers of the foregoing essay may gather the impression that here is a positivist who at long last has seen the light and turned realist. My record tells a different story. Except for articles in which I primarily reported on the development of Logical Positivism and Logical Empiricism (esp. [Feigl and Blumberg, 1931a], [Feigl, 1934b and 1943a]), I have throughout twenty years attempted to formulate and to vindicate an Empirical Realism ([Feigl, 1929a, 1936a, 1945a and to some extent also 1943a]). It is especially gratifying to me that Carnap, in his progress from a brilliant phenomenalistic reconstruction of Knowledge [Carnap, 1928c] to a sounder Physicalism and its criterion of confirmability [Carnap 1936, 1945b], holds now in much more perfected form a view that I was insufficiently equipped effectively to defend in the early days (1926–1930) of the Vienna Circle. Even if, as amply indicated above, I cannot accept the *probabilistic* justification of Reichenbach's Realism, I have always been highly appreciative of Reichenbach's realistic emphasis in his conception of Scientific Empiricism.

Among the major realistic philosophies of a generation ago there is much that seems to me still vital enough for reconsideration, especially in the works of R. B. Perry, R. W. Sellars, A. O. Lovejoy in America; also in the earlier Schlick, whose magnificent epistemology [Schlick, 1925] unfortunately never received in the English speaking countries the attention it deserved; and whose critical realism was unfortunately supplanted by a phenomenalistic positivism (not lacking, however, some qualifications in the direction of an empirical realism) [Schlick, 1938], mainly under the influence of Carnap and Wittgenstein during the late twenties. I have had reference already to the ever thought-provoking analyses of the British realists, especially B. Russell and C. D. Broad.

I wish to acknowledge with sincere gratitude also the stimulation and aid received from my friends R. Carnap, C. G. Hempel, E. Nagel, A. Pap and especially Wilfrid Sellars (son of Roy W. Sellars). In a series of brilliant essays, Wilfrid Sellars has outlined a new version of a realistic epistemology on the basis of pure syntax, semantics and pragmatics. What I had only dimly perceived, particularly in connection with the distinction between evidential base and factual reference (which I expounded in an unpublished lecture on 'Meaning, Meaningfulness, Reference and Epistemic Reduction' at the Sixth International Congress for the Unity of Science, Chicago, 1941) has since been independently and much more systematically elaborated by Wilfrid Sellars. In my estimation, he holds greater promise than any other contemporary thinker for doing justice to, and to provide a new synthesis of, the justifiable claims of realism and positivism, as well as of rationalism and empiricism.

Philosophical analysis, not entirely unlike science, progresses through dialectical oscillations between equally unacceptable extremes and reaches higher levels of sophistication and enlightenment in steps of successive approximation toward the horizon of complete clarity. The story of the controversy between phenomenalism and realism thus construed forms an exciting chapter in the history of thought.

The present essay, I fully realize, leaves much to be desired along the lines of further more detailed and painstaking analysis. The precise forms of the 'definitions' or meaning rules for existential constructs; the application of pure pragmatics and semantics in the reconstruction of scientific language etc., are tasks which I could scarcely begin to outline in this essay. No matter how inadequate my own constructive suggestions may be – I should be satisfied if I succeeded in revitalizing an issue that has been threshed out *ad nauseam* and that has too often been dismissed as hopeless, meaningless or dead. I shall be delighted to learn what others may have to contribute by way of destructive and especially constructive criticism and clarification.

NOTES

* For the precise meaning of 'existential hypothesis' cf. p. 201 below.

[1] For the sake of simplicity we disregard here the phenomena (involving high velocities, etc.) which call for the kinematics and dynamics of the theory of relativity.

[2] Permitting a rough identification of such substances as glass, leather, gold, aluminum, moisture, etc.

[3] This purely formal parsimony might be disputed if the causal relations are construed as material (or else as general) implications. It does seem to hold, however, if they are construed as modal implications.

[4] For the present purpose the otherwise extremely important distinction as to the qualitative semi-quantitative (topological) or metrical form; as well as concerning the deterministic or statistical character of the laws may be disregarded.

[5] Similarly: The width of spectral lines (or the character of the Brownian motion) as evidence for molecular velocities; the motion of oil drops (Millikan) as evidence for the electric charge of individual electrons; Geiger-counter effects as evidence of the impact of single electrons: Bragg and Laue diffraction patterns as evidence for the atomistic structure of crystals, etc. – For the reasons previously given, we shall, on the whole, use the simpler illustrations from magnetics.

[6] To regard verification of the conjunction 'p.q.' as a verification of the implication 'if p then q' obviously does not render adequately what is intended here. As to whether the introduction of causal modalities (e.g., Reichenbach's 'nomological implication' or some

other modal conditional) satisfactorily explicates the meaning of lawful connection and of counterfactual conditionals, is one of the controversial issues of present day logic. Cf. Hempel [1945], Hempel and Oppenheim [1948], Chisholm [1946], Goodman [1947], Reichenbach [1947], Lewis [1946], W. Sellars [1948a], Popper [1949].

[7] Just as (Va) represents the syntactical refinement of (V), so (VIIa) may be taken as a semantical refinement of (VII). Cf. Wilfrid Sellars [1947a, 1947b, 1948b].

[8] We trust that this very loose and picturesque formulation will be allowed here.

[9] No matter to what extent and in whatever specific form such a coherent scheme may be attainable.

[10] The historical adequacy of the labels is questionable. My primary concern was not to display the strife of 'schools of thought' but to survey and explore the typical gambits in a well-known problem.

[11] Cf. *The Philosophy of G. E. Moore*, ed. Schilpp, Liberty of Living Philosophers, Volume 4: especially the articles by Malcolm, Lazerowitz, Ambrose, Wisdom; also the more searching articles by Marhenke and Langford.

[12] Comments on Professor Feigl's paper by Philipp Frank, C. G. Hempel, E. Nagel, A. G. Ramsperger and C. W. Churchman appeared in the April, 1950 issue of *Philosophy of Science*, together with a reply by Feigl.

13. LOGICAL RECONSTRUCTION, REALISM AND PURE SEMIOTIC

[1950c]

In this rejoinder to the critical comments elicited by my essay 'Existential Hypotheses', I propose to deal first with the challenge coming from the avowedly different philosophical outlook of Professor Churchman. My other critics, Professors Frank, Hempel, Nagel and Ramsperger, on the whole, share my basic conception of the tasks of philosophy of science and epistemology, even if they dissent in one important respect or another from the special solution I suggested. But since I discern even in Professor Nagel's remarks (and possibly also between the lines of Professor Ramsperger's comments) a pragmatist or instrumentalist strain akin to the major contentions of Professor Churchman, it will be well to begin with a defense and further clarification of my underlying point of view. Only after this restatement of my platform will I undertake to defend semantic realism against the specific criticisms advanced by the last four authors.

Professor Churchman wants to know what is the use of logical reconstruction. Does it help science in its own progress? Will the methods and techniques of science benefit from metalinguistic analyses of its concepts and assertions? My answer is unhesitatingly in the affirmative. If logical reconstruction were no more than an idle parlor game invented solely for the delight of those who care to play it, I should not for a moment admit it as a legitimate task for philosophy. Ultimately I too share the value judgments of the pragmatists. But I think the history of science shows most convincingly that new levels of reflection, even when they seemed remote from direct practical application, have either indirectly or in the long run helped in the main concerns of the scientific enterprise. Modern symbolic logic at first was widely condemned as a play thing; it was considered sterile by as eminent a thinker as Poincaré. Even if its present day utilizations in the axiomatics of mathematics, probability theory, theoretical physics, biology, psychology, etc., may not be as impressive as the utilization of non-Euclidean geometry or of group theory, etc., there is no doubt that we could no longer do without it. The highly important and practical results of cybernetics (of N. Wiener and others) to which Churchman refers with such high appreciation may well be mentioned as in part dependent on the developments in mathematical logic. And who can conceive of contemporary mathematical logic without making use of the reconstruction that such metalinguistic studies as pure syntax and pure semantics have made possible? – But I need not rely on parallels, analogies or vague promises for my argument. Even if these be questioned I would maintain that the complexity of the logical structure of present day science requires analysis, at least for the reason that we attain the clarity requisite for the avoidance of inconsistencies and confusion of various kinds in our thinking about science. If, for example, experimental physics is regarded as a tool for the acquisition of knowledge applied in engineering (and so ultimately, we hope, for human welfare), then theoretical physics and pure mathematics, each in turn are further tools in the same enterprise; and finally, the operational, syntactical and semantical analysis of the symbolisms used on these various levels is a further instrument, a further technique designed to help in the progress of science. Does Churchman need to be reminded that studies, e.g., in the foundations of probability and statistics (in which he himself has shown a strong interest) are of the nature of a logical reconstruction? Does he, the esteemed editor of the *Philosophy of Science* journal, need to be refreshed on the difference between statements in science and statements about science? Doesn't he too, for the sake of clarity, find it imperative to distinguish among the latter very sharply between statements concerning the psycho-socio-historical aspects of science (i.e., the context of discovery) and statements concerning the logical aspects (i.e., the context of justification[1])? If Churchman is at all willing and able to distinguish an account of the genesis or development from an account of the validating reasons of scientific knowledge claims, would he not then wish to make each of these accounts as explicit, articulate and efficient as possible? If so, then I think he will have to admit that the philosopher of science, or in any case the *logician* of science (in contradistinction to the sociologist of science) is charged with the task of ascertaining the validating grounds and the validating principles of scientific assertions. What specific forms the fulfilment of this task may take; what specific tools may prove the most useful in its pursuit, these are of course further questions regarding which there can be legitimate controversy.

Turning more specifically to the issue under discussion I would urge the following consideration upon Professor Churchman, and perhaps upon pragmatists in general:

(1) Knowledge-claims in the natural and the social sciences are legitimate only if they are based on specifiable evidence. If we are not to be bogged down with the well-known troubles of a pure coherence view of confirmation we must at any given stage of science be able to quote observations that serve

as evidence, at least until further observations impel us to discredit such erstwhile evidence and replace it by other observations.

(2) While I have not only admitted but indeed stressed that there is no completely sharp division between the directly observable and the indirectly confirmable, I would most definitely oppose any attempts to make more of this 'fuzziness' than it warrants. A. C. Benjamin, for example, possibly under the influence of Whitehead, seems to me to have blurred important issues by the too facile device of expanding his distinction (of degree) of the "clearly given" and the "obscurely given" beyond the very narrow zone of border-line vagueness in perception [Benjamin, A. C., 1937, esp. pp. 131–134]. The indirectness of confirmation in the case of existential hypotheses of my type B (e.g., as regards past events, electric fields, nuclear processes, etc., etc.) is radical and irremediable. Visible spectral lines are evidence for the dynamics of atoms. Inscriptions on tombstones are evidence for events in history. Only by distorting the ordinary meaning of the word 'evidence' beyond recognition could one reverse these relationships. In the explication of the justification of our knowledge claims we must pay attention to the (epistemic) primacy of the observable, even if, as I have stressed, this primacy becomes irrelevant in the ('realistic') account of the finished scientific theory. In which other way could one possibly state on what grounds the more highly theoretical assertions of science can be warranted? If Churchman's outlook does not embrace this basic minimum of empiricism, then I have either not even begun to understand him or he is not entitled to classify himself as a pragmatist or as an experimentalist.

(3) The term 'observable', as Churchman maintains, may indeed be taken to refer to a dispositional property, and as such it involves some of the relativities which Professor Ramsperger stresses. But it must be remembered that in the (indeed customary and legitimate) sense in which both commentators use the term, it is a scientific term and therefore presupposes the frame of the scientific account of the world that enables us to speak of observers and observed objects, of organisms and their environment, of the conditions and the consequences of perceptual processes, and the like. In the context of logical reconstruction however the scientific characterization of observability must be understood as an extrasystematic, as it were, marginal, didactic or elucidatory hint. In the reconstruction in terms of pure pragmatics certain predicates are distinguished from others by purely formal features. In the earlier phase of logical positivism we used to speak of primitive or undefined predicates which served as the basis for the introduction (by explicit or contextual definitions) of derived predicates. Although we have changed our views

on some basic features of the forms of reconstruction we still maintain as radically as ever the distinction between the logical analysis of scientific language from the account of the world given in terms of the scientific language. That the logical analysis must, in order to be adequate, finally disclose a certain congruence with the scientific account of knowledge, is precisely one of the major points of my essay, and I regret that it was not accepted as an olive branch by the pragmatists.

(4) To the teleological or purposive character of the scientific (as well as of the logico-analytic!) enterprise I attribute the same importance as does Churchman. All goal directed behavior depends on drives or needs (primitive or derivative) and is characterized by docility (analyzable in terms of feedback mechanisms). Far from condemning teleological concepts I would be eager to see them completely purged of their metaphysical connotations and clarified in terms of causal-statistical mechanisms. The new discipline of cybernetics strikes me as very hopeful in this connection. But again, this pertains to science (for our issue, the bio-psycho-sociology of knowledge) and not directly to the philosophy of science. Teleological considerations become relevant only if we step outside the context of logical reconstruction and ask for a (pragmatic) justification or vindication of the very principles that are presupposed in the (cognitive) justification (validation) of our knowledge claims. I have tried to disentangle this delicate and complex problem in a special essay on the meaning and the limits of justification [Feigl, 1950a].

Turning to Professor Ramsperger's suggestions in favor of Objective Relativism or Contextualistic Realism, I admit that I have dismissed this view all too briefly without discussion. But even with the helpful illustrations in Ramsperger's comments (as well as the fuller presentation in his book *Philosophies of Science*) I am still unable to see how his position can obviate the issue of realism vs. phenomenalism. Granting the relativity to context of the facts of perception, I would first of all ask: how is this relativity itself confirmed? Ultimately by observations, is it not? And this for the simple reason that any statement of the dependence of perceptual fact upon the conditions of observation is in the nature of an empirical law (psychological, psycho-physical, psycho-physiological, physical, socio-psychological, etc.). Such laws can however be established only by confirming instances, and the description of these instances must, at that stage of inquiry and until further notice, be taken as data with no 'if' or 'provided' attached to them. Availing myself of one of Ramsperger's own examples I would advance the following further considerations and questions: In order to predict what color impression a given observer will receive (and report) when exposed to light rays we must know (or assume) a good many facts about the observer as well as about the radiation. Is it not a conclusion of scientific research, rather than one of its basic epistemological presuppositions, that the color impression is dependent upon the conditions of the observer (his location, adaptation, accommodation, retinal and neurophysiological, psychological, etc., characteristics) and the conditions of the radiation (frequency, intensity, polarization, etc.)? Quite generally, do not all statements of contextuality or relativity occur *within* the frame of scientific concept formation? In this age of relativistic and quantum physics I am as fully impressed with the relational character of 'reality' as is Professor Ramsperger. But I do not see how the very statement of any relational or relativistic situation can be significant (let alone fruitful) unless it is made in terms of some *invariants*. The program of scientific knowledge proceeds unmistakably from narrow and local contexts to wider and (ultimately or ideally) universal contexts. Of course we can never be sure that the basic constants (such as those of contemporary physics: *c, e, m, h*, etc.) are not themselves relative to as yet unrecognized contexts. But this reservation is only the indispensable 'valid until further notice' clause, the warning call of caution, insisted upon by any empiricist aware of the self-correcting nature of scientific research. Fully granting all this, I still maintain that any statement of relativity to context can serve in scientific explanation and prediction only if it is formulated in terms of functional relations which, at least for the time being, are regarded as invariant. I trust that I shall not be grossly misunderstood as advocating some metaphysical absolutes. Now the natural laws stating some of the more pervasive invariancies of relationships contain concepts of the hypothetical-construct type. In the above example we may make use of Maxwell's concept of the electro-magnetic field. If it were maintained that the total meaning of statements containing such concepts consists in the (infinite) set of directly verifiable statements describing observable results in observable contexts, then I would characterize this position as (contextualistic) phenomenalism. However, Ramsperger, does allow for something more; namely counter-factual conditionals. And if I may suggest a few important distinctions, Ramsperger requires counterfactuals not only of the ordinary type, i.e. those that specify what would be observed under (a) actually unrealized and (b) technically unrealizable conditions, but also, and this is notable: (c) physically unrealizable conditions. I shall not elaborate the obvious objection here that the actual procedures of scientists do not involve considerations of the last sort; for example, no atomic physicist seriously depends upon the fictional conditional concerning how he would perceive a hydrogen atom if his organism were reduced to comparable size. Rather I should urge Ramsperger to realize, that whatever he can state in the form of the last type (c) of counterfactual hypotheticals he actually deduces from a theoretical system which he *presupposes*. That this theoretical system in turn is arrived at by inductive and analogical reasoning, or in any case can be justified only by inductive logic, will be granted. But neither the meaning nor the validity of the system depends upon the mentioned fictitious conditionals.

This brings me, finally, to the searching questions and criticisms of Professors Frank, Hempel and Nagel. (Since their comments have a good deal in common I shall address myself to these three critics simultaneously.) Their major doubts concern the precise nature of the *surplus of meaning* which according to my view attaches to existential hypotheses which are *in principle* only indirectly confirmable. Much as I have tried in my essay to make myself clear on the significance of these crucial (italicized) phrases it seems I have not fully succeeded. Let me tackle first the last point once again. In the scientific account of the cognitive process (which Churchman mistakes for a philosophical account, and which Ramsperger presupposes for the formulation of his counterfactuals) we trace the adaptations of the organism (human being) to its environment (physical and social) in terms of the psychology of learning. The organism, being a spatio-temporally minute part and quite recent arrival in the vast setting of the processes of the universe, acquires habits of action and of expectation. In this task the symbolic function of language is of the greatest importance. The 'mapping' of the universe is carried out by means of the reference of linguistic, or in any case rule-governed, symbols. Only some symbols (or rather individual tokens thereof) actually confront their designata within human experience. The vast majority has what some realists are fond of calling 'transcendent reference.' It was my concern to show that this transcendence is completely unobjectionable in contradistinction to the transcendence invoked in metaphysical speculation. The manner in which the knowing organism is embedded in the world of which it is a part simply precludes *in principle* direct experience or confrontation of all but a minute portion of that world. This 'impossibility in principle' is not a logical impossibility in the sense of self-contradiction. It is a physical impossibility in the sense that it involves incompatibility with acknowledged basic features and laws of the universe. From a philosophical point of view it is important to differentiate those physical impossibilities that involve specifically the cognitive processes of organisms from those that don't. In the latter class we find for example the various types of *perpetuum mobile* of thermodynamics. But the 'egocentric' and 'present-moment' predicaments of epistemology involve the knowing subject (organism). As I indicated in my essay, the impossibility of a return to the past (or the impossibility of reducing one's size to atomic dimensions, or of making one's retina sensitive to radio-waves) are matters either of the basic structure or of special laws of nature. In the context of logical reconstruction we reflect these 'predicaments' simply by the choice of the *basis* of reconstruction. The evidential basis thus understood can therefore never provide for sets of statements that would be equivalent to statements whose factual reference transcends the physically possible direct evidence. The counterexample contrived by Hempel is specious in that the logical equivalence of the two statements depends (according to his own presuppositions) upon the non-factual character of the statements: that the 1st of June (1949) falls on a Wednesday is either an analytic proposition and therefore irrelevant for what the example is to prove, or else factual (socio-linguistic), but then it is not *logically* equivalent with the other statement (that the 2nd of June falls on a Thursday). – Even if we allow, as Nagel does, for infinite sets of statements, capable of extension in unforeseen directions, the epistemic predicaments will 'in principle' preclude strict logical equivalence of statements only indirectly confirmable with statements directly confirmable. This is my reason for ascribing a 'surplus meaning' to the former.

Professor Frank's comments seem to imply a denial of this surplus meaning. Indeed, if he identifies the meaning of scientific statements with their truth conditions (this latter term understood in the sense of evidential basis), then this is precisely the phenomenalistic position I was concerned to refute. But if we take the term 'truth-conditions' in its recent semantic usage, then it coincides completely with the sense in which I used the term 'factual reference.' In this (semantical) sense, and in this sense only, can we say of an existential hypothesis that it means precisely what it says. The intention of my essay was to avoid both the metaphysical excesses of traditional physical realism, e.g., that of M. Planck and in much of American Critical Realism as well as the reductive fallacies of phenomenalistic positivism (Mach, the Vienna Circle, etc.). The slogans of operationism and of the first phase of logical positivism were: 'A concept is identical with the set of operations that determine its application'; 'the meaning of a statement is the method of its verification.' I agree with Frank that these slogans were excellent devices for the elimination of metaphysics. But as is so frequently the case in the history of ideas, these extreme measures, this all too radical handling of Occam's razor, went too far in the other direction. Schlick's memorable essay on 'Positivism and Realism' [Schlick, 1938a] attempted to do justice to both sides but unfortunately remained vague and vacillating just in the most crucial points. At the time of writing that essay, Schlick was reacting against his earlier realistic position. This accounts for the decidedly phenomenalistic trend in all his later work. The proper synthesis, I shall maintain, could be found and formulated only in terms of pure semiotic which became fully available only after Schlick's premature death in 1936. Nevertheless, Schlick's early realism, expounded in his *Allgemeine Erkenntnislehre* of 1918 and 1925, was an admirable informal anticipation of the sort of realism toward which Carnap (ever since 'Testability and Meaning' and his recent work on inductive logic) has been modifying his earlier positivism.

Twentieth century physics, I should like to suggest against even such an authority as Frank, does not lend more than a superficially convincing support to phenomenalistic positivism. The evolution of recent physics of course represents a powerful argument against any sort of apriorism. But it is consonant only with a sufficiently broadminded empiricism. As I indicated already in my remarks to Ramsperger's comments, there is no difficulty in combining the idea of factual reference with whatever relativities need to be taken into consideration. In regard to spatial or temporal determinations we can take a certain frame of reference for granted, and express in the coordinate language thus provided the measurable (or inferable) quantitative values of lengths, durations, masses, etc. relative to that frame of reference. Or else, we decide upon Minkowski's representation in which case the four-dimensional intervals between space-time-points (events) are among the objects of factual reference. Although I cannot possibly enter here into a discussion of quantum mechanics, I should like to anticipate at least one challenging question that arises out of the interpretation of Schrödinger's equation. I might be asked about the factual reference of the wave and the particle concepts. In agreement with the generally accepted interpretation by Max Born I would of course consider the values of ψ^2 as statistical frequencies. But I would insist that the frequencies concern micro-events which according to only indirectly confirmable existential hypotheses have *some but by no means all* the characteristics of (classical) particles in motion, collision, etc. Semantic realism as I should like to see it understood, is free from the dangers of metaphysics precisely because it does *not* prescribe anything at all about the *nature* of the designata of our theoretical constructs. It is concerned only with the most abstract and formal features of the semiotic situation. There is no danger that the wish for picturization, so strong in the older, metaphysical forms of realism, will dictate the application of the categories of commonsense to domains where they are notoriously out of place. Things are and will always be – as far as we can meaningfully talk about them – what they are confirmably *knowable* as; and it is up to the advance of science, not to logical or semiotic analysis, to tell us what things are 'really' like. But it is the task of logical analysis to tell us by means of what rules of our language we describe the objects of our knowledge, and (this was our major concern) what we mean by the surplus of the *knowable* over the *known*.

The exact explication of this surplus meaning is a further task which I have indeed only sketched in outline. I readily concede that pure pragmatics has not been developed to the extent that its indispensability or fruitfulness is as obvious as is (to my mind at any rate) the value of pure syntax and pure semantics. Fortunately I can here again refer to the articles by Wilfrid Sellars [1947a, 1947b, 1948a, 1948b] in which the basic ideas of a pure pragmatics are set forth. The work of W. Sellars has impressed upon me the perfect analogy of all three branches of pure semiotic: syntax, semantics and pragmatics. Ironically, the general resistance against recognition of the clarifying power of these three disciplines appears to be inversely related to their philosophical importance and must be overcome one by one in the chronological order of their development. As I see it, Frank and Nagel allow for syntactical studies of the language of science and supplement them by methodological or operational analyses. But those latter analyses are still mixtures of the descriptive pragmatics as pursued in the history of science and pure pragmatics which is a formal discipline that deals with the norms of meaning, meaningfulness, verification, confirmation, verifiability and confirmability. In his concluding remarks, Hempel concedes that a purely syntactical account of science must be supplemented by a "semantical interpretation of at least some of its terms." I suspect that Hempel has here in mind only the predicates whose designata are observable thing-properties and the proper names which designate the objects of direct acquaintance. The various arguments that I adduced against this syntactical positivism and in favor of a semantic (or perhaps, as I had better call it, 'pragmatic') realism simply amount to the claim that when we fully and justly explicate the way in which we use the language of science (or homologously, the language of commonsense) we cannot do without a set of designata that are in principle beyond the reach of direct experience. I maintain that a good many statements concerning theoretical constructs and hypotheses made by Frank, Hempel and Nagel are *de facto* statements in the pure pragmatics of science. Any surprise of my good friends at having, at best implicitly, utilized the metalanguage of pure pragmatics would be no better justified than the surprise some of us had some fifteen years ago when we learned from Tarski that any statements about the truth of sentences or the designation of terms (with which ordinary conversation and certainly logical discussions abound) belong to the metalanguage of semantics.

(Monsieur Jourdain was surprised that he had been speaking prose all his life.)

A more serious and difficult question pertains to the probability of existential hypotheses. This is a highly involved issue because neither the frequency theory nor the logical theory of probability have as yet provided a full and satisfactory account of the probability of complex scientific hypotheses. Moreover the issue between the frequency and the logical interpretations of probability is still controversial. It seems to me, however, that on either interpretation we can define inductive probability only if we have first of all clearly settled the vocabulary and the rules of the language in which both the hypotheses and their supporting evidence are formulated. The ratio of ranges which defines the degree of confirmation of an hypothesis cannot be determined unless we presuppose a definite set of particulars, predicates and relations. It was my contention that the language of science employs terms whose designata extend far beyond the scope of the phenomenal data. The temperature of a body, for example, is not to be identified with any or all the possible operational indications of that temperature. It is a state of that body to which we can refer only after the language has been sufficiently extended to include, besides the predicates needed to describe the various indications, also the states indicated. I must admit that I cannot at present furnish an accurate reconstruction of the meaning of inductive probability for existential hypotheses (Type B). But it seems obvious, especially considering the inadequacies of phenomenalistic interpretations granted by Nagel, that we cannot identify the probability of an existential hypothesis, e.g., regarding the surface temperature of the sun, with the probability of the outcome of any one (or all) of the various indirect indications and measurements that would confirm that hypothesis. Since Hempel, in criticizing my statement, refers to Carnap's article on 'The Two Concepts of Probability', I may in turn refer him to footnote 20 in that same article in which Carnap explicitly endorses my 'empirical realism.' The interpretation of the language in which we can meaningfully speak either of limits of statistical frequency or of ranges of propositions seems to me to be precisely the one of semantic realism.

At the risk of provoking intense controversy I might suggest an argument that goes beyond the considerations of my essay. This argument would require a good many qualifications to safeguard it against misinterpretation. Since there is no space here to do this I shall state my point quite bluntly, but would not wish to insist on either its cogency in its present form or its indispensability for my point of view. In brief, I contend that there is a specific kind of difference that *makes* a difference between Syntactical

Positivism and Semantic Realism. I still maintain of course (as before) that this difference is not of the kind that we so often encounter in the case of rival scientific hypotheses or theories. Differences between rival theories, if they consist in discrepancies of their factual content and not merely in their logico-mathematical formulations (i.e., in their respective degrees of formal simplicity) can indeed be determined by empirical tests. The difference between Syntactical Positivism and Semantic Realism lies in their different semantical interpretation of one and the same theory. The kinetic theory of heat, to take a simple example, from 19th century physics, is interpreted by syntactical positivists as merely a convenient formal device designed to correlate and unify the various empirical laws of thermodynamics. A phenomenalist like Mach would admit as much as this only in his more tolerant moments. In view of the triumphant success of the molecular, atomic and quantum theories during the last eighty years, more recent phenomenalists (such as P. Frank, R. von Mises, N. R. Campbell, H. Dingle, G. Bergmann, a.o.) do not in the least deny the fruitfulness of those 'constructions.' But would physicists have pursued this type of theory construction and attained their goals with such remarkable success if they had really held the phenomenalistic interpretation and not merely paid lip-service to it (as did some of them, e.g., Heisenberg and Dirac)? Now this question might be dismissed in the familiar manner as a purely psychological and historical one, concerned with the development of scientific ideas, and the heuristic efficacy of pictorial models. I hasten to assure the reader that my argument is intended in a *logical* sense, concerned with the semantic interpretation, not with the heuristic value of the picturization of theoretical systems. Here, then, is what I suggest: The difference that makes a difference can be explicated by the differing inductive probabilities of concrete predictions. In the example of the kinetic theory, a consistent phenomenalist would say (and did say) that Maxwell's theorem concerning the distribution of velocities among the molecules of a gas is merely part of the mathematical model whose exclusive task is to integrate into an expedient deductive structure the various experimental laws which state the relations between such observables as pressure, volume, temperature, concentration, rate of diffusion, viscosity, etc. Since 'constructs' like the mass and the velocity of individual molecules are expressly viewed (by the phenomenalists) as 'nothing but' parameters in an abstract model, he could not *on this interpretation* have predicted with any appreciable probability the outcome of such experiments as that of Born and Stern. In this experiment it became possible, by a simple but most ingenious device, to measure the speeds of individual molecules. In order to derive this outcome with the high probability that physicists in general attach to such predictions[2] the macro-observable setup of the experiments must be interpreted in terms of micro-existential hypotheses. This, however, involves the abandonment of the phenomenalistic interpretation of the theory in question. I should like to ask syntactical positivists, and phenomenalists generally, to provide a plausible reconstruction of this striking feature of modern science: the high objective probability of the results of experiments of the kind mentioned. A purely syntactical interpretation of the postulates of the theories in question does not seem to me at all adequate for the explication of this feature.

My point is simply this: The customary probabilistic realism in trying to justify 'transcendent' hypotheses on the basis of experimental findings has put the cart before the horse. Only after the introduction of the realistic frame can we legitimately argue inductively either from the theory to the outcome of as yet unperformed experiments; or vice versa from the results of experiments to *specific* postulates of the theory. But the presupposed introduction of the realistic frame, i.e., the semantic-realistic interpretation of the theory, is a step that can be justified only instrumentally: It furnishes the very possibility of a theory that is inductively fruitful.

Looking back to the realism-positivism controversy of two generations ago (Boltzmann and Planck vs. Mach and Ostwald) we may say that the subsequent developments in epistemology and especially in pure semiotic have enabled us to eliminate the metaphysics from realism by utilizing the positivistic warnings against picture thinking. On the other hand we have preserved the sound element of realism in the idea of the factual reference of (some of the) hypothetical constructs. A positivism freed from the confines of a narrow phenomenalism can yet retain its most vital safeguard: the confirmability criterion of meaning. The resulting synthesis, empirical realism reconstructed in terms of pure semiotic, should help in avoiding wasteful controversies in the development of science. Thus, even if such logical reconstruction bakes no bread and builds no bridges; even if in and by itself it does not yield new techniques of empirical research; it may yet fulfill a function that even pragmatists might recognize as quite useful.

NOTES

[1] Cf. Reichenbach [1938], Section 1, for a clear statement of this difference.

[2] Similar cases in point are: the outcome of the von Laue and Bragg X-ray diffraction patterns revealing the atomic structure of crystals; the cloud chamber tracks and Geiger

counter indications in many other experiments; the Stern-Gerlach results on the magnetic moments of atoms; and countless other results in recent experimental atomic and nuclear physics. – Genetics and bacteriology furnish analogous illustrations.

14. DE PRINCIPIIS NON DISPUTANDUM . . . ?

On the Meaning and the Limits of Justification

[1950a]

Arguments purporting to justify beliefs or evaluations often proceed from specific to more general issues. Opposition and challenge tend to provoke critical reflection; through various dialectical moves higher levels of justification are reached and made explicit. Argument usually terminates with appeals to principles which are considered indisputable, at least by those who invoke them. But, notoriously, initial disagreements cannot always be removed by what is called 'rational argument.' Frequently enough, initial disagreement can be traced back to disagreement in basic presuppositions. It is a characteristic of those modern cultures which endorse freedom of thought that they countenance divergencies in religious, political, or economic positions. 'It is all a matter of one's ultimate presuppositions' – this phrase and its variants indicate that enlightened common sense is aware of the limits of argument and justification. But on the other hand there is also the deep-rooted wish to be *right*, absolutely right, in one's basic outlook. When the disagreement concerns mere gastronomical matters, we are quite willing to reconcile ourselves with the saying, '*De gustibus non est disputandum.*' Art critics and aestheticians, however, do not unreservedly extend such tolerance to all issues of aesthetic evaluation. Most people, including the majority of philosophers, are still more reluctant to grant any relativity to the basic standards of moral evaluation. There is, at least in this age of science, almost complete unanimity as regards the criteria by which we judge the claims of ordinary factual knowledge. And perhaps genuine indisputability is attributed to the principles of formal logic. At least the simplest canons of deductive reasoning, as they are exemplified, e.g., in some of the syllogisms or in elementary arithmetic, are quite generally accepted as indispensable presuppositions of any sort of argument.

While there is no intention here to cast doubt upon the particular gradation just sketched, it need not be taken for granted either. What we do wish to clarify is the status of the very principles which in each of these various fields constitute the standards of validity or the bases of justification. The question mark attached to the title of this essay is not only to indicate that I am going to raise more questions than I shall be able to answer, but also to stress the deeply troublesome and controversial character of the main issues of justification.

The present essay aims at the illumination and at least a partial resolution of the following puzzling questions:

(1) What are the meanings of the term 'justification' and what are the logical structures of the corresponding procedures of reasoning?

(2) If justification consists in the stating of reasons, and if the fallacies of the *petitio principii* and of the infinite regress are to be avoided, what are the limits to which justification can legitimately be pursued? By what criteria do we know that we have reached the limits of justification?

(3) What is the nature of the 'ultimate presuppositions' which serve as the uppermost principles of justification?

(4) Can disagreement with respect to these principles be settled only by such nonrational procedures as persuasion, indoctrination, propaganda, therapeutic influence, or coercion? Or else, *if* rational argument concerning first principles *is* possible, what are *its* standards of justification?

(5) How is the issue regarding the primacy of 'theoretical reason' vs. the primacy of 'practical reason' to be resolved?

In order to approach these intriguing questions with any hope for clarification we shall first have to make sure that we understand what we mean by 'justification' and what major types of justification are employed in various contexts.

The search for justification, the capacity for critical reflection, are among the marks of the much vaunted rationality of man. He is sometimes able, and often willing, to state the reasons for accepting or repudiating knowledge claims and evaluations. The procedure of justification is here taken to be precisely this stating of reasons. More fully explicated, justification consists in the disclosure (exhibition, demonstration) of a conformity of that which is to be justified (the *justificandum*) with a certain principle or a set of principles which do the justifying (the *justificans*). We justify claims of factual knowledge by means of empirical confirmation. We cite evidence. But the facts that constitute what we call 'evidence' have a bearing on our knowledge claim only by virtue of some principles of confirmation (or induction). We justify claims of mathematical truth by proof. But the validity of a proof depends upon conformity with the principles of deduction. We justify moral approvals or condemnations by reference to ethical principles, and so on. Justification as here understood thus invariably involves at least an implicit reference to some standards or norms which serve, in the given context, as principles of justification. When challenged to justify any one of these principles in turn, people are apt to get impatient or "probably blow up right in your face, because you have put your finger on one of [their] absolute presuppositions, and people are apt to be ticklish in their absolute presuppositions" [Collingwood, 1940, p. 31; cf. also p. 44]. Indeed, if we ask a typical laboratory scientist what justifies him in his unquestioning acceptance of arithmetic or of the principle of empirical induction, he will, at the very best, tell us that he takes these principles for granted and that it is not part of his business to validate or justify them. We are apt to get an analogous answer from, say, a democratically minded statesman engaged in promoting some measure of social reform, if we ask him for a justification of the principle of justice for all. However, all this is psychology. The facts mentioned may be taken as symptoms of the ultimacy of the principles in the given context. A symptom of a distinctly logical character (but not decisive either, as regards logical ultimacy) is the circularities that are apt to arise on this level of argument. Requests for justification, if complied with at all, tend to elicit answers which are more or less disguised forms of question begging.

It is generally recognized that one of the major tasks of philosophical analysis consists in making explicit (i.e., formulating articulately) the more basic justifying principles. It is almost equally well recognized that *giving reasons* for our beliefs is something altogether different from pointing out their causes. He who has not grasped this difference has not even begun to understand what philosophy is all about. For philosophical analysis endeavors to reconstruct (explicate, clarify) the procedures of justification.

The uses of the word 'reason' suffer unfortunately from a good deal of ambiguity. Besides naming a capacity of the human mind (part of which may be the ability to state reasons), it is used to designate causes and purposes, as well as grounds of validation. What a rich source of confusion lies in the little phrases 'the reason why', 'because', and 'since'! Aristotle and Schopenhauer and many thinkers between and after have struggled to disentangle these and other meanings of 'reason.' Kant stated what concerns us here very clearly by distinguishing between the questions *quid facti* and *quid juris*. Husserl, at the beginning of this century, opposed most explicitly the psychologistic confusion of the two questions. But, even like Kant and Husserl, many other philosophers have not been free from serious relapses into the very confusion they had recognized and severely criticized.

The word 'justification' shares some of the ambiguities of the word 'reason' (as used in phrases like 'giving reasons'). As we proceed we shall find it not only indispensable but also highly clarifying to distinguish between justification in the sense of *validation* and justification in the more usual sense of an argument concerning *means* with respect to *ends*. The type of justification which we wish to distinguish from validation may be called 'pragmatic' or 'instrumental' justification (*justificatio actionis* as contrasted with *justificatio cognitionis*). In what follows we shall take the terminological liberty of using the term 'vindication' as a short expression for this second meaning.

Other terms related to 'justification' are 'criticism' and 'appraisal.' In their ordinary meanings these terms stand for reflective acts which are directed toward other acts or attitudes, rather than upon the cognitive content of propositions. If it is cognition that is under critical examination it is usually the process of reasoning, or the acceptance or rejection of knowledge claims, rather than the validity of the relevant propositions that is being 'criticized.' But here again the two meanings are usually so intimately intertwined that it takes a special analytic effort to separate them neatly from one another. The distinctions that we just indicated will be applied presently in the comparative study of four domains of justification; and it will be more systematically elaborated in the concluding part of this essay.

ILLUSTRATIVE DISCUSSIONS OF FOUR MAJOR DOMAINS OF JUSTIFICATION

Material for the analysis and rational reconstruction of justification abounds in the countless varieties of argument. A review of the interplay of dogmatic theses, skeptical antitheses and critical syntheses in some of the major domains of argument will help in avoiding the dangers of pale abstraction and sterile formalism that unavoidably impair discussions of justification in general and *in vacuo*. We are now going to discuss (1) logical, (2) methodological, (3) epistemological, and (4) ethical contexts of justification.

Justification involving appeal to principles of formal logic

Justification or criticism of the processes or the results of reasoning may involve, *inter alia*, questions of meaningfulness, of truth (or of reliability), of consistency, and of formal correctness (or conclusiveness). We shall concentrate on the two last mentioned criteria in the present section.

In appraising the correctness of a deductive argument we confront it with the rules of deductive logic. Conformity with these rules establishes the correctness, violation establishes the incorrectness of our reasoning. In traditional logic we may prove the legitimacy or expose the illegitimacy of some argument by reference to the rules of the syllogism. In the more generalized disciplines of modern logic a much greater wealth of forms and standards becomes available.

A specific deductive argument such as a valid syllogism may then be justified by reference to the well-known rules of the syllogism. And these rules in turn may be justified by reference to definitions and more fundamental principles such as the *dictum de omni et nullo* or (in modern logic) the rules of substitution and inference. Here the logician qua logician usually rests his case. But if he is *philosophically* curious in regard to the justification or justifiability of those basic rules of formal deduction, he will involve himself in peculiar perplexities. If he assumes that the laws of logic are the most general laws of nature, he ascribes a factual content to them that, no matter how thin, would require inductive justification. But inductive justification, while irreducible to deductive justification, presupposes the rules of deductive logic and is therefore impossible without reliance upon them. It may be urged that those most general laws of the universe are known by *a priori* intuition and, thus being self-evident, are neither in need nor capable of validation. This reply, however, is unacceptable for at least three reasons: The difficulties (Kant's heroic efforts notwithstanding) of accounting for the possibility of synthetic *a priori* knowledge are notoriously insurmountable. The reference to self-evidence involves us in the confusions of psychologism. Finally, closer analysis reveals a difference of kind (not merely of degree) between the laws of logic and the laws of nature.

This view (espoused by Mill and others) which construes the laws of logic as psychological laws of thought is merely a variant of the just criticized factualistic interpretation. Thought, as a matter of fact, does not generally conform to the rules of logic. Even if it were impossible to *think* a simple self-contradiction, it is only too painfully obvious that even a slight degree of complexity in argument often conceals just such an inconsistency to the thinker who then blithely asserts self-contradictions at least by way of implication.

A more promising view construes the rules of logic as the *norms* of correct reasoning. Leaving the question of the nature and status of rules or norms aside for the moment, we may say that the *rules* of logic in their totality *define what we mean by correct reasoning*. This view presupposes that we possess, at least implicitly, a criterion (or a set of criteria) by which we can tell whether reasoning is correct or incorrect. The formulation of the rules then merely explicates these criteria.

But this position provokes the question: What assures us of the adequacy of our explicandum, i.e., our pre-analytic notion of correct reasoning? The most widely accepted answer here refers us to the analytic character of all implication relations upon which correct deductive inference must be based. Deducibility, logical (or 'strict') implication, entailment, or whatever else we may call it can be accounted for by reference to the *meaning* of the terms employed in deductive argument. We are then likely to say that whatever follows from propositions by virtue of the meaning of the terms they contain and by virtue of the meaning the propositions have as structured wholes follows *with necessity*. But how are we to decide what (if anything) follows from a given set of premises or is implied by given meanings unless we utilize the very rules of logic which we were going to justify? The emergence of circularity, here as well as elsewhere, is symptomatic of the fact that we have reached the limits of justification, that we are at least in the neighborhood of what are called 'ultimate presuppositions.'

More specifically, what is it that makes a presupposition ultimate? One well-known reply to this question maintains that any attempt to deny, reject (or replace by alternatives) an ultimate principle involves its reaffirmation. This view appears indeed most plausible in the domain of formal logic; it is much less convincing in other domains of justification. Yet, even in regard to the laws of deductive logic, the argument, if intended as a *validation*, is specious. The denial of the law of noncontradiction leads to contradiction, that is, if we use all terms ('contradiction', 'denial', etc.) in their customary sense. This sense is the 'customary' one precisely because it conforms with the basic rules of ordinary (two-valued) logic. The argument therefore demonstrates only that a denial (in this sense!) of the laws of logic involves us in inconsistencies. These inconsistencies, however, are such only within the frame of the rules that determine the logic from which we expect to deviate and define the meaning of 'denial' by means of which we attempt to deviate.

Let us then examine the widely held claims as to the legitimacy of *alternative logics*. The three-valued systems of Lukasiewicz and Tarski, of Brouwer and Heyting, or the logic for quantum mechanics of von Neumann, etc., are called systems of *logic* not only because they, like the two-valued systems of Aristotle or of Whitehead and Russell, are capable of axiomatic deductive presentation, but also because (and this is much more important) they too provide us with rules of inference. We shall pass over in silence the rather confused claims of the disciples of Hegel, Marx, or Engels – not to mention Korzybski – in favor of a dialectical logic. If there are any tenable insights in these trends of thought, they would have to be first separated from a great deal of outright nonsense or egregious equivocation. Only by the most charitable interpretation can those tenable elements be assimilated to the aforementioned three-valued (or many-valued) systems.

Is it a matter of 'arbitrary' decision whether or not to bestow the title of 'logic' upon such alternative systems? An adequate discussion of this controversial question would take us too far afield. I can here only sketchily indicate what seems to me to be involved. First there is the question whether 'deduction' or 'deducibility' as defined in various alternative logics despite vague analogies is not something so radically different from what these terms signify in two-valued logic that the use of these terms without proper qualification is bound to lead to confusion. Secondly, there is the question whether the rules according to which we manipulate the symbols in a three- (or many-) valued calculus must not themselves be applied according to yes-or-no principles which in turn would impose two-valued character upon their metalinguistic formulation. No matter what the structure of a language or of a calculus, if we are to proceed according to constant rules at all, if we are to be able to answer questions, solve problems, etc., in a responsible manner, must we not at some level introduce the *definiteness* which has throughout the ages been regarded as the very essence of the logical? Is not the requirement of unambiguous designation the very root of the regulative principles of semantics? And is not *logic* as we customarily understand it predicated upon adherence to rules of univocal designation? Can communication from person to person, as well as with oneself, dispense with the principles that ensure self-consistency? Are we not continually trying to remove ambiguity and vagueness from our concepts precisely in order to safeguard ourselves against inconsistency?

These questions and their obvious answers seem also to imply a repudiation of a view of logic which has lately gained some currency especially among mathematically oriented philosophers. These thinkers take their cue from the conventionalism of Poincaré and from Duhem's views on scientific method. According to this outlook there is no way of justifying laws or principles in isolation. Only the total set of laws, hypotheses, and principles is capable of test by experience. The principles of logic are simply the ones we are unwilling to surrender or modify, except as the last resort if our total system proves inadequate. The principles of logic are thus considered as in no way sharply distinguishable from those that we ordinarily would call 'empirical.' The advocates of this view then deny that there is a sharp distinction between the analytic and the synthetic types of propositions, and they deny accordingly also the sharp distinction between the *a priori* and the *a posteriori* type of validity. While it is difficult to see how such a position can be maintained wherever analyticity depends upon explicit definitions (as in 'All roans are horses'), it may be granted that the distinction between analytic and synthetic propositions within such systems as theoretical physics is more problematic. A given formula may represent an analytic or a synthetic proposition depending upon the context of inquiry. Or, more precisely speaking, it depends upon the specific interpretation (by way of co-ordinating definitions, or semantical designation rules) to which the given formula (or a whole postulate system) is subjected. But the admitted last-ditch-surrender policy in regard to logical principles would seem to indicate that it is the data of experience which have jurisdiction over the factual content of a theory, whereas criteria of a very different kind are relevant for the adoption (or rejection) of the logical principles. In the customary view of the theories of the factual sciences, the principles of logic and pure mathematics are silent partners, presupposed but not explicitly listed among the postulates of the given theory (geometry, mechanics, electrodynamics, quantum mechanics, etc.). They are, to use a Kantian phrase with greater propriety than we find it used by Kant himself, the necessary conditions for the very possibility of any theory whatsoever. To ensure definiteness of meaning for our symbols, to ensure (the related) conclusiveness of deductive inference we have no choice but to conform to the principles of identity, of noncontradiction, and of the excluded middle. No matter whether we understand these principles as tautologies of the object language (as we do in propositional and functional logic) or whether we understand them as semantical precepts (formulated in the metalanguage), it is impossible to abandon, e.g., the law of the excluded middle, without at the same time abandoning the other two principles (as well as the principle of double negation, or the principles '$p \vee p \equiv p$,' '$p \cdot p \equiv p$,' etc.). Only if we allow ourselves to tamper with the implicitly understood meanings of 'proposition,' 'negation,' 'equivalence,' 'disjunction,' 'conjunction,' etc., can we responsibly deviate from one principle without affecting the others. And even if, upon modifying some of these meanings, we arrive at an 'alternative logic,' we shall yet have achieved no more than if, for example, we had perversely decided to replace the numeral '4' by the numeral '5' in arithmetic. That under such conditions '2 + 2 = 5' becomes a true statement is not in the least surprising. Actually, the parallel with arithmetic is (as everybody should have realized, at least since Frege's contributions) not merely a superficial analogy but genuinely a consequence of the fact that arithmetic (in ordinary interpretation) is a branch of logic (in ordinary, i.e., in Frege-Russell-Whitehead interpretation). Even if in some other world putting two and two objects together resulted invariably in a total of five objects, we should need ordinary (good old) arithmetic in order to formulate the rather peculiar natural laws of that world. Thought experiments of this kind reveal that the fundamental principles of logic are indeed independent of the data of experience. They show that these principles are indispensable not because of some pervasive features of the world that is to be symbolized, but because of the requirements of the process of symbolization itself. Every symbolic system has its tautological equivalences, based on conventional synonymities.

What is it then that accounts for the unique and ineluctable character of the principles of logic? If it is not the data of experience or the facts of the universe at large, what is it that makes compliance with them imperative and their violation a sin against the spirit of Reason? Neither the Ten Commandments nor the Law of the Land prescribe any rules of logic or language. What then is the 'authority,' what are the 'sanctions' that dictate conformity with these principles? It will be suggested that at least part of what we mean by 'mental sanity' consists precisely in compliance with those laws. Assuredly so, but this merely shows that 'mental sanity' is (in part) defined by conformity with logic. The appeal to sanity therefore amounts to begging the question.

It is obvious that we have reached the limits of justification. Justification in the sense of validation involves reliance upon the principles of logic and can thus not provide their validation. Justification in the sense of a pragmatic vindication of the adoption of, and compliance with, the rules of logic would then seem to offer itself as a further opening, if we insist on pursuing our quest to the bitter (because trivial) end. If we wish to give such a justification, it too must be given by reasoning and will thus have to rely upon the very standards whose adoption is now the issue at stake. Will this involve us in a vicious circle? It will not, if we are perfectly clear that we are not seeking a *validation* of the principles of logic. If pragmatic *vindication* is sharply distinguished from validation, then all it can provide amounts to a recommendation of a certain type of behavior with respect to certain ends. We may say to ourselves: If we wish to avoid the perplexities and discomforts that arise out of ambiguity and inconsistency, *then* we have to comply with the rules of semantics and logic. If we wish to derive true propositions from true premises then we must conform to the rules of inference and the rules of substitution. The reasoning concerning these means-ends relations utilizes, as any such reasoning must, the forms of deductive and inductive inference. Perhaps in the extreme case – the degenerate case, as it were – only deductive inference is required.

There is only one more question that the dialectical process will finally bring forth: Why should we accept the ends which we took for granted in the vindication of compliance with logical rules? If the end is a means to further ends, then the question merely shifts to those further ends. And just as in the reconstruction of validation we disclose ultimate validating principles, so in the reconstruction of vindication we encounter ultimate ends or purposes. Whether definiteness and conclusiveness of reasoning are to be viewed as ultimate ends or as means to certain other values is a question of psychology. The well-known transformation process of means into ends has brought it about that for many persons, especially of the scientific type of mind, the virtues of logicality have become ends in themselves. But for the vast majority of mankind logicality is primarily a means in the struggle for existence and in the pursuit of more satisfactory ways of living. These ends we pursue as a matter of stark fact; they are part of our human nature. The question whether this ought to be our nature, if not altogether preposterous, falls at any rate outside the domain of logical justification. If there is a meaningful question here at all, it is one for *ethical* considerations to decide.

Justification involving appeal to principles of methodology

It is generally granted that consistency and conclusiveness of reasoning are necessary conditions for arguments purporting to substantiate knowledge claims. It is almost equally well agreed that while consistency and conclusiveness may be sufficient in the purely formal disciplines, they are not sufficient in the realm of factual knowledge. The sort of justification that the claims of factual knowledge require leads us then to a consideration of principles outside the domain of formal logic. These principles belong to the field of inductive logic.

One qualification may be in order here. Inductive logic is not required for a justification of factual-knowledge claims that involve no transcendence beyond the completely and directly given. Perhaps a better way to state this contention is in the subjunctive mood: If there were knowledge claims susceptible to complete and direct verification (or refutation), their justification would involve appeal only to immediate data, to designation rules, to definitions, and to the principles of formal logic. We shall not attempt here to clean up this particular corner of the Augean stables of philosophy. In any case the doctrine of immediate knowledge seems to me highly dubious. The very terms in which we formulate observation statements are used according to rules which involve reference beyond the occasion of direct experience to which they are applied. If they involve no such reference then they are not terms of a language as we usually conceive languages. Terms referring exclusively to the data of the present moment of a stream of experience could not fulfill the function of symbols in observation statements that are connected with symbols in other statements (of laws and/or other observation statements). If we are not to be reduced to mere signals indicating the individual occurrences of direct experience, we must formulate our statements regarding these individual occurrences in such a manner that they are capable of revision on the basis of other observation statements and of laws that are confirmable by observation.

Appeal to the justifying principles of inductive logic is inevitably made if the dialectic question 'How do you know?' is pursued to the limit. An engineer may rest satisfied with reference to specific physical laws when he justifies his claims as to the efficiency (or inefficiency) of a particular machine. A physicist in turn may justify those specific laws by deduction from very general and basic laws, such as those of thermodynamics or electromagnetics. But when pressed for the reasons of his acceptance of those more (or most) general laws, he will invariably begin to speak of verification, of generalization, or of hypothetico-deductive confirmation. A researcher in medicine will proceed similarly. In order to justify a particular hypothesis, e.g., one according to which a certain disease is caused by a virus, he will quote evidence and/or will reason by analogy and induction. Justification of knowledge claims in the historical disciplines (natural as well as social) conforms to the same pattern.

A given item of observation is evidence not in and by itself, but only if viewed in the light of principles of inductive inference. On the level of common sense and on the more 'empirical' levels of scientific inquiry those principles are simply the accepted laws of the relevant field. Utilizing some items of evidence and some pertinent laws we justify our assertions concerning past events in history, concerning the causes of diseases in pathology, concerning the existence of as yet unobserved heavenly bodies in astronomy, concerning motivations or learning-processes in psychology, and so on. The search for causes quite generally presupposes the assumption that events do have causes. The investigator of a crime would give us a queer look (if nothing worse) if we asked him how he could be so sure that the death under investigation must have had *some* cause or causes. He takes this for granted and it is not *his* business to justify the principle of causality. Philosophers however have felt that it is their business to justify the belief in causality.

We need not review in too much detail the variety of attempts that have been made in its behalf. The assimilation of causal to logical necessity was definitely refuted by Hume. Kant's transcendental deduction of a causal order depends on the premise that the human understanding impresses this order upon the data of the senses. Part of this premise is the tacit assumption that reason will remain constant throughout time and that it therefore can be relied upon invariably to bestow (the same) causal order upon the data. In this psychologistic version of Kant's epistemology we thus find that the lawfulness of the world is demonstrable only at the price of an assumption concerning the lawfulness of Reason. (We shall not press any further questions concerning the credibility of this ingenious but phantastic account of the nature of cognition.) Turning to the presuppositional version of the *Critique* we gladly acknowledge that Kant, more incisively than any of his predecessors, disclosed the frame of justifying principles within which the questions of natural science are raised as well as answered. But the elevation of strict determinism and of Newtonian mechanics to the rank of synthetic *a priori* principles proved to be a mistake. The development of recent science provides, if not a fully cogent refutation, at least a most serious counterargument against any such attempt at a rationalistic petrifaction of the laws of science of a given epoch. More crucial yet, Kant did not achieve what he proposed to do: overcome Hume's skepticism. The presuppositional analysis furnishes no more than this: 'Knowledge' as we understand this term connotes explanation and prediction. Therefore, if knowledge of nature is to be possible, nature must be predictable. This is true enough. But do we establish in this fashion a synthetic *a priori* guarantee for the order of nature? Not in the least. All we have attained is an *analytic* proposition drawn from the definition of 'knowledge.'

Can *intuition* justify the belief in causality? Even if we granted that we have an intuitive acquaintance with causal necessity in some of its instantiations, how can we assure ourselves without *inductive* leap that the intuited samples are representative of the structure of the world at large?

It should scarcely be necessary to point out that the again fashionable attempts to rehabilitate the concept of Real Connections will not advance the problem of the justification of induction. We grant that problems such as those regarding the meaning of contrary-to-fact conditionals show that Hume's (and generally the radical empiricist) analysis of causal propositions is in need of emendation. Indeed, it seems inevitable to establish and clarify a meaning of 'causal connection' that is stronger than Hume's constant conjunction and weaker than entailment or deducibility. Perhaps the distinction between laws (nomological statements) and initial conditions will help here. Our world, to the extent that it is lawful at all, is characterized by both: its laws and its initial condition. Counterfactual conditionals tamper with initial conditions. The frame of the laws is left intact in asking *this* kind of hypothetical question. The 'real possibilities' which are correlative to the 'real connections' are precisely the class of initial conditions compatible with the laws of our world (or of some other fancied world). These considerations show that the laws of a given world may be viewed as something stronger than can be formulated by means of general implication. They are the very principles of confirmation of any singular descriptive statement that is not susceptible to complete and direct verification. They are therefore constitutive principles of the conception of a given world, or rather of a class or family of worlds which are all characterized by the same laws [cf. Sellars, 1948a, p. 287].

But the reconstruction of laws in terms of modal implications does not alter one whit their status in the methodology of science. Clearly, besides counterfactual hypotheticals we can equally easily formulate counternomological ones. Here we tamper with the laws. And the question of which *family* of worlds our world is a member can be answered only on grounds of empirical evidence and according to the usual rules of inductive procedure. Upon return from this excursion we are then confronted, just as before, with the problem of induction.

Obviously the next step in the dialectic must be the lowering of our level of aspiration. We are told that it is the quest for certainty that makes the justification of induction an insoluble problem. But we are promised a solution if only we content ourselves with probabilities. Let us see. 'Probability' in the sense of a degree of expectation will not do. This is the psychological concept to which Hume resorted in his account of belief, or that Santayana has in mind when he speaks of 'animal faith.' What we want is a justifiable degree of expectation. And how do we justify the assignment of probability ratios to predictions and hypotheses? That depends on how we explicate the *objective* concept of probability. But here we encounter the strife of two schools of thought. According to the frequency interpretation there is no other meaning of 'probability' than that of the limit of relative frequency. According to the logical interpretation 'probability' (in the sense of strength of evidence, weight, degree of confirmation) consists in a logical relation between the evidence which bestows and the proposition upon which there is bestowed a certain degree of credibility. The adherents of this logical interpretation urge that the statistical concept of probability presupposes the logical one. For the ascription of a limit (within a certain interval) to a sequence of frequency ratios is itself an hypothesis and must therefore be judged according to the degree of confirmation that such hypotheses possess in the light of their evidence. Contrariwise, the frequency interpretation urges that locutions such as 'degree of confirmation' or 'weight,' whether applied to predictions of single events or to hypotheses of all sorts, are merely *façons de parler*. Basically they all amount to stating frequency ratios which are generalized from statistical findings regarding the events concerned (the occurrence of successful predictions).

We shall not attempt to resolve the issue between these two schools of thought. Our concern is with the justification of induction. And here perhaps the divergence of interpretations makes no fundamental difference. Inductive probability in the sense of a degree of confirmation is a concept whose definition renders analytic every one of its specific applications. If we use this concept as our guide, that is if we believe that it will give us a maximum of successes in inductive guessing, then this could be explicated as an assertion about the (statistical) structure of the world. We are thus led to essentially the same rule of induction which the frequentists propose: Generalize on the basis of the broadest background of available evidence with a minimum of arbitrariness. This principle of straightforward extrapolation in sequences of frequency ratios applies to a world which need not display a deterministic order. Statistical regularity is sufficient. Does the application of the rule guarantee success? Of course not. Do the past successes of procedures according to this rule indicate, at least with probability, further successes? No again. Hume's arguments refuted this question-begging argument before Mill and others fell victims to the fallacy. Only if we utilize the *logical* concept of probability can we achieve a semblance of plausibility in the criticized argument. But for the reasons stated before, mere definitions cannot settle the issue as to whether our world will be good enough to continue to supply patterns of events in the future which will support this definition as a 'useful' one.

The presuppositional analysis sketched thus far has merely disclosed one of the ultimate principles of all empirical inference. Any attempts to validate the principle itself involve question-begging arguments. Its ultimate and apparently ineluctable character can be forcefully brought out by considering how we would behave in a world that is so utterly chaotic and unpredictable that any anticipation of the future on the basis of past experience is doomed to failure. Even in such a world after countless efforts at inductive extrapolation had been frustrated we would (if by some miracle we managed to survive) abandon all further attempts to attain foresight. But would not even this be yet another inductive inference, viz., to the effect that the disorder will continue? The inductive principle thus is ultimately presupposed but in turn does not presuppose any further assumptions.

The various attempts (by Keynes, Broad, Nicod, Russell, and others) to deduce or render probable the principle of induction on the basis of some very general assumption concerning the structure of the world seem to me, if not metaphysical and hence irrelevant, merely to beg the question at issue. Assumptions of Permanent Kinds and of Limited Variety, provided they are genuine assertions regarding the constitution of the universe, themselves require inductive validation. To assign to such vast hypotheses a finite initial (or 'antecedent') probability makes sense only if 'probability' means subjective confidence. But nothing is gained in this manner. Any objective probability (logical or statistical) would presuppose a principle of induction by means of which we could ascertain the probability of such world hypotheses in comparison with the (infinite) range of their alternatives.

We have reached the limit of justification in the sense of validation. Can we then in any fashion provide a 'reason' for this acknowledged principle of 'reasonability'? Obviously not, if 'reason' is meant in the sense of validating grounds. What we mean (at least in part, possibly as the most prominent part) by 'reasonability' in practical life as well as in science consists precisely in the conformance of our beliefs with the probabilities assigned to them by a rule of induction or by a definition of degree of confirmation. We call expectations (hopes or fears) irrational if they markedly deviate from the best inductive estimates. The attempt to validate one of the major principles of all validation, it must be amply obvious by now, is bound to fail. We would be trying to lift ourselves by our own bootstraps.

The only further question that can be raised here concerns a justification of the *adoption* of the rule of induction (or rather of one of the various rules of induction, or one of the various definitions of inductive probability or of degree of confirmation). Such a justification must have the character of a vindication, a *justificatio actionis*. Our question then concerns the choice of means for the attainment of an end. Our end here is clearly successful prediction, more generally, true conclusions of nondemonstrative inference. No deductively necessary guarantee can be given for the success of such inferences even if we follow some rule of induction. The probability of success can be proved, but that is trivial because we here utilize the concept of probability which our rule of induction implicitly defines. This probability cannot be construed as an estimate of the limit of frequency. We do not know whether such a limit exists. Only if we grant hypothetically that there is such a limit can we assign weights to its various values (i.e., to intervals into which the limit may fall) on the basis of (always finite) statistical samples. What then justifies our optimistic belief in the convergence of statistical sequences toward a limit? Since any attempt at validation would inevitably be circular, we can only ask for a vindication of the rule according to which we posit the existence of limits. Reichenbach's well-known but widely misunderstood justification of induction [Reichenbach, 1938; see also 1940, p. 101] consists, as I see it, in a vindication of the adoption of that rule. It amounts to the *deductive* proof that no method of attaining foresight could conceivably be successful if every sort of induction were bound to fail. Perhaps there are alternative techniques of foresight that might even be more efficient or reliable than the laborious method of scientific generalization. But such alternative methods (let us none too seriously mention crystal gazing, clairvoyance, premonitions, etc.) would themselves have to be appraised by their success; i.e., they would have to be accepted or rejected on the basis of statistical studies of the frequency ratio of correct predictions achieved by them. And our confidence in such 'alternative' techniques of foresight would therefore ultimately be justifiable only on the basis of normal induction. *If* there is an order of nature at all (i.e., at least a statistical regularity), not too complex or deeply hidden, then a consistent application of the rule of induction will reveal it. This statement is of course a tautology. It should not be confused with such bolder and undemonstrable factual assertions as that the inductive procedure is the only reliable one, or that it is our best bet. Reliability and optimal wagering presuppose inductive probabilities and thus cannot be invoked for their justification. We cannot even say that straightforward generalization is a necessary condition (never known to be sufficient) for the success of predictions. The air of plausibility that this statement shares with its close relative, the common-sense slogan 'Nothing ventured, nothing gained' arises only if we disregard (logically conceivable) alternative routes to predictive success, such as sheer inspiration, capricious guessing, intuition, premonition, etc. The unique character that the inductive procedure possesses in contrast with those alternatives rests exclusively in this: The method of induction is *the only one for which it can be proved* (deductively!) that it leads to successful predictions *if* there is an order of nature, i.e., *if* at least some sequences of frequencies do converge in a manner not too difficult to ascertain for human beings with limited experience, patience, and ingenuity. This is the tautology over again, in expanded form, but just as obvious and trivial as before. In the more ordinary contexts of pragmatic justification the validity of induction is invariably presupposed. If we want to attain a certain end, we must make 'sure' (i.e., probable) that the means to be chosen will achieve that end. But if we ask for a vindication of the adoption of the very *principium* of all induction, we deal, so to speak, with a degenerate case of justification. We have no assurance that inductive probabilities will prove a useful guide for our lives beyond the present moment. But equally we have no reason to believe that they will fail us. We know furthermore (as a matter of logical necessity or tautology) that *if* success can be had at all, in any manner whatsoever, it can certainly be attained by the inductive method. For this method is according to its very definition so designed as to disclose whatever order or regularity may be present to be disclosed. Furthermore, since the inductive method is self-corrective, it is the most flexible device conceivable for the adaptation and readaptation of our expectations.

The conclusion reached may seem only infinitesimally removed from Hume's skepticism. Philosophers do not seem grateful for small mercies. In their rationalist quest for certainty many still hope for a justification of a principle of uniformity of nature. We could offer merely a deductive (and trivial) vindication of the use of the pragmatic rule of induction. But for anyone who has freed himself from the wishful dreams of rationalism the result may nevertheless be helpful and clarifying. It is the final point which a consistent empiricist must add to his outlook. We refuse to countenance such synthetic *a priori* postulates as Russell (perhaps not with the best intellectual conscience) lately found necessary to stipulate regarding the structure of the universe. We insist that no matter how general or pervasive the assumptions, as long as they are about the *universe*, they fall under the jurisdiction of the rule of induction. This rule itself is not then a factual assertion but the maxim of a procedure or what is tantamount, a definition of inductive probability. In regard to rules or definitions we cannot raise the sort of doubt that is sensibly applicable to factual assertions. In order to settle doubts of the usual sort we must rely on some principles without which neither doubt nor the settlement of doubt makes sense. The maxim of induction is just such a principle.

Justification involving appeal to epistemological principles: the criteria of factual meaningfulness

The slogan '*De principiis non est disputandum*' is perhaps most emphatically invoked in the justification of theological and metaphysical doctrines. But since the rise of pragmatism, operationism, and logical positivism, those issues have been undercut. Prior to the examination of the validity of theological or metaphysical assertions is the question of the type of meaning with which they can justifiably be credited. It is generally admitted that the hypotheses of inductive metaphysics, e.g., cosmological speculations, may be merely more problematic than the kind of hypotheses that we call 'scientific.' The distinction is one of degree and concerns only the probability or strength of evidential support of those conjectures rather than their factual meaningfulness. It may likewise be granted that such theological arguments as the teleological, or the inferences from the facts of religious experience, can be so formulated that they conform to the inductive or hypothetico-deductive patterns of scientific reasoning. Empiricists like Hume or Mill were critical of such arguments because they felt that the supporting evidence was extremely weak. But, notoriously, some theologians and metaphysicians have thought they could protect their knowledge claims against such criticisms by severing their assertions from observational evidence altogether. This much more orthodox reliance on pure faith or intuition involves, however, a radical shift. The assertions are now completely transcendent, i.e., transempirical. They have lost whatever intersubjective testability was possessed by the previous 'empirical' approaches. According to the logical positivists' analysis the significance of linguistic expressions that lack confirmability in principle may consist in pictorial, emotional, and/or motivational appeals. But the presumption of factual reference is (according to this analysis) erroneous or illusory. This is perhaps most strikingly obvious in connection with the uses of the word 'belief.' We say that we believe that the earth is a sphere or that the law of the conservation of energy holds for atomic processes or that unemployment causes social unrest. But we also say that we believe in the dignity of man or in equal rights for everybody. Clearly, the latter usage of 'belief' is utterly different from the former. It formulates a commitment to certain values, while 'belief' in the former (empirical) sense applies to confirmable hypotheses.

There is no space for more than just a few hints regarding some of the issues of traditional metaphysics and epistemology. The problems of the existence of an external world, of the past, of other minds, of substance, of causal necessity, etc., appear *prima facie* as factually meaningful because they are often assimilated to the problems of the existence of *specific* physical objects, past events, other persons' particular mental states, physiochemically characterized substances, concrete instances of causal influence, etc. Problems of this latter type are indeed meaningful precisely because we raise them within the frame of a language and of (inductive or hypothetico-deductive) procedures that makes responsible answers possible. But the aforementioned problems of metaphysics are in principle insoluble because they confuse questions of fact with questions regarding the semiotic frame which is *presupposed*. The justification of this presupposed frame must therefore be radically different. The only sort of justification we can give for this ('realistic') frame of our language consists in showing the indispensability and the adequacy of the language required for the purpose of such sciences as physics, psychology, or history. [For fuller analysis of these issues see Feigl, 1950b; and W. Sellars, 1948b.] Since we deny that the data of religious experience require a wider frame than the one sufficient for science, we do not admit that any alleged theological presuppositions have a status coordinate with, or analogous to, that of the presuppositions of science.

The criterion of factual meaningfulness has been the issue of intense disputes for more than twenty years. We need not review the well-known arguments. The issue as it concerns us here turns on the so-called, 'weaker verifiability criterion,' i.e., the condition of (at least) incomplete and indirect verifiability or refutability. Logical empiricists recognize today that this criterion formulated as a principle is a proposal and not a proposition. It could be expressed as an analytic proposition only in a metalanguage that in addition to syntactical and semantical concepts contains also such pragmatic concepts as verification and confirmation. Given such a sufficiently rich metalanguage (viz., of *pure pragmatics*) the term 'factually meaningful sentence' can be explicitly defined in terms of confirmability.

Two questions seem quite generally pertinent in regard to proposed definitions or stipulations of this sort: (1) Does the definition explicate adequately what is, no matter how vaguely, intended by the term (the *explicandum*) in the language of common sense and of science? (2) Does the definition of a term that has an emotive halo ('meaningful' is certainly such a term) succeed in stipulating a meaning that, when consistently employed, will be fruitful in its application?

There is a great deal of opportunity for dispute on the first point. I am inclined to contend, however, that once the distinction between emotive appeals and cognitive meanings is accepted, there is much that can be said in favor of the adequacy of the confirmability criterion. It explicates what is quite commonly regarded as the distinction between what 'makes sense' and what doesn't. The second point is at any rate much more important. It raises the question of a vindication of the proposed meaning criterion. The criterion may then be viewed as a rule for the delimitation of factually meaningful from factually meaningless sentences. The purpose of the rule is obviously to distinguish discourse that can justifiably claim to embody factually true or false statements from discourse that does not fulfill this function – even if it serves other, viz., noncognitive purposes. The vindication of the criterion must then consist in showing that its adoption will produce the sort of clarity that we seek when we realize that confusion of the various functions of language leads only to endless perplexity and vexation with pseudoproblems. In other words, if we do not wish to open the floodgates to countless questions which by their very construction are in principle unanswerable, then the adoption of the confirmability criterion is indispensable.

Justification involving appeal to ethical principles

Moral approvals and disapprovals are formulated in normative judgments. These judgments need not be, and usually are not, expressed in sentences that are explicitly in the imperative form. 'This is a case of fraudulence,' 'You are cruel to your brother,' 'It is right to be loyal to your country,' 'All men have equal and inalienable rights' – these and countless other sentences contain words that have a normative significance in addition to a component of factual meaning. There is even a kind of tautology (and correspondingly a kind of contradiction) that is exclusively based on the normative significance of the terms used. For example: 'Wanton cruelty is condemnable' is a tautology, and its denial a contradiction.

The justification of specific normative judgments may be sought along the lines of their validation. Suppose we have made sure of all the relevant empirical facts and we can state that someone charges 25 per cent interest on debts owed to him. This is then condemned as 'usury.' The use of this term is a specific way of classifying the previously described fact and combining this classification with moral disapproval. How do we justify the moral disapproval? We may do this deductively (syllogistically in this simple case): 'All usury is morally wrong: this is a case of usury; therefore . . . ' We may next try to validate the major premise of the preceding argument by deducing it from two universal premises such as: 'All types of action in which one person benefits himself by harming another are morally wrong.' 'Usury is a type of action, etc. . . . ' Perhaps we can succeed in deducing the major premise of this last argument from a still more general moral principle, such as the golden rule, or some principle of justice, or the like. This kind of regressive reasoning will of course terminate with some premise which cannot plausibly or fruitfully be deduced from any still more general or fundamental principle. It is generally conceded, however, that Kant's categorical imperative in its first or second formulation is vacuous and that therefore his deductions of specific moral percepts were fallacious. But let us suppose that we could formulate some principle such as Kant intended (i.e., a principle of justice, impartiality, or of 'no special privileges') in a form that is not vacuous and hence would yield specific moral judgments when applied to empirically characterized actions or attitudes. What can we say in answer to skeptical questions regarding such a supreme principle? Derivation from theological or metaphysical principles will not do. Apart from the epistemological criticisms (discussed in the previous section), the question of the goodness or rightness of divine commandments or of metaphysically founded imperatives can only be silenced but not answered in any intelligible or enlightening fashion. Appeal to intuitive self-evidence is equally fruitless, if not suspicious, in this age of cultural anthropology and social psychology. Obviously we have reached the limits of validation. Just as in the other domains of justification, we may disclose the ultimate presuppositions; we may explicate the principles of validation of moral judgments. We may say that these principles define the moral terms ('right', 'wrong', 'ought') and delimit the moral universe of discourse, just as the principle of induction defines the methodological terms ('evidence', 'probability') and delimits the universe of discourse of empirical knowledge.

We could rest more easily satisfied with this conclusion if the idea of alternative ethical systems were as plausibly refutable as that of alternative deductive or inductive logics. The vindication of the principles of meaning and knowledge is so trivial precisely because, given the purposes of language and knowledge, there are no genuine alternatives for fulfilling them. But we do know of alternative systems of moral norms. An aristocratic ethics such as Nietzsche's and a democratic one such as Jefferson's are clearly incompatible with each other. The ethics of capitalism and the ethics of socialism may serve as a (related) further example.[1] Even if there are areas in which such ethical systems may have elements of agreement, there are others in which they irreconcilably diverge.

It is of no avail to criticize one ethical system on the basis of another. The issue is *logically* symmetrical here. Such criticisms would merely utilize implicit persuasive definitions of such terms of 'true morality', 'the really right attitude', etc. They would beg the question at issue. From a more rigorous analytic viewpoint it may be remarked that ethical criticism in the usual sense applies only to actions and attitudes, not to propositions but to belief in them ('belief' in the sense of disposition toward action). Moral criticism makes sense only *within* the frame of a set of basic moral standards. Any incompatible alternative basic standards are as much beyond criticism as are the axioms of a non-Euclidean geometry within the frame of the axioms of Euclidean geometry. Quite generally we may conclude that, granting our rational reconstruction of the hierarchy of levels of justification in ethics, the acceptance or rejection of the supreme principles can be only a matter of pragmatic justification. But this, in a sense, has been obvious to all those who approach ethics from the point of view of the interest theory of value, or from that of anthropology, or from the 'emotive-meaning' school of analysis. This meaning of moral judgments has been such a troublesome and enigmatic issue because these judgments in their protean ways simulate (1) the analytic truths of logic, in that they may be construed as necessary implications holding between the general value criteria and their specific applications; (2) factual propositions, in that they must always refer to empirically characterized classes of actions or attitudes; (3) purely emotional or motivational expressions and appeals, in that they evince attitudes and/or contain an imperative component. No wonder that moral norms have been viewed as synthetic *a priori* propositions, as altogether *sui generis*, endowed with 'normative meaning.' But normative meaning is just as analyzable here as it is in logic or epistemology. Criticism presupposes norms, and norms (like any *rules*) differ from (descriptive) laws in being *prescriptive*. To the extent that a purely factual ('naturalistic') interpretation of norms falls (inevitably) short of an adequate analysis, it must be amended by proper attention to their directive or motivational components. If the word 'ethics' is not to be used for a descriptive socio-psychological study of human conduct and its actually occurring evaluations, if it is to stand for something that is not a mere account of past and present moral codes, their origins and evolutions, then 'ethics' is used in the sense of a system of norms which makes criticism and justification possible. At least the supreme justifying principles in every domain of validation must have the status of norms which implicitly define the *critical terms* of the domain in question: 'correct', 'conclusive', 'warranted' in logic; 'meaningful' in epistemology; 'right' in ethics. All these terms have directive significance, just as all (nominal) definitions have a motivational appeal, in addition to whatever cognitive meaning may likewise be present. Now according to our reconstruction, moral judgments are valid if they are in accordance with the relevant ethical norms. The supreme norms define the standards of morality of a given system. The terms 'right' and 'wrong,' if not rigorously scrutinized, are apt to be applied in a doubly persuasive manner: first, *in accordance with the norms* of the given system; and secondly, *to the norms* of the system. But we have already warned against this confusion. It is just as nonsensical to approve of a definition of 'right' as right, as it is to approve of a definition of 'probable' by saying that it is probable. The mistake in the first case is not as obvious as in the second because of the greater ambiguity of the term 'right'. The other meanings of 'right' that are apt to interfere here are (a) the adequacy of the explication achieved by the reconstruction of the system of norms and (b) approval (endorsement) of the supreme norm in the sense of expressing personal agreement with it. It is this latter sense that would result from a vindication of the norms.

Now just what does a vindication here amount to? It consists in showing that adoption of the norms of a given moral system fulfills a purpose. Well, then, what is the purpose that is fulfilled by adoption of, e.g., the golden rule, or a principle of impartiality? The answer clearly depends on the individual's personality. Perhaps he obeys the golden rule because of sheer prudence and 'enlightened egoism.' Perhaps he has by training, education, experience, or reflection developed genuinely altruistic interests and thus holds the ideal of the greatest satisfaction for the greatest number. Generally, no vindication will prove convincing unless it appeals to the needs, interests, or aspirations of the individual concerned. If there are fundamentally incompatible purposes, unmitigated by any purpose to eliminate divergence of purposes, then only segregation or, in the extreme case, coercion will be able to settle such disagreement in attitudes. But given an interest in avoiding conflict there are the techniques of (unilateral) persuasion or of (bilateral) compromise.

No matter what factual ('naturalistic') content we associate with the (otherwise emotively significant) value terms 'good', 'right', 'ought', etc., and their contraries, it is of the utmost importance to distinguish the *rules* that serve as *justificantia cognitionis* in the validation of moral judgments from the *goals* that serve as *justificantia actionis* in the vindication of the adoption of such rules. This distinction enables us to see more clearly what is involved in the quarrel between 'deontological' and 'teleological' moral philosophies. This quarrel can be adjudicated by allotting to validation and vindication their proper roles. The usual formulations of utilitarianism, for example, are logically questionable because they attempt to combine validation and vindication by telescoping together rules (such as 'Only kindly acts are right') with goals (such as 'the greatest happiness of the greatest number').

An obvious objection to this analysis urges that the adoption of moral rules requires a justification that goes beyond a mere sanction by purposes. In other words, it may be asked whether those purposes are morally good. But clearly this question presupposes moral standards and without them remains unanswerable. If the moral standards drawn upon are those that formulate the system whose vindication is under discussion, then (given complete logical consistency) we obtain a validation of the value judgment concerning the adoption of its standards that is bound to be analytically true. If the standards are taken from a system that is incompatible with the one under discussion, we obtain an invalidation resulting from logical contradiction.

Another related but more serious question concerns the ethical relativism which the preceding analysis seems to support. Now 'ethical relativism' is a phrase which exerts a strong negative emotive appeal because it is taken to imply that there are no grounds for preferring one ethical system to another.

This is often exaggerated in the charge that for a relativist moral standards can be no more than a matter of arbitrary decision, of whim and caprice. But these are gross caricatures of a position which merely combines a sociological conclusion with the results of a logical analysis of the structure of justification. A judgment of indifference is still a value judgment and thus justifiable only within its own frame of reference. More significantly yet, the purposes that we adduce in the vindication of ethical standards are not a matter of personal caprice but are (usually) the resultants of age-long experience in the harmonization of intra- and inter-individual needs and interests, of experience, personal and social, guided by the adaptive and integrative influence of intelligence. Far from being 'arbitrary' or 'capricious' in the usual sense of these words, our terminal purposes are usually held with the most serious and profound conviction. The only sense in which the misnomer 'arbitrary' could be sensibly interpreted here is in the sense of 'ultimate,' i.e., nonvindicable and resting on the (logically) contingent traits of human nature. Such standards[2] as those of justice and kindliness, as well as of self-perfection, are the counterpart of goals such as those of a harmonious, peaceful, and progressive humanity. The goals or purposes are in turn resultants of the nature of man and his needs and interests in ever widening and ever more interdependent social contexts. The evolution of a global code of morality out of its tribalistic precursors offers in many ways a striking parallel to the development of the norms of scientific method out of its magical, animistic, and metaphysical origins. In a continuous process means and ends underwent selections, revisions, replacements, focussings, diversifications, and harmonizations. Experience of ever widening scope teaches the lessons of cautious conservation and daring innovation, of sweeping synthesis and attention to specific detail. Well-established knowledge and clarity of meaning are indispensable. But only if they are in the service of purposes which emerge from the broadest experience will their utilization be regarded as justifiable. In this sense, and perhaps in this sense only, can the etymological association of wisdom with philosophy be supported.

SOME OBSERVATIONS BY WAY OF CONCLUSION

The conclusions we have reached are neither new nor should they be startling. Justification is a form of argument which requires some platform of basic agreement on one level, even if on a different level there is doubt or disagreement. (Argument need not involve two persons. One may try to justify some belief for one's own acceptance.) In order to resolve doubt or disagreement we must not, at least in the given context and until further notice, call into question the very means by which such doubt or disagreement is to be resolved. The status of the validating principles (in logical reconstruction) is that of stipulations, definitions, or conventions. They differ from other less fundamental and less consequential conventions, in that they determine whole domains of justification. That is the reason why so many rationalistically inclined thinkers feel tempted to view them as synthetic *a priori*. But like all definitions or conventions the *justificantia* are *a priori* precisely and only because they are analytic. Since they are definitional in character, we can not ask whether they are in any sense true to fact. Their virtue lies in fulfilling a purpose. And purposes are resultants of the very needs of our lives. The key terms which are defined by the justifying principles (e.g., 'correct', 'valid', 'true', 'confirmed', 'good', 'right', 'meaningful', etc.) carry emotional prestige and therefore lend themselves to persuasive redefinitions. This indicates that the content of these terms depends on the purposes we pursue. It is diagnostic of the purposes and ideals of our (Western) civilization and of this age of science that the term '*reasonableness*' has come to embrace (at least) the following five connotations: (1) logical consistency; (2) inductive plausibility; (3) reflective clarity; (4) impartiality; (5) abstention from violence in the settlement of disagreements [cf. Dennes, 1939 and 1946].

There are two familiar phrases that serve as a last resort to philosophers when challenged as to the grounds for their reasons: 'ultimate presupposition' (or 'basic postulate') and 'pragmatic justification.' The use of these phrases is often regrettably glib. They are often employed as rhetorical devices designed to intimidate the inquirer and to put a stop to further argument and questions. They serve only too often as verbal sedatives for the philosopher himself. Our analysis has given us a clearer idea of what is involved in the responsible use of these two modes of justification hinted at by those expressions.

We must also guard ourselves against slipshod notions of *presuppositions*. The term 'presupposition' is far from clear and univocal. Sometimes it refers to premises that imply. At other times it refers to consequences that are implied. The latter sense would be in keeping with the idea of necessary or indispensable conditions. Combining both we have a third meaning, namely, sufficient and necessary condition (and this would amount to logical equivalence or mutual deducibility of that which presupposes and the relevant presupposition). None of these explications in purely logical terms will quite meet the intended meaning of 'presupposition.' The class of premises that imply and the class of consequences that are implied by a given proposition are indefinitely wide in scope. We mean something much more restricted than all that.

Let us consider some examples. We say that the rules of the categorical syllogism (or what is tantamount in symbolic logic: the rules of the propositional and the lower functional calculi) presuppose the laws of identity, noncontradiction, and of the excluded middle. This might quite correctly mean that these laws are logical consequences of those rules. As is well known, the reverse does not hold here. We have here a case of necessary condition. But since an indefinite number of other consequences follow equally rigorously from the rules mentioned above, the prominence and distinction traditionally attributed to the three (so-called) laws of thought cannot thus be explained or defended.

Our difficulty might be resolved by distinguishing between necessary conditions (logical consequences) within a system of statements and necessary conditions in the sense of requirements imposed upon the system as a whole. These requirements would have to be stated in a metalanguage and when so stated could still be read as necessary conditions, but with this difference: The *conditio sine qua non* concerns the semantical (and/or syntactical) structure of the language (i.e., the object language) in which we customarily formulate the laws of logic. For example, we might say: Only if the language used in syllogistic reasoning is constructed in accordance with the *semantical* rule of identity (univocality of symbols, i.e., unambiguous designation rules) will we be able to validate the rules of the syllogism. Likewise we could say: Only if the definition of 'degree of confirmation' is so chosen that it incorporates what we usually call the principle of induction (i.e., the regulative maxim of simplest extrapolation or straightforward posits) can we justify the customary rules for the assignments of inductive probabilities. The same sort of analysis may be applied to epistemological presuppositions. The stipulation of criteria of meaningfulness is an indispensable prerequisite for the justification of the intended distinction between genuine and pseudo-problems, and this precisely for the reason that only an object language that complies with those criteria will not contain sentences that are in principle unconfirmable.

We found it tempting to think that these considerations may equally well apply to the presuppositions of moral evaluations. Unfortunately very little has as yet been achieved in the formalization of ethical systems, so that our conclusions must here remain tentative. It does seem plausible, however, that the ultimate validating principles of a given system of moral evaluations are incorporated by definition (stipulation in the very meaning of the basic terms of that system. In any case it is clear that whenever we are engaged in ethical evaluations we are, so to speak, operating within a system. This is of course an idealization, because in practice we are rarely aware of the logical structure of the system, nor do we ever approach anything like the consistency of a postulate system. (The situation is still more amorphous and unstable in aesthetics.) The important point, however, is this: Any doubt raised with respect to the presuppositions of a given ethical system or any comparative evaluation of different ethical systems requires a further frame of reference (in the ideal case an alternative system) for the responsible settlement of such doubt, or for a justifiable preference. It is one thing to compare ethical systems in the value-neutral manner of a logician. It is another thing to criticize them ethically. The latter endeavor relies on fresh presuppositions. This is analogous to the fundamental shift that is required when we criticize certain, now obsolete, criteria of truth (e.g., revelation, authority, intuition, self-evidence) in the light of our modern criteria. The presuppositions of the older justifications of knowledge claims may indeed have been the necessary conditions of justification as it was then conceived. Historically and psychologically speaking, we might say (cf. Collingwood) that the ultimate or absolute presuppositions (criteria, standards) vary from epoch to epoch. They remain unscrutinized and uncriticized within the given epoch of thought. As long as no alternative frame of justification is envisaged, we are seldom fully aware of the one that functions jurisdictively at the moment. But while the historian may legitimately establish the cultural relativity of basic presuppositions, the philosophical analyst will be interested in (1) their explicit formulation and the recognition of their logical function and (2) the criteria that justify their criticism and revision. It is precisely the possibility of revision of erstwhile ultimate standards that urges us to amend the view according to which '*de principiis non est disputandum*.' This view is unassailable as long as it maintains that the criteria which are definitive of a certain mode of justification can not themselves be justified *within that mode*. Attempts in this direction are bound to be fallacies of the vicious circle, either ordinary *petitiones principii* or else violations of a semantical rule of types.

The ultimate principles of logic, semantics, methodology, and axiology are (as *justificantia*) not susceptible to cognitive justification. But if the question be raised *why we should adopt* those, rather than some alternative principles, then this obviously concerns not the validity of the principles but the justifiability of our attitude toward them. We do not wish to elicit the trivial answer that the reason for adoption of those principles is that, once adopted, they enable us to carry out such justifications that it is the very business of the principles to make possible. This answer would merely amount to a restatement of the *conditio-sine-qua-non* character of the principles, which was just explicated. Nor are we here concerned with the justification of any particular explication or formulation of the justifying principles. What we are looking for is, instead, a pragmatic justification of our adoption (choice, preference) of the given principle.

This question is best understood as the formulation of a highly artificially generated doubt. The quest for the ultimate justifying principles already exemplified this Cartesian procedure. We are not even one step further removed from the type of doubt that would arise on the common-sense level. We are asking what reasons we have for embracing the very principles whose function it is to help in removing the more familiar types of doubt. The often implausible, unconvincing character of the results of philosophical analysis does not detract from its value. In the case of explication it is the unfamiliar character of the *explicatum*. In the case of justification it is the remote and unconvincing character of the 'reasons.' Convincingness is at its strongest where it dispels the kind of doubts we are apt to raise against the background of the currently accepted beliefs. But where doubt (artificially pretended) concerns those beliefs themselves, reasons given to reestablish those beliefs cannot possibly convince more strongly than do reasons embodying those accepted beliefs.

What we want to know then are the practical reasons that justify our choice of an entire mode of justification. Now, a pragmatic justification amounts to showing that something serves as a means toward an end. It thus requires a prior agreement (1) as to the desiredness of the end and (2) as to the method or type of reasoning by which the appropriateness of the means to the end is to be shown. As to the first point it may be said that the ends which we here acknowledge (without questioning) are simply taken as objects of certain interests. In regard to the field of cognition we bluntly acknowledge that the multifarious experiences (successes and failures) in the enterprise of knowledge have gradually given rise to a strong interest in the following desiderata: (1) *clarity*, i.e., freedom from confusion as to types of significance; (2) *definiteness*, i.e., univocality of meaning, possibly enhanced by quantitative precision; (3) *consistency and conclusiveness*, i.e., absence of self-contradiction (by means of rules of inference that ensure the truth of conclusions derived from true premises); (4) *warranted* (*reliable*) *assertibility*, i.e., availability of evidence that confers a high degree of confirmation on our knowledge claims; and (5) *maximum scope*, i.e., as complete and detailed a coverage of fact as is compatible with the foregoing conditions. Now, if it be granted (as to our previous point) that we are entitled to employ deductive logic in showing that the listed virtues of knowledge can be attained (or approximated) *only if* we adopt certain principles as the norms of cognitive procedures, then we may manage to offer a pragmatic justification of the principles of formal logic and semantics, of the meaning criteria, and of the principle of induction. The ideals of cognition are, for our purposes, to be conceived and formulated in terms of pure syntax, pure semantics, and pure pragmatics. If so formulated it can obviously be shown deductively (and asserted as an analytic statement) that conformity with the principles is a necessary condition for the attainment of those ideals. But, as already hinted, such a demonstration relies in turn on the principles of deductive logic. Does this involve a vicious circle? I think not. Our argument was not to establish the validity of logical principles. It was to show that their acceptance as regulative standards is an indispensable prerequisite for the fulfillment of certain ideal requirements. And since such 'showing' is intended as an argument (or demonstration), it could not possibly abstain from the utilization of logical procedures.

Just how far removed from triviality must a deductive argument be in order to escape the criticism of circularity? We suggest that an inference is *strictly* circular only if its conclusion appears *literally* among the premises. Many arguments that are free from strict circularity are of course extremely trivial nevertheless. Since triviality is a psychological feature, there can be no universal rule as to just what complexity an argument must have in order to be enlightening or helpful to a person of given intelligence and curiosity. Our Cartesian quest for justification may have to be satisfied with demonstrations that have the tang of utter obviousness. The extent to which these demonstrations may nevertheless prove clarifying depends in part on the measure of psychological novelty that attaches to the explications of the key terms of the given context. For example, to say that only confirmable statements make sense (have factual meaning) is not just an arbitrary definition but really explicates what in some fashion we have known all along – implicitly. To bring it into full focus is the merit of the explication. The vindication of the meaning criterion is therefore neither strictly circular nor so trivial as to be unenlightening.

The preceding analyses, especially the remarks on pragmatic justification, would be grossly misunderstood if they were projected upon (i.e., translated into) the customary reasoning and language of common sense. Common sense operates *within* a frame of presuppositions and purposes; it never raises questions concerning this frame. If philosophical analysis had no other task than that of a hygiene and therapy of language, it might well restrict itself to the procedures utilized by G. E. Moore and L. Wittgenstein (and their disciples). The extravagances by which we are apt to deviate from common good sense into uncommonly strange and bad sense may indeed be successfully attacked by those methods. But full awareness of the basic principles of knowledge and evaluation can be attained only by a systematic analysis of the 'rational reconstruction' type. The language of rational reconstruction is legitimate in a strict sense only if it is governed by a set of metalinguistic rules. But in order to introduce this apparatus of analysis and to give it its proper points of application a few didactic fictions and deviations from common-sense language are practically very helpful. Unlike the metaphysician who seriously proposes statements concerning what things really are, we use those didactic fictions (like the famous Wittgensteinian ladder) merely as a makeshift which can be discarded after it has fulfilled its purpose. When we spoke of choosing the principles of an entire mode of justification, we did not wish to assimilate such a choice to the choice situations of ordinary life, any more than, for example, when we speak in semiotic of 'conventional' in contrast to 'natural' signs. No one who uses this phrase intends to suggest seriously that some time in the remote past primitive men convened around stone tables and decided upon the meanings of the words they were going to use. Even if, as we claimed, there are no genuine rivals or alternatives to the principles of deductive logic (nor, in a weaker sense, with respect to inductive logic and epistemology), it is instructive and enlightening to ask what from the point of view of common sense must indeed appear like silly or foolish questions, concerning the reasons for their adoption. Such inquiries are as clarifying in philosophy as are thought experiments with 'outlandish' possibilities in science. They bring out with distinct prominence the role of the presuppositions which would otherwise never be made fully explicit.

A SUMMARY ON CIRCULARITY AND PRIMACY

The following diagram traces the characteristic tangle of circularities which inevitably results if the distinction between cognitive and pragmatic justification is not properly attended to.

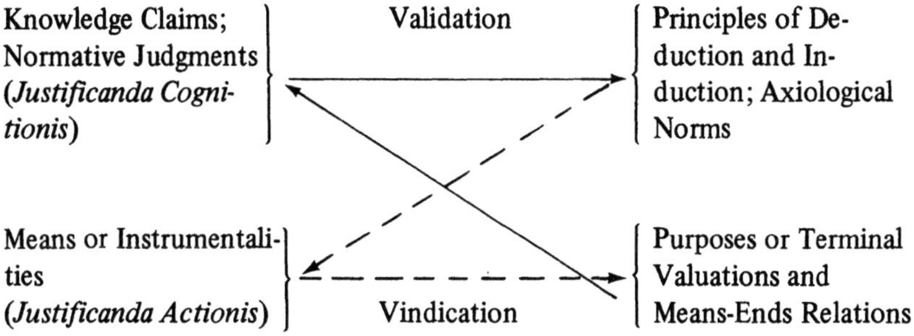

Knowledge claims are justified (validated) by (reference to data and) appeal to the principles of deductive and/or inductive inference. Moral (normative) judgments are validated by (reference to empirical facts and) the axiological principles (norms) of the given moral code. The adoption of these principles themselves may be justified pragmatically (vindicated) by reference to purposes (terminal valuations) and the (usually inductive or, in the degenerate case, deductive) relations between means and ends. These relations themselves may be asserted as knowledge claims and thus require in turn cognitive justification. And so we are apt to keep going merrily around, wherever we start with our quest for justification. The first step toward disentanglement consists of course in the realization of the fundamental difference between cognitive and pragmatic justifications. (The fully drawn arrows represent the former, the broken arrows the latter type of justification.) Our purposes as formulated in terminal valuations as well as the means-ends relations must be viewed as the unquestioned frame of pragmatic justifications. However, since 'terminal' is not intended in the sense of 'absolute,' the terminal valuations may indeed, in a fresh context, be revised or replaced by alternative valuations. Obviously also our knowledge of means-ends relations (if inductive) is subject to revision. Correspondingly, the principles of deduction and induction form the unquestioned frame of cognitive justification. But this frame, too, as the history of logic, mathematics, and the empirical sciences amply demonstrates, is capable of revision and modification. Circularity and infinite regress can be avoided only if we reconstruct our justifications in such a manner that only one type is considered at a time and its corresponding frame taken for granted in that context and until further notice. The old puzzle of *primacy* thus resolves itself very simply in that we may realize that primacy is relative to the context of justification. Kant's insistence upon the primacy of practical reason is plausible enough within the context of pragmatic justification. But, clearly, in the context of cognitive justification 'pure reason', i.e., the principles of deduction and induction, reigns supreme and functions as unquestioned presupposition. Hence, in arguments such as those of the present essay (and quite generally in philosophical analysis) the primacy of 'pure reason' is a necessary and obvious consequence of the nature of the questions proposed for discussion. The structural analogies of the limits of justification in logic and in ethics must not obscure the equally important differences between the two fields. Violation of the logical principles vitiates the very purposes of reasoning and communication; a deliberate repudiation of the law of noncontradiction results in paralysis. But alternative moral codes are just as self-consistently conceivable as are alternative geometries. Clearly, there can be no alternative logics *in the sense* in which there are alternative systems of geometry or ethics. (For further analysis of related issues see W. Sellars [1950]; cf. Feigl [1950c], pp. 186–195.)

NOTES

[1] We assume here that capitalist and socialistic ethics differ in their respective conceptions of social justice. This is of course debatable. (The difference between the two ideologies may consist merely in a disagreement in belief concerning economic and sociological facts.)

[2] Perhaps we should remind the reader that standards are ideal norms. We are under no illusion to what little extent actual conduct conforms with them. But the degree of universality of the ideals embodied in the 'moral sense' of people all over the earth is remarkable.

15. EMPIRICISM AT BAY?

[1971e]

REVISIONS AND A NEW DEFENSE

At the risk of being regarded a 'square' and 'reactionary' I wish to defend, explicate, reformulate and 'salvage' whatever seems valuable to me in the empiricist tradition in the philosophy of science. I am fully aware of the almost hostile vogue of 'beating-up' on the empiricists, a fashion that is still in full swing. My only consolation is the remark of a well-known British philosopher. When he visited us at the University of Minnesota a few years ago, he said to me: "Don't worry Feigl, if you don't catch the bus of philosophy the first time, just wait a while, it'll come 'round again!" Since I was lucky enough to catch the bus of philosophical fashion some forty years ago, I hope to be lucky once again during the rest of my 'natural life'.

Of course I know full well that some drastic revisions – in several respects – are in order. What I shall criticize are mainly some highly questionable exaggerations and distortions that the critics of empiricism have mistaken for genuine objections against the central tenets of classical as well as of modern (or 'logical') empiricism. My major aims then are criticism and clarification of misunderstandings. Since my topic is large and since it seems wise to limit my speech to – possibly – less than an hour, I am forced to restrict myself to a number of terse remarks. I trust you will excuse the somewhat dogmatic tone thereby necessitated. I regret that I can think of no alternative manner of presentation.

Modern empiricism has been, and is still being attacked on many fronts. I shall try to deal with what seem to me the most important and incisive issues.

The late, great Hans Reichenbach [1938; and Feigl, 1970c] distinguished – I think most fruitfully and cogently – between two types of analyses, viz.: in the 'context of discovery' and in the 'context of justification'. Historians of science recognize something closely related to this distinction (in their own way) in that they speak of 'external' and 'internal' history of science. *External* history of science concerns the psychological, social-political-economic contexts and conditions of the developments in the sciences. *Internal* history attends to the reasoning at a given stage of science, i.e., the matters of evidence and inference, of presuppositions or basic assumptions (be they scientific or philosophical), the conceptual, definitional and postulational aspects.

This basic distinction has been attacked by a great many outstanding thinkers. I was astonished that such brilliant and knowledgeable scholars as N. R. Hanson, Thomas Kuhn, Michael Polanyi, Paul Feyerabend, Sigmund Koch *et al.* consider the distinction invalid or at least as misleading. N. R. Hanson complained[1] about the 'bifocal' views of science as presented by the logical empiricists (such as Carnap, Reichenbach, Hempel *et al.*). Hanson criticized the distinction (separation was intended by no one!) between the logical and the empirical aspects of scientific theories. But he (along with others) was attacking a straw man of his own making. None of the Logical Empiricists ever maintained that in an account of the origination and development of scientific theories those two aspects can be sharply kept apart. Hanson may have been quite right in his view that in many a scientific revolution, i.e., in many a drastic change of scientific *theories* something like a '*Gestalt-switch*' has been (and still is) operative. But interesting as that is, it pertains to the psychology of productive, creative thinking and problem solving. It is in this respect that the affinities between great science and great art – stressed recently, for example by Bronowski, Polanyi (and should I mention Koestler?) – are especially significant. But this does not begin to cast any doubt on the need for the 'bifocal' approach. It is clearly one thing to appraise the logical validity of a mathematical derivation (be it in pure mathematics, or in the mathematically formulated empirical sciences) – and it is quite another thing to examine the evidential support of the premises of such derivations in the empirical sciences. Even in the most elementary of Aristotelian syllogisms the distinction is glaringly obvious. 'Is there a formal fallacy?', and what is the evidence for the major and minor premises?' are wholly different questions and require entirely different considerations in order to be responsibly answered.

To be sure, W. V. O. Quine has raised serious doubts about the traditional Kantian (but also empiricist) distinction of analytic and synthetic proposition. It must be admitted that both in ordinary language, and especially in scientific theories the distinction can often be drawn only by arbitrary definitional decisions. Scientific theories, consisting of networks of nomological postulates (plus explicit definitions of derived concepts; derivations; and operational definitions of low-level empirical concepts) contain propositions which are difficult to classify as either analytic – or synthetic. Carnap [1963, pp. 915–923; cf. also 1966, esp. Chapters 27 and 28] has, I think, fairly successfully dealt with this very question. In any case, outstanding logicians like Quine or

Tarski (their occasional protestations to the contrary notwithstanding) should not have any difficulty with, or resistance to, acknowledging the radical difference between pure mathematics and applied mathematics (as e.g. in physical theories). The observations, measurements and experiments that are indispensable in the confirmation or refutation of physical theories function in a role *toto coelo* different from whatever the observation of written symbols does for the appraisal of mathematical proofs. (I shall not digress here into a discussion of the role of electronic computers. Their philosophical significance, in this respect, is not different from that of algorithms in written or printed form.)

There is one way of reconstructing physical theories (take as a relatively simple example that of classical mechanics), in which I have – long before N. R. Hanson or Ernest Nagel – in my seminars on philosophy of physics for some thirty-five years suggested that the analytic-synthetic distinction can be retained in that a given formula, e.g., Hooke's law or Newton's second law of motion, may be given alternatively (and systematically interchangeably) the status of definitions or of laws, such that the total set of feasible alternative reconstructions helps in explicating what is at stake in various given contexts of scientific inquiry. I don't insist that this mode of reconstruction is always the most felicitous, faithful or fruitful one. We may have to get used to the idea that every single specific reconstruction has some demerits along with whatever merits it may possess. So much on the reproach of 'bifocality.'

Now I am turning to the core doctrine of classical empiricism (from Hume to Carnap and Nagel) according to which there are no synthetic *a priori* truths in the factual sciences, or in factual knowledge generally. I am not much impressed (though perhaps just a little bothered) by the examples of the phenomenologists (Husserl and his disciples) and those of Langford and others. It is claimed that once we have confronted a case of the phenomenological *a priori* we shall unfailingly know (by 'intuitive induction') the universal and necessary truth of the proposition that formulates such 'facts' as e.g., orange is between red and yellow; whatever has shape, has size (and *vice versa*); that a three-dimensional body with six squares forming its surface (i.e., a cube) has twelve edges; that a given spot cannot be red and green simultaneously, etc. Even if these examples are at first puzzling, they have been subjected to various analyses, completely consistent with the nonexistence of synthetic *a priori* knowledge. If the worse comes to the worst, I would be willing, if the joke be permitted, to classify cases of the mentioned sort as (*horribile dictu*) analytic *a posteriori* or as synthetic *a priori*. In any case, examples of this sort represent the 'puny *a priori*' compared with the 'grandiose *a priori*' of the classical rationalists and of Kant, viz. regarding space, time, substance and causality. The refutation of the grandiose *a priori* through the development and application (in physics and astronomy) of the non-Euclidean geometries, the elimination of the notion of absolute substance (e.g., of the ether by relativity theory) and the extremely serious doubts regarding strictly deterministic causality in quantum physics (if not since 1900, then certainly since 1926!); I say these refutations, to put it mildly, have punctured rationalistic philosophies in their most important contentions. If historians and philosophers of science refuse to see this, then I wonder whether it is not a waste of time even to discuss these matters with them any more.

It should hardly be necessary to emphasize that the new empiricism is neutral in regard to the nativism issue that agitates the psycholinguists. Noam Chomsky and Jerry Fodor would be the first to admit that only *empirical* evidence can ultimately decide what, if any, of the 'deep-language structures' are both universal and innate. Hence the main contention of logical empiricism thus remains unscathed. To be sure, Chomsky's work is of great interest not only psychologically and linguistically, but also philosophically. But as a refutation of empiricism it can be taken as relevant only to the older, seventeenth century empiricism (e.g., parts of Locke's epistemology in contrast to doctrines of innate ideas such as those of Plato, Descartes or Leibniz). The question of the *origin* of our knowledge and competences (abilities) is the *old* issue of rationalism vs. empiricism. Even though neither Hume nor Kant were consistent on the salient points, the *new* issue of rationalism vs. empiricism was clearly seen by them and has ever since been the battleground of the Kantians and the modern empiricists. This new issue concerns exclusively the grounds of the validity of our knowledge claims. Hence, no amount of evidence advanced by Michael Polanyi in favor of the 'tacit dimension' is relevant to this issue. In order to prevent possible misunderstandings here, I gladly admit that as regards 'knowing how' (in Gilbert Ryle's sense), Polanyi and Gerald Holton may well be right in thinking that Einstein in 1905 did not know (or consider) the Michelson-Morley experiment. From my scanty knowledge of the historical data of the period I am inclined to believe that Einstein – through his acquaintance with the work of H. A. Lorentz – must have been at least dimly aware of the negative outcome of that famous experiment. But be that as it may, Einstein's genius consisted (in part) in his extraordinary capacity for correct conjectures. Six years before the remarkable observations on double stars by W. de Sitter, Einstein quite cavalierly assumed 'on general electromagnetic grounds' that the velocity of light coming from those stars is not influenced by their radial motions (velocities) relative to the terrestrial observers. Einstein's reliance on the general proportionality of the gravitational and inertial masses of physical bodies is another example of his ingenious intuition. A fact already clearly known to (but left uninterpreted by) Galileo and Newton thus became the empirical cornerstone of the highly speculative general theory of relativity.

Our new empiricism can thus fully agree with Leibniz' famous and brilliant saying *Nihil est in intellectu quod non prius fuerat in sensu, nisi intellectus ipse*. As we also realize from recent computer science, unless a computer has the requisite 'competence', it cannot be programmed for the corresponding tasks of calculation or problem-solving. The human brain (or, if you prefer, mind) has an amazingly rich organization which no doubt is the basis of our learning capacities and of our productive thinking. Questions of embryological development aside, surely there are innate features. They are beginning to be understood through neurophysiological and microbiological research.

Empiricism has often been accused of leading to skepticism, if not to solipsism. Philosophers should know that reproaching them for solipsism is psychologically akin (if you pardon a little psychoanalytic jargon) to the diagnosis of narcissism, if not an outright schizophrenia or paranoia. But most forms of classical as well as modern empiricism can easily be shown to be free of these vices or diseases. Skepticism and solipsism thought through consistently to the bitter end are self-refuting – in addition to being self-stultifying. As is well known, the type of reasoning that leads to ordinary solipsism relentlessly pursued ends up with a solipsism of the present moment. This would condemn the philosopher to complete and permanent silence – and what worse fate could befall his loquacious profession?

More significant for the issues regarding the status of scientific theories is the fact that modern empiricism is no longer to be identified with positivism or operationism. In their search for a secure foundation of knowledge (really engendered by the rationalist Descartes) empiricists like Berkeley and Hume, positivists like Comte, Mill, and Mach, and operationalists like Bridgman attempted to ascribe this role to sense impressions, the immediately given, sense data (neutral elements in Mach, elementary mensurational and/or experimental procedures in Bridgman). Carnap in his early impressive *Logical Structure of the World* [1928b, Eng. tr. 1969] (stimulated by Mach and B. Russell) chose a neutral basis of elementary direct experience and the relation of remembered similarity as the basis of his rational reconstruction of all empirical concepts.

After decades of criticism, these 'sense-data' and 'pointer-readings' doctrines have been largely abandoned. Some philosophers of science again flirt with Otto Neurath's suggestion that logical-reconstruction – from scratch – (i.e., from an ultimate and indubitable basis) is chimerical. Neurath himself, in order to illustrate the lines of scientific progress, used the analogy of rebuilding a ship on the high seas. Translated into what is common to Quine and Popper, this means that scientific progress (or if you prefer a more neutral designation, scientific change) consists in working from some tentatively accepted background knowledge, and by (often bold) conjectures introducing hypotheses or theories which then have to be severely scrutinized both for logical consistency and conclusiveness as well as especially their evidential (observational-experimental) corroboration. But what precisely is to be taken as empirical evidence? Popper's basic propositions are perhaps still not so very different from Wittgenstein's elementary propositions or Carnap's experiential ones.

In the course of developments the 'incorrigibility' (i.e., infallibility) of these foundations was completely repudiated. Feyerabend, in a long sequence of essays,[1] has thoroughly undermined all attempts at reconstruction on a phenomenal basis. And going far beyond this he has – buttressing his arguments with a multitude of examples from the history of astronomy and physics – energetically and indefatigably argued that all (or most) observation statements are 'theory laden', some of them even contaminated by the very theories which they are evidentially to support. Perhaps Goethe had something like that in mind when (in one of his aphorisms) he remarked: "The most important thing to remember is that whatever we call 'fact' already contains some theory."

In what follows I beg to differ sharply with this point of view. Of course, I grant – and would even insist – that there are hardly any important statements in the sciences that do not in some way presuppose a conceptual frame-work, if not some theoretical presuppositions. But I wish to point out that there are 'theories' and 'theories.' Our common sense assumptions of the relations of observer to observed events or processes, i.e., of the manner in which the observing subject (person) is embedded in the world observed are indeed taken for granted in the testing of scientific theories. But these common sense 'theories' are, as a rule, not under scrutiny when some high level theory is examined in regard to its empirical support.

By way of a brief digression I do want to state my opinion that in an *epistemological* reconstruction of the common sense view of the world (i.e., roughly what my friend Wilfrid Sellars calls the 'manifest image') something akin to a sense data analysis is indispensable. But sense data, far from being 'given', let alone being 'constructs', are 'reducts' (or 'destructs'), i.e., artificial reconstructions of the items of direct experience which can serve as substantiating evidence for our everyday knowledge of matters of fact. Consider, for example, the acquisition (i.e., the learning) of what the late, great psychologist Edward C. Tolman called a 'cognitive map' of our environment. My scanty knowledge of the topography of parts of Boston; the location of the University buildings, the surrounding streets, the River Charles, other public buildings like post offices, hotels, restaurants, banks, etc., is picked up by my visual perceptions – which in the serial order in which they came to me, form a chronological sequence, something like a movie film. A German philosopher, I remember only his last name, Gerhards (in *Die Naturwissensch.*, 1922) called such sequences 'phenograms.' Moreover it not merely the temporal sequence of impressions (as represented in the 'phenograms') but rather – and more decisively – the repeated and repeatable observed conditions – observed consequences among our impressions that form the testing ground and the supporting (or refuting) evidence for our 'cognitive maps'. C. I. Lewis, I think, was emphasizing this in his conception of 'terminating' judgments. It is the more or less lawful relationships among our impressions, rather than the impressions (sense data) themselves that furnish the corroborating evidence for our knowledge of things, topographies, and the regularity of events on the level of the common life conception of the world.

Now, even if some critics of empiricism are not ready to accept the foregoing sketch of an epistemology, I maintain – with much greater confidence – that the situation in the philosophy of science is largely analogous – (or perhaps 'homologous' is a more apt expression). To put it briefly, I think that a relatively stable and approximately accurate basis – in the sense of testing ground – for the theories of the factual sciences is to be located not in individual observations, impressions, sense-data or the like, but rather in the empirical, experimental *laws* (not necessarily in metricized form). Infallibility ('incorrigibility') is *not* claimed for these empirical laws. They, of course, are revisable not only in view of further empirical evidence, but also in the light of theories. In the well-known level structures of descriptions, empirical laws and theories (sometimes of two or more layers of theories), corrections 'from above' are the rule, although there are exceptions as well. By 'corrections from above' I mean such revisions as that of the Kepler laws in the light of Newtonian (and Laplacean) perturbation theory, or the incisive re-interpretation of he second law of thermodynamics in view of the kinetic theory of heat and of classical statistical mechanics, or the modification of Newton's laws of mechanics and gravitation in Einstein's special and general theories of relativity. These modifications, it is true, become significant observationally only for the rather extreme ranges of such relevant variables as velocities, masses, and (in cosmology) for the topology and metrics of the entire universe. The development of the theories of the chemical bond, of the genetic theory of evolution, and the corrections of Mendel's laws in the light of the new molecular biology, furnish further examples.

Despite these admissions, I maintain, especially against Feyerabend, Lakatos – and to some extent their teacher, Sir Karl Popper, that

(1) There is an important practical difference between empirical laws and theories. This difference is epistemic, not ontological.

(2) While it may well be the case that all theories were (or are) 'born false' – i.e., that they all suffer from empirically demonstrable anomalies, there are thousands of empirical laws that – at least within a certain range of the relevant variables – have *not* required any revision or corrections for decades – some even for centuries of scientific development.

(3) While I admit that 'theories come and go' (but nevertheless favor a realist over an instrumentalist philosophy of theories), I insist that the growth of scientific knowledge depends upon the relative (comparative) stability of empirical laws.

(4) That this is so, seems to follow from the methods by means of which scientific theories are successively 'secured' (Reichenbach's term) and often successfully established – the 'until further notice clause' of course always understood. Similarly understood is the promise of possible improvements in quantitative exactitude.

(5) The successive securing of theoretical knowledge-claims rests upon the (tentative!) reliance upon the (approximate) correctness, within the pertinent range of the relevant variables, of the empirical laws which characterize the functioning of the instruments of observation, experiment, measurement (or statistical designs).

To illustrate briefly by a few typical examples: Such theoretical principles of physical chemistry as the one of Guldberg and Waage according to which the rate of chemical reactions is proportional to the concentration of the reacting substances, may ultimately be tested by weighing the amounts of chemical elements or compounds before they are brought together in aqueous (or alcoholic, etc.) solution. The weighing itself is usually done by the use of chemical balances. The (nowadays very obvious) principle of the balance is the ancient law of the lever discovered by Archimedes. And while this law was much later recognized as a consequence of the conservation of energy principle, Archimedes' law has not been in need of any revisions for about 2200 years. The laws of light reflection and of refraction (precisely known at least since Snell's discoveries in 1621), and of geometrical optics generally (certain corrections on the 'fringes' of the relevant phenomena to the contrary notwithstanding), are still used in reflector and refractor telescopes, and in optical microscopes (and on a simpler everyday level in magnifying glasses, spectacles, etc.). Spectroscopes since Fraunhofer, Bunsen, and Kirchhoff exhibit the spectral lines of various chemical elements (or of compounds). All those optical devices are utilized in modern astronomy (or biology); they are often of crucial importance in the testing of far-flung astrophysical and cosmological (and microbiological) theories. It should be clear that in the context of such testing the empirical ('low level') laws of optics are hardly ever questioned. Of course, I grant, that it is in principle conceivable that astrophysical theories may some day suggest revisions of optics. But I am not impressed with such purely speculative possibilities which the opponents of empiricism indefatigably keep inventing with shockingly abstruse super-sophistication! My point is very simply that thousands of physical and chemical ('low-level') constants figure in amazingly stable empirical laws. The refraction indices of countless transparent substances (such as the various types of glass, quartz, water, alcohols, etc., etc.), the specific weights, specific heats, thermal and electric conductivities of myriads of substances, the regularities of chemical compounding (or disassociation), the inverse square laws for the propagation of sound and light, similarly the Coulomb laws for magnetic and electric interaction and (after the experimental confirmations by Cavendish, Jolly, and Krigar-Menzel!) yes, even Newton's inverse square laws for gravitational forces, Ohm's, Ampère's, Biot-Savart's, Faraday's, Kirchhoff's, and so on, laws of electricity, etc., etc. are all intact, and are needed for the testing of higher level theories. I could continue with the empirical regularities of the thermal expansion of mercury, alcohol or of various gases which have been used for a long time as thermometric substances – and (again with certain corrections 'coming from above') have served in the confirmation of important theoretical assumptions. Thus the (approximate) correctness of the empirical Boyle-Charles law – or for that matter, of Newton's law of cooling is well confirmed, and relied upon in the scrutiny of thermodynamic theories. Or I could mention that the variability of g – a consequence of Newton's theory that has been tested with the help of the empirical law of the (physical) pendulum; or that the empirically discovered formula of Balmer (in 1885) for the order of lines in the hydrogen spectrum still holds, and thirty years later became an important piece of evidence for Bohr's quantum theory of the atom.

It is obvious that examples of this sort could be indefinitely multiplied. But I trust the tedium need not be prolonged. My main point, I repeat, is that while there are of course genuine revolutions in the realm of theory (shifts from one paradigm to a new one, if we speak Thomas Kuhn's language) or of research programs, be they 'progressive' or 'degenerative', to use the terminology introduced by Imre Lakatos, the empirical laws have remained amazingly stable. But I disagree with Kuhn and Feyerabend in that differences in the conceptual framework and the presuppositions of scientific theories, do not make them logically incommensurable. Not only is it always (in principle) possible to compare the logical structures and the evidential bases of differing theories, but we can also appraise their respective explanatory powers in the light of *which* and *how many* empirical laws (often qualitatively quite heterogeneous ones) can be derived from the postulates of the competing theories.

Sub specie aeternitatis, I admit, of course, that Feyerabend has some excellent arguments for the 'theory-laden' character of (at least some) empirical laws. But since both he and I wish to do justice to the history and the practice of the scientific enterprise, it is imperative to give a faithful account of the actual procedures in the growth of scientific knowledge. Anyone who has worked in physical or chemical laboratories (I *have* for six years) can not help being impressed with the (relative) constancy and stability of the empirical laws, and their indispensability in the testing and stepwise corroboration of theories.

As I know only too well from my many discussions with Popper, and his disciples (most of them now renegades from the original position of Sir Karl), if confronted with my 'benighted' inductivism, they roll out the heavy artillery ('Big Bertha' I have come to call it) of Humean skepticism in regard to inductive inference or inductively based beliefs. Popper keeps saying that he does not 'believe in belief' [Popper, 1971]. But he has never deigned to answer *the* crucial question of his critics as to what justifies him to trust (or 'place his bets') on theories, hypotheses or laws that are – in *his* sense – highly corroborated. If he responds at all (and I have often pestered him with that query [Feigl, 1964a]) he says either – that is a matter of practical action (and I think this is quite compatible with Hume's psychological view of belief and 'custom'), or he says this is a piece of 'good metaphysics.' I wholeheartedly agree with the pragmatic answer and will elaborate on it shortly. But when it comes to metaphysics, Popper and I (along with most logical empiricists) have a common 'enemy' there which we both have vigorously and relentlessly criticized. I know of no criterion that Popper could use in order to demarcate good from bad metaphysics, except the one by which he distinguishes between science and non-science. In that case 'good metaphysics' would have to be continuous with science, perhaps mostly more general –

but still in principle testable. But then the problem of induction returns with a vengeance to good metaphysics. As I know from many personal conversations with my good friend Karl Popper, he regards the general assumption of the lawfulness (not necessarily deterministic, according to Popper's opinion, actually indeterministic – but not chaotic, i.e., with essentially statistical regularities) of the universe as a 'good' metaphysical presupposition of the scientific quest. But this makes him vulnerable to the same sort of criticism which applies to the early 'world hypotheses' of C. D. Broad, J. M. Keynes, and the later (1948) B. Russell.

A side remark may be in order here. Popper's *modus tollendo tollens* view of falsification is highly persuasive in the case of simple syllogisms. I doubt that the much vaunted asymmetry of falsification and verification obtains in the case of scientific theories. With the late Bela Juhos I think that empirical laws in science are refuted only by better (inductively!) established alternative laws that are logically incompatible with the previous ones. Thus, for example, the regularities of the speed of light (discovered by Foucault in optically denser media) are incompatible with Newton's assumption regarding the speed of light (for instance in water or glass). This is the usual way in which assumed empirical laws are refuted, and this casts serious doubts on Popper's asymmetry thesis.

Has Nelson Goodman (ever since his article on the 'Infirmities of Confirmation' and his book *Fact, Fiction and Forecast*) contributed to the solution of Hume's problem? I doubt it. The 'pathological' predicates like 'grue' and 'bleen' that have been discussed *ad nauseam* seem to be (after a suggestion by Paul Teller – in conversation) merely the analogon – on the qualitative-predicate level – of the well-known interpolation-extrapolation problems long familiar in connection with 'curve-fitting.' With a limited supply of discrete mensurational data there is a non-denumerably infinite number of curves (or functions) that fit the given array of points. Einstein and Popper are surely right in saying that there is no 'straight' (unique) path that would lead from the data of observation to the laws (let alone the theories!). Kurt Gödel, in an unpublished appendix to the doctoral thesis (Vienna, 1928) of our dear late friend Marcel Natkin, proposed an elegant definition of the formally simplest curve (for interpolation purposes). It singles out (using variation calculus) that curve by the criterion of the minimum integral of the curvature of the curve within the given interval of the relevant variables. But, of course, as to whether 'nature' is kind enough to display that sort of simplicity is again a matter of empirical evidence. Hence this does not solve the problem of induction.

Nelson Goodman's own attempt at a constructive solution, recommends that we use only 'entrenched' predicates in our inductive projections. But 'entrenched' predicates in his sense are of the kind that have thus far proved fruitful in inductive generalizations. Hence, Goodman has at least re-stated the problem of induction but he has hardly advanced its solution.

Jerrold J. Katz, in his ambitious book *The Problem of Induction and Its Solution* [1962] tries to show by highly polemical – but partly defective – criticisms that there is no and there cannot be a justification of induction. In other words, Katz's solution is: there is no solution!

Rudolf Carnap in more than thirty years of impressive work on an inductive logic, had to revise, fix and mend various axiom systems in order to adapt them to the elementary requirement that they must be in keeping with the fact that we learn from experience and that 'probability is our guide in life.' Moreover, even if Carnap's inductive logic were otherwise impeccable, it is applicable to only extremely simple worlds of individuals and a finite number of predicates. Carnap hoped optimistically that his inductive logic could be generalized and extended to the quantitatively (metrically) formulated theories of science. But his much lamented death (at the age of 78) in September of 1970 terminated his work abruptly.

Imre Lakatos, perhaps Popper's keenest disciple (though nowadays a renegade) has in recent years advocated a solution in terms of what he calls 'research programs' [I. Lakatos, 1968; I. Lakatos and A. Musgrave, 1970]. As anyone even superficially familiar with the history of science knows very well, certain theoretical frames (I am tempted to say ideologies) have been extremely fruitful in solving problems, explaining a variety of phenomena (i.e., empirical regularities), and in suggesting new problems whose solution may be plausibly and auspiciously pursued within the same frame. On the other hand there have been theories (or ideologies) in the sciences which even despite some initial plausibility needed so much 'fixing' and 'mending' with *ad hoc* hypotheses, that they became suspect, and in the light of a new research program were displaced by a different more fruitful theory. Without entering into the otherwise important differences between Lakatos and Kuhn, such 'shifts' (of research programs) are what Kuhn characterizes as scientific revolutions or as transitions from one 'paradigm' to another. Now consider Lakatos' claims. He does not wish merely to chronicle these changes in scientific outlook. He presents his views as instruments of critique and advice to scientists. He encourages progressive problem shifts and research programs and warns against degenerative ones. But doesn't this make him (*horribile dictu!*) an inductivist after all? Doesn't he have to watch the development of scientific theories and place his bets, or (since probabilities may be unobtainable here) in accordance with how his 'race horses' (i.e., research programs) *run*? If Lakatos thus turns out to be a second level inductivist, I wonder whether he could not be persuaded to become as well a first level inductivist in regard to the validity of empirical laws.

To be sure, such justifications of induction (or inductivism) as have been presented by R. B. Braithwaite [1953] and Max Black [1962] are – despite their sophistication – clearly circular and thus beg the question at issue. Admittedly it is tempting to justify inductive inference by all its past successes, but David Hume, and most empiricists after him, have criticized this grievous fallacy. It is conceivable, I submit tentatively, that there is logically speaking, such a thing as 'virtuous circularity'. As C. I. Lewis so nicely put it (in *Mind and the World Order*) – a circle is the less vicious the bigger it is. So, I consider it possible, that in an all encompassing reconstruction of scientific theories together with their supporting evidence, a network of propositions may be so exhibited that the postulates justify the observational propositions *and vice versa*. A. S. Eddington may have had this sort of thing in mind in his masterly (and unjustifiably maligned) three philosophical books [A. S. Eddington, 1928; 1959; 1958; cf. also J. Witt-Hansen, 1958; and the positivistically based critique in L. S. Stebbing, 1937]. But no exact reconstruction of this type has thus far been worked out in any detail. ('The thing to do, is do it!') In any case certain constraints imposed by whatever be the observational evidence would seem indispensable for the adequacy of any such reconstruction.

Carnap had some limited sympathy with the approach of D. C. Williams in *The Ground of Induction* [1947]. The essential point here (originally suggested by C. S. Peirce) is the inference from sample to population. Williams thought he could derive inductive probabilities converging toward practical certainty in this manner. But he overlooked one essential precondition. Without some uniformity assumption (something like permanent kinds and limited variety à la Broad or Keynes) no probabilities can be derived. Nevertheless I think there is a kernel of truth in Williams's ideas. I think the much discussed pragmatic justification of the adoption of an inductive rule, the reasoning that Hans Reichenbach,[2] and I have advocated (my term 'vindication' of inductive inference has been widely used, e.g., by Wesley Salmon, Jerrold Katz *et al.*) can be given an alternative formulation in terms of 'sampling'. What I have in mind is the following simple thought: Consider observations, measurements and experimentation (and, of course, statistical designs) as a way of obtaining samples of the regularities of nature. Now of course without further assumptions (or prior probabilities – as used by the

Bayesians and the 'personalists') there is absolutely no way of estimating the probability of the 'fairness' (representative character) of the sample. But consider how '*ir*rational' it would be to assume without any reason whatever that the samples are unrepresentative. (Of course they may well be unrepresentative in fact, in one way or another.) Yet, to *assert* that the samples *are un*representative opens the floodgates to limitless possibilities. It seems to me that assuming (until further notice!) that a given sample *is* representative is tantamount to one of the more important formulations of the requirement of simplicity. It secures the uniqueness – with respect to previous experience – which in many discussions has been seriously questioned.

Let me candidly state that this sort of vindication (just as in Reichenbach's or my own earlier formulations) must appear weak to anyone who expected a 'genuine solution' of Hume's 'grand problem' of the justification of induction. Compared with the subjective practical certainty that we attach to the inductive and analogical inferences on which we base the countless expectations of everyday life, and the best established empirical laws in the sciences, the pragmatic justification looks very pale indeed. But here I am inclined to use, for once, some 'Oxbridge' type philosophy: If this sort of justification (of the adoption of a principle of induction) seems disappointing to you, what else on earth (heaven or hell) would you possibly suggest? Even if we have a guarantee from God (or from Laplace's demon) that the world is strictly deterministic (and no such guarantee will be forthcoming!), scientists will always realize (or should at least be willing to admit in principle), that there may be as yet unknown parameters in our galactic region of space-time which are absent or completely different in other parts of the universe. No assurance can be given that the regularities of the world are unchanging. Entirely different empirical laws may have to be adopted tomorrow (or even beginning with the next second!).

If it is the major aim of the factual sciences to understand – in the sense of explaining the facts, events, processes (and their regularities) of nature, and as long as scientific explanation consists of the hypothetico-inferential reasoning (from deterministic or statistical) premises, then we had better hold on to whatever regularities have been empirically established.

The one concession I (still as an empiricist!) gladly make to Kant (and some of the neo-Kantians) is that our understanding of nature is predicated upon the well-known patterns of explanation, i.e., of derivation – deductive or probabilistic, as the situation may require – from (tentatively) accepted premises. We have not a glimmer of an idea what else understanding or explaining could consist of. It seems plausible that this basic feature of human reason (mind or brain) is thus a logically contingent, but extremely important feature of our world. The forms of intuition and of the understanding that Kant himself considered necessary conditions for *Erkenntnis überhaupt*, have had to be loosened up tremendously with the advances of recent science. Nevertheless, it is just conceivable that there are limits of our understanding, owing to the structure of our minds (or brains). There is nothing sacrosanct about the empirical laws of science. But I have tried to point out that very many of them have stood the test of time and of many searching criticisms. If the data of observation should surprise us (and of course I admit they have frequently done just that; think, for example, of the discovery of radioactivity or of the isotopes), then it is time to revise our laws and theories.

Popper, Feyerabend, Lakatos and many other brilliant philosophers of science, have understandably fixed their attention upon such dazzling and exciting high level theories as those of Newton, Einstein or of the quantum physicists (Bohr, Heisenberg, Schrödinger, Born, Dirac, von Neumann *et al.*). They have rarely condescended to the 'low' level of experimental laws. Of course, I agree with them, high-level theories are much more fascinating. But if Popper thinks that his critical approach helps us to understand how science can get 'nearer to the truth', the neglect of the empirical laws prevents him from giving a convincing account of the progress of knowledge. Popper relies on the well-known Tarski-Carnap semantic definition of truth. But this definition merely clarifies what we *mean* by the word 'truth'; as Popper admits, it does not provide him with a criterion of truth or of an explication of 'nearer to the truth.'

Close attention to the actual procedures of the scientists does show that they do use very simple, almost homely, criteria for the truth and the approximation of the truth by laws or theories. A theory is 'near to the truth' if it enables us to derive a large number of well-established empirical laws. And empirical laws are well established or 'close to the truth' if repeated experiments and measurements, if possible performed under varying conditions, yield practically convergent series of values of the natural constants that are pivotal in the empirical laws. To take at random one of thousands of examples: the speed of sound in air of normal temperature and pressure (on the surface of the earth) is approximately 1087 feet per second. Repeated measurements will show a practical convergence toward this (non-mathematical!) limit – once the theory of the errors of measurement has made the necessary corrections.

Finally, just a few words about Popper's criterion of demarcation. Empiricists can be quite happy with his distinction of scientific and nonscientific questions. The logical positivists took testability as a criterion of meaningfulness, but as Carnap has pointed out already in 1934 this was a mere proposal, to be judged by its fruitfulness. What is not so clear in Popper is an additional distinction that the positivists *did* make, namely between scientific and *un*scientific knowledge claims. Though I am extremely wary and skeptical of the claims of Psychical Research (ESP), I would certainly not reject them as meaningless. But I can afford to wait for more convincing evidence before I am persuaded that we have here a genuinely scientific enterprise. In other words, until further notice I go at least through the 'motions of an open mind'. Astrology, palmistry, alchemy, and phrenology, however, are, I think, rightly regarded as *un*scientific because all (or most) of the available evidence speaks harshly against these fake disciplines. In regard to some recent hypotheses in astrophysics, cosmology and of nuclear theory, however, the evidence is still too weak to decide whether or not they will end up on the scrap heap of failures of – however highly respectable – approaches toward honest science.

Psychology and the social sciences present problems and difficulties of their own. It is my considered opinion that nothing like a Newtonian or Einsteinian synthesis is here to be expected. I am inclined to agree with B. F. Skinner, whose search for reliable empirical laws (mostly statistical ones) of learning and motivation, and studies of schedules of reinforcements, have been remarkably successful. He does not, like his predecessor in the behavioral sciences Clark L. Hull, look for all-embracing psychological theories in molar-behavioral terms. As Sigmund Koch has shown in his devastating critique of Hull's work, it is both logically and empirically extremely questionable. I am, however, disappointed by B. F. Skinner's ignoring (if not denying) the possibility of progress in psychology through advances in neurophysiology. (Some of the contributions in that field already have considerable explanatory power.) And I am even more gravely disappointed and dismayed with S. Koch (my former brilliant student of ca. 1938 at the University of Iowa) who has lately adopted what I can only call the obscurantist stance of Polanyi, Kuhn and Feyerabend. An unprejudiced closer look at the actual research in psychology seems to me to show that we do not need any 'holistic' psycho-social-economic theories of the relation of observers to observed organisms, for fruitful theory construction. How a rat turns in a T-maze, and with what frequency, how, and at what rate, it presses the lever on a Skinner box, or the words uttered by a person in psychoanalytic interviews, etc., etc. can be recorded by reliable machines. But this is a large topic and it is getting late.

Let me say by way of conclusion that empiricism, though in need of renovation, will remain a fruitful and adequate philosophy of science.[3]

Postscript. In order to forestall misunderstanding – such as involved in the reproach that I cannot 'have it both ways' – let me briefly clear up the essential point. Of course I *can* have it both ways: Thousands of empirical laws in physics and chemistry have *not* been in need of revision (except for minor adjustments in quantitative exactitude). Nevertheless, I admit, of course, that in quite a few (perhaps a few dozen) cases, empirical laws received incisive 'corrections from above', i.e., in the light of new and well-established theories. The empirical laws of Newtonian mechanics, and the empirical laws of classical electrodynamics *have* been drastically modified. But the thousands of physical or chemical constants that figure in the vast majority of the 'low-level' empirical laws, and are listed in such tables (as in Landoldt-Boernstein years ago) are pivotally characteristic of the empirical laws which remained essentially unchanged!

NOTES

[1] N. R. Hanson [1969].

Even more critical of the empiricist and positivist philosophy of science: P. K. Feyerabend [1965]; [1970a]; and [1970b]. Equally challenging, P. K. Feyerabend [1970c].

[2] H. Reichenbach [1935, Eng. tr. 1949b] (see especially the exciting last chapter of this book). More recent statements regarding this 'vindication' of induction may be found in H. Feigl [1950a] and Wesley C. Salmon [1957]. But cf. also his later excellent and extensive, penetrating discussion and scrutiny in Salmon [1966].

[3] For a modified statement of modern empiricism in the philosophy of science, cf. C. G. Hempel [1970], and a qualified argument in favor of traditional empiricism, H. Feigl [1970b]. Also, H. Feigl [1969a].

For a brilliant and 'unorthodox' critique of traditional empiricism, as well as of some aspects of K. R. Popper's views on the logical of science, cf. Grover Maxwell [1974].

A very lucid and judicious discussion (with ample references) is contained in J. J. C. Smart [1968].

The literature on this subject is growing so vast that I must stop giving references. There are extremely sane and sound papers by Ernest Nagel [1971], and by Dudley Shapere [1966].

Quite recently (August, 1972) a most brilliantly sound article came to my attention: Carl R. Kordig [1971]. This, I think, is the very best and concise critique of the ideas of Feyerabend, Hanson, Kuhn, and Toulmin. I wish I had known about this article sooner.

For an excellent and very thorough analysis and discussion of this and closely related issues, cf. also Adolf Grünbaum [1971].

16. THE MIND-BODY PROBLEM IN THE DEVELOPMENT OF LOGICAL EMPIRICISM

[1950d]

The cluster of puzzles and perplexities that constitute the Mind-Body-Problem of modern philosophy owes its origin to a great variety of motives and considerations. The central issue, however, may justly be located in the disputes between Dualism and Monism. The dualistic doctrines have a twofold root: Firstly, there are the mythological, animistic, theological, and religious-moral contentions as to the sharp distinction, if not actual separability, of the mental and the physical. The deeper-seated and culturally fairly widespread wishful belief in some form of survival after bodily death, as well as the exaltation of the spirit and the deprecation of the flesh in so many Eastern and Western religions and moral codes may be regarded as the emotional root of dualism. The other, scientific, root of dualism may be found in the rise of science, most prominently beginning with the seventeenth century, although at least adumbrated in ancient thought. The striking success of the method of the physical sciences was, at least historically, contingent upon a clear-cut division of the physical and the mental, and the relegation of the latter to the limbo of a sort of secondary or epiphenomenal existence. But the development of modern psycho-physics and psycho-physiology from the nineteenth century on, culminating in present-day neuro-physiology, Gestalt-psychology, psycho-somatic medicine, and cybernetics, has revived the interest in monistic interpretations. One discrepant tendency may of course be seen in the dualistic claims of the researchers in the still highly questionable fields of Parapsychology (extra-sensory perception, psychokinesis, etc.). Another and very different kind of opposition comes from philosophers of various schools who either on the basis of their metaphysical commitments or simply in the name of clear thinking insist that the physical and the mental are *toto genere* and irreconcilably distinct and different, so that any monistic attempts at their identification must be rejected on purely logical grounds.

This is not the place to review even in outline the history of dualistic and monistic arguments and systems from Descartes and Spinoza down to our time. Two notable conclusions seem to emerge from a study of this history:

1. The clarification of the badly tangled issues requires as an indispensable first step the discrimination between the factual and the logical questions involved in the mind-body-problem. The factual questions depend for their

solution on the progress of scientific research, such as in psycho-physiology. The philosopher *qua* logical analyst has no business either imaginatively to anticipate or dogmatically to endorse hypotheses that can be established only by painstaking empirical investigations. Since the philosopher is concerned with the analysis of meanings, he can at best examine the consistency of various hypotheses and clarify their precise content by an examination of their logical implications.

2. It is evident that different thinkers have been impressed with different aspects of the very complex problem of the relations of the mental and the physical. Descartes was puzzled with the question how something of the nature of a non-spatial substance (thinking) could be causally related with a spatial substance (matter). Some philosophers of the nineteenth and twentieth centuries tried to tackle another 'spatial' problem: the location of sense data. Still others have tried to account for the difference of the mental and the physical in terms of the distinctions of the qualitative and the quantitative or of content and structure. Some were intrigued with the 'private' character of consciousness and the 'public' character of behavior and of neurophysiological processes. Others again, found in the 'meaningful', 'intentional', 'referential' nature of mental states an insuperable obstacle to the attempted identification with 'blind' brain-states. Similar objections arose out of the considerations of 'purpose', 'free choice', 'reason' on the mental side as juxtaposed with 'mechanism', 'determinism', 'cause' on the physical side. Normative and critical predications (like 'correct', and 'incorrect', 'success' and 'failure', 'responsible' and 'irresponsible', 'justified' and 'unjustified', (morally) 'right' and 'wrong', etc.,) seem to apply meaningfully only to minds, mental states, attitudes or functions but not to physical things, processes or events.

This list of juxtapositions, which could easily be expanded, may serve as a reminder that any present-day advocate of monism (in the sense of an identity-theory) is confronted with a considerable task. Recent naturalistic philosophical and psychological movements, such as positivism, pragmatism, neo-realism, critical realism, behaviorism and some phases of analytic philosophy, have in one way or another attempted various resolutions of the puzzles posed by the apparent incompatibilities of the essential features of the mental and the physical. A good many of the traditional questions in the total complex of the problem have fairly generally been recognized as pseudo-problems, arising out of conceptual confusions. This may be asserted with assurance in the case of the free choice vs determinism perplexity. Almost equally definite seem to me the clarifications of the problems of spatial localization, of emergent novelty and of teleology. The proper view of the referential, normative and critical functions of 'mind' or 'reason' depends on an adequate formulation of rule-guided behavior. Although a good deal of work along these lines is still required, it is evident even now that some of these questions pertain not so much to the distinction of the mental and the physical, but rather to that of logical structure to psychological (or behavioral) fact. Common to all these issues, however, is the irrepressible and most controversial question: In which sense is the identification of the mental and the physical to be understood? It is interesting to note that Logical Empiricism in the twenty-five years of its career since its beginnings in the Vienna Circle has in succession embraced three different monistic views and has temporarily countenanced also a more agnostic (parallelistic) form of dualism. In recent years Logical Empiricists have prepared a return to their first monistic position, however, reformulated in a more cautious and therefore more auspicious manner. In connection with the very brief review of the four previous positions that I am now going to present, it must be kept in mind that the affinities these positions display with the more traditional metaphysical doctrines are, on the whole, more of the nature of historical analogies than genuine identities of theoretical import. Logical empiricists have from the beginning disclaimed any intention of pronouncing ontological truths. The sole concern has been the analysis of language and meaning. It was precisely on the basis of such reflections that ontologies of *all* sorts were declared as devoid of factual meaning. The metaphysicians, understandably hurt in their pride and unconvinced by the negativism of the positivists, kept reading into the logical analysis of the latter all the traditional tenets and categories. As already admitted, the flavor of the traditional monisms (or of parallelism) was there, but only historically-culturally speaking. The first position, for example, can easily be regarded as a double-aspect, or double knowledge, view of the type held by critical realism. This was Schlick's [1918, 1925] outlook before the formation of the Vienna Circle, i.e., before the impact of the ideas of Carnap and Wittgenstein.[1] However, even anticipating the later emphases of logical positivism Schlick regarded the difference of the mental and the physical as a difference between two conceptual systems, of which the physical, as a matter of fundamental empirical fact, is universally applicable, whereas the psychological pertains only to a small part of the total realm of reality. This early point of view is therefore more appropriately characterized as a 'double-language' theory.

With the first phase of logical positivism, most markedly represented by Carnap's *Der logische Aufbau der Welt*, the rational reconstruction of empirical knowledge was pursued on a phenomenalistic basis. It is therefore not surprising that metaphysicians misinterpreted this approach as a revival of a Berkeleyan subjective idealism. While Carnap explicitly disavowed any claims regarding the ultimate reality-problems of the mental and the physical, he shared of course with Berkeley, Hume, Condillac, Mill, Mach, and Avenarius the conviction that there is no ontological mind-body problem that could be legitimately formulated. The only genuine problem, Carnap claimed, was one of logical analysis, i.e., the question of the formal relations between the concepts that describe the data of first-person-experience, the concepts of physics, and those of (behavioristic) psychology. The 'basic situation' of the mind-body-relation was identified with the parallelism of data that a person would experience if he were to observe by means of some 'cerebroscope', his own cerebral processes alongside with the stream of images or feelings which 'correspond' to those brain processes. But the internal difficulties of a strictly phenomenalistic reconstruction were soon recognized. The translatability of statements about physical objects into statements about phenomenal data could no longer be held to obtain in the sense of mutual deducibility. And the absurdities of a metaphysical solipsism were parallelled by the absurdities of a phenomenal language that was doomed to be 'private', 'soliloquistic', 'incommunicable.'

The second phase of logical positivism arose largely out of a reaction against the phenomenalism (experientialism) of the first phase. Under the influence of O. Neurath's and K. Popper's critical suggestions, Carnap [1934, 1932b, 1935b, 1938] formulated his *physicalism*. It was easy again for metaphysically-minded opponents to misconstrue this position as a variant of ontological materialism. But Carnap's aim was, just as in the previous phrase, merely that of an analysis of language. He outlined a logical reconstruction of factual knowledge on the basis of an intersubjective (physicalistic) thing-language. This position, though independently arrived at, was generally akin to the methodological behaviorism that had been formulated even somewhat earlier but with much less formal precision by E. A. Singer [1924] and K. S. Lashley [1923]. It is important to distinguish two phases in the development of physicalism. The first phase was rather rash in its claim of the translatability of the statements of physics and those of psychology into those of the thing-language. Availing ourselves of the material idioms (realistic language), this radical and crude form of physicalism may be said to amount to an identification of mental states with overt behavior. Early behaviorism (especially that of J. B. Watson) has been rightly accused of just this fallacious reduction. This view was essentially revised and corrected in the later formulations

[Carnap, 1936a]. Strict translatability depends of course on explicit definitions. But no explicit definitions that would serve the purpose could plausibly be constructed. The concepts of physics and psychology could perhaps be *introduced* by means of test-condition-test-result-conditionals but not in any way be regarded as synonymous with concepts of the thing-language (or purely logical compounds thereof). Carnap [*ibid.*] advanced his reduction sentences as a possible formulation of those conditionals. While it has become increasingly doubtful that this formulation is logically adequate, the underlying and related ideas of confirmability and degree of confirmation are now quite generally accepted. No statement of physics nor of (intersubjective) psychology can be considered as completely and directly verifiable (or refutable) by the observations as formulated in the protocol-statements of the thing-language. These protocol-statements confer only a degree of confirmation upon the statements in the scientific languages of physics and psychology.

Reichenbach's version of scientific empiricism [1938, 1936] had for many years opposed the narrow verifiability criterion of the Viennese positivists. His emphasis on probability and induction led him to advocate a more inclusive confirmability criterion, amounting approximately to the same delimitation of factual meaning as Carnap's criterion (in the second phase of physicalism). Reichenbach's account of the mind-body problem, based on his empirical realism, represents in many ways a position similar to that of Schlick in his early realistic approach. Before we turn to a fuller discussion of this view we must briefly mention a more agnostic position which arose out of a reaction against the earlier, rather immature arguments in favor of mind-body identity.

Felix Kaufmann [1944], and similarly also Norman Jacobs, generally in sympathy with the principles of Logical Empiricism, insisted that strict identity would have to be tantamount to *logical* equivalence of phenomenal (introspective) descriptions of mental states with the descriptions of the 'correlated' neurophysiological processes. But it seems obvious, so Kaufmann argued essentially, that the investigations of psycho-physiology are of a factual-empirical character. Which mental state is correlated with which neural processes can be determined only by experimental investigations. The statement of the correlation is therefore synthetic and the 'equivalence' of the two descriptions thus can at best be only of (universal) *empirical* character. Reading this conclusion again in terms of traditional metaphysics it may be taken as a formulation of dualistic parallelism. Wolfgang Köhler in one of his later works [1938], and other thinkers trying to be cautious in such delicate matters, have essentially retreated to this obviously safer (because less daring) position. If anyone (like, e.g., E. G. Boring [1933]) wanted to account for the parallelism by means of a supposedly more fundamental identity, he usually availed himself of the help of the principle of parsimony.

The principle of parsimony itself needs careful analysis. Occam's razor has really, as it were, *three* blades. The simplicity it advocates may be the descriptive or purely formal (or logico-mathematical) expediency that distinguishes, e.g., the heliocentric from the geocentric description of the planetary system. It may be the factual (or inductive) simplicity that arises from a reduction of the number of independent empirical hypotheses. This is probably the purport of Newton's first *regula philosophandi*. But finally, Occam's razor may be used to cut away metaphysical entities. In what follows I shall contend that this third blade, the confirmability criterion of Carnap and Reichenbach, if properly applied, removes the metaphysical surplus, without cutting into the flesh of knowledge. I shall contend also that this new point of view involves (1) a fundamental revision of phenomenalistic positivism and radical operationism (and behaviorism); (2) a re-instatement of a clarified critical realism on the basis of pure semantics and pure pragmatics; (3) a return to a reinterpreted identity (or double-language) view of mind and body.

1. The slogan of Vienna Logical Positivism: 'The meaning of a statement is the method of its verification' [M. Schlick, 1936]; and the slogan of Bridgman's operationism [Bridgman, 1928]: 'A concept is synonymous with the set of operations' [which determine its applications] were excellent preventives of the transcendent type of metaphysical speculations. They have had a most salutary purifying effect. Logical empiricism in its later development, however, had to replace these radical principles by more conservative ones. As already indicated, the meaning of scientific statements cannot in general be identified with their confirming evidence. This is obvious in all those cases in which the evidence must in principle be indirect. Historical statements concerning past events, predictions of future events; existential hypotheses concerning radiations, subatomic processes in physics; genes, filterpassing viruses in biology; unconscious motivations in psychology; etc., are only some of the more striking types of assertions whose meanings (i.e., the states of affairs to which they refer) cannot be identified with the states of affairs that can conceivably serve as evidence for them. For a more specific but very simple example we may refer to the concept of the temperature of a body. As ordinary and scientific commonsense (untouched by ultra-positivistic reductionism) would put it, thermometer (or pyrometer) readings,

spectroscopic findings, and other types of measurement merely indicate something about the body in question, namely the intensity of heat which is a state of that body. No matter whether this heat intensity is construed in terms of classical (macro-) thermodynamics or in terms of statistical (micro- or molecular) thermodynamics, it is in any case only *evidenced by but not identical with* those indications. Similarly for psychology: The overt symptoms and behavior that indicate an emotion like, e.g., anxiety, are confirmable and measurable in terms of skin-temperature, endocrine secretions, psychogalvanic reflexes, verbal responses, etc. but must not be confused with the emotion itself. Generally, the 'theoretical constructs', i.e., the hypothetically assumed entities of the sciences cannot be identified with (i.e., explicitly defined in terms of) concepts which apply to the directly perceptible facts as they are manifest in the contexts of ordinary observation or of experimental operations.

2. The required correction and emendation of the phenomenalistic phase of positivism and operationism can best be achieved by means of a reconstruction in terms of pure semantics and pure pragmatics. Semantics as developed primarily by Tarski and Carnap enables us in a precise way to speak, in a metalanguage, about the relation of designation that holds between the symbols of a given language (the object language) and the objects, properties, relations and states of affairs they symbolize. The required metalanguage must of course have a sufficiently rich vocabulary to allow for this. It is in the field of pure pragmatics (thus far only sketched in outline by Wilfrid Sellars) that the rules and the scope of the metalanguage are determined. The pragmatic prerequisites of a workable scientific language extend far beyond the conditions that must be fulfilled for the sake of logical consistency and for the purposes of deductive inference. They also include the condition of confirmability, with all that this implies: a set of proper names (or co-ordinates) and of predicates only some of which correspond to directly confrontable items of immediate experience; a set of relationships that connect the directly verifiable with the only (indirectly) confirmable predicates and statements. With such a reconstruction a distinction necessarily neglected by phenomenalism can be reinstated. It is the important distinction between the evidential basis and the factual reference of terms and statements. In acknowledging this distinction we retain the empiricist conditions for meaningfulness and for factual adequacy: Only if our terms are nomologically related to terms that designate items or aspects of what is directly observable can they be factually meaningful; and only if statements are supported at least by incomplete and/or indirect evidence can they be justifiably asserted.

But in the recognition of the incompleteness and indirectness of the verification of practically all scientific statements we implicitly allow for a genuinely critical realism. This new version of realism is free from the objectionable metaphysical elements in the older forms of realism. Much of the perplexities in the time-honored reality-problems arose out of a confusion of the intuitive, experiential idea of reality with the cognitive, objective concept of reality. The agonies that attend all attempts to solve the 'problem of transcendence' can be avoided once it is realized that this is a pseudo-problem. The solution that had been sought involved plainly an inconsistency: The non-given was to be proved just as real as the given. But if by 'real' one means *given*, then obviously the wish for a demonstration is doomed because of the self-contradiction. If however one wishes to connect with the word 'real' not an ineffable but a cognitively expressible significance then the usage of this term in common life and in science may profitably be taken as a standard. 'Real' and 'unreal' are of course ambiguous and often emotively tinged words. But in the context of the traditional realism-phenomenalism controversy it is clear that the distinction connoted by these terms cannot be intended to achieve a division among things, events or processes. Once anything is at all classified under one of these three headings it is *eo ipso* considered real. Dreams and delusions are (even according to common-sense) real enough as occurrent events. What is not real are the referents (designata) of certain terms or assertions that we sometimes formulate on the basis of certain *interpretations* of dream or delusion-experiences.

The realistic correction of positivism consists in the identification of meaning with factual rêference. This conforms well with customary usage according to which a statement *means* a state of affairs; and is *true* if that state of affairs is fulfilled ('is real', 'exists'). This is the obvious grammar of 'meaning', 'truth', and 'reality.' Metaphysical problems cannot arise as long as we combine those definitions with the empiricist requirement that in order to be meaning*ful*, a statement must in principle be confirmable. The confirmation rules which formulate the connections between the evidential basis and the factual referents of statements are the metalinguistic correlate of those laws without which inference of specific unobserved or unobservable states of affairs would be impossible. Just which network of laws and existential assumptions will most adequately and parsimoniously serve for a comprehensive and predictively fruitful organization of the data çan of course not be settled in any *a priori* fashion. Nevertheless, only within the frame of a language that makes such a network possible can we legitimately assign probabilities to hypotheses on the basis of relevant evidence. The ('realistic') frame itself however cannot be justified by considerations of inductive probability. The adoption of this frame can be vindicated only by its fruitfulness for the purposes which it helps to fulfill. Like other principles which rationalists mistake for synthetic *a priori* presuppositions this is, from the viewpoint of logical reconstruction, a basic convention, capable only of pragmatic but not of cognitive justification.[2]

3. We are now ready to develop the implications of the just outlined clarified empirical realism for the mind-body problem. There are three demonstrably mistaken reductions by means of which monistic solutions have been attempted. There is firstly the crude and simple-minded identification of the stimulus-aspects with the mental qualities. Obviously we cannot say that a color sensation is identical with the radiation (of a certain intensity and frequency-pattern) which, under certain conditions merely elicits that sensation. Secondly, in our critique of phenomenalism we have also refuted the identification of physical bodies with complexes or configurations of elements of direct perception. Thirdly, the behavioristic identification of mental states with the responses (including linguistic utterances) of organisms is equally fallacious. It is of course granted that the confirmation of objective statements concerning 'physical' bodies is possible only on the basis of the evidence of direct experience. Similarly, intersubjectively meaningful statements concerning mental states are confirmable only on the basis of behavioral evidence. If we are to avoid the errors of phenomenalist reduction and quite generally of the negativism of orthodox positivism then all the relationships mentioned are not identities, but – at best – lawful (causal) connections between distinguishable states or events. The equivalence of statements about each pair of states or events can therefore be only of the empirical type. The precipitous assertion of a logical equivalence was of course based on the phenomenalistic claims of the explicit definability of the entities in one realm in terms of the entities of the corresponding other realm. This, as we have tried to point out, was completely unwarranted.

Curiously enough, the same sort of critique has been applied also to the identification of mental states with processes inside the organism, i.e., neurophysiological processes. It seemed quite incredible how a color sensation, a remembrance of things past, an act of thought concerning mathematical relations or a feeling of indignation, could in any sense whatsoever be 'the same' as some brain-process or other. Here again it was urged that the relation can be no other than, at best, that of a lawful correspondence or parallelism of simultaneous events. The many arguments in favor of this view are well known. One of the more important among these arguments contends that the attributes of mental states and events and the attributes of the corresponding neurophysiological processes are so different that the respective predicates characterizing each of the two types of processes can stand only in the relation of general (empirical) equivalence but never in that of a logical equivalence. Hume argued that statements of specific causal relations are synthetic *a posteriori* because alternatives are always conceivable without self-contradiction. Similarly, it is contended that a brain process which a future neurophysiology might characterize as of a definite type could conceivably be associated with a phenomenologically described immediate experience of a type radically different from that with which, as a matter of empirical regularity it is actually associated (say a sentiment of nostalgia). Eddington once argued that even the most detailed physiological and physical knowledge of the behavior and the nervous processes in the human organisms occurring on some November 11th at 11 A.M. in London could not possibly indicate to a Martian super-scientist unfamiliar with human history and unendowed with human sentiments that these events 'mean' a commemoration of the armistice. This fascinating argument, however, rests on two fallacies. Correcting these errors, it may be said, firstly, such a utopian knowledge of the neurophysiological processes would enable the Martian to derive the actual and potential verbal behavior of the Londoners; it would also enable him to reconstruct the physical account of the origin of the ritual (two minutes' silence, etc.) and thus to know, in principle, everything that can be known about those events in an intersubjective manner. Secondly, this can be achieved even if the Martian, because of the differences or limitations in his repertoire of emotions, cannot empathize, let alone share, the sentiments in question. A congenitally blind man, equipped with modern physical devices, could investigate not only the physics of colored surfaces, of light radiations reflected by them, etc., but also the (behavioristic) psychology of color sensation, discrimination, and perception (on the part of subjects equipped with eye-sight). Similarly, a Martian could *know* all *about* human feelings and emotions without having *knowledge of* them, i.e., without directly experiencing them or being *acquainted* with them by intuition or imagination.

Quite generally, one of the difficulties that are so frequently adduced in the critique of the identity-theory of mind and body rests on a confusion of *acquaintance* with *knowledge*. No one denies that the *image* of a brain, as perceived by a surgeon or as pictured in terms of an atomic model has totally different properties from a melody-as-heard or a sentiment-of-elation-as-actually-lived-through. But images or other directly experienced acts or data are not in and by themselves concepts. Knowledge proper is always conceptual. This insight is an important point of agreement between such otherwise divergent recent philosophers as Poincaré, Bergson, James, Dewey, Russell, Eddington, R. W. Sellars, C. I. Lewis, Schlick, Wittgenstein, and Carnap. What then is meant by 'conceptual knowledge'? What is meant by 'concept'? The best answer we can give today rests on a repudiation of psychologism and upon the results of pure semiotic. Concepts are meanings (intensions) of symbols constituted by the syntactical, semantical, and pragmatic rules which determine the relations of those symbols to one another, to their designata and to their evidential basis. The crucial question then concerns the conditions of the identity of concepts. What is the criterion for identity? We can safely follow Leibniz' *principium identitatis indiscernibilium*, here as elsewhere. If two terms are defined by the same set of rules, they are merely different symbols for the same meaning, they represent the same concept. Such synonymy however may arise in various ways. The most obvious and trivial case is that of explicit definition in which we arbitrarily stipulate the unrestricted mutual substitutibility of symbols. More interesting and more relevant for our problem is the case of epistemic (or 'systemic') synonymy. We may determine certain meanings uniquely by different definitions of the type known as 'definite descriptions' (Russell). Thus two explorers may unwittingly have observed the same mountain from different directions, and only after comparing notes come to realize that it was really identically the same mountain. This is a systemic identity in that it can be established only if the system of empirical geometry and optics is presupposed. Quite analogously, the identity of the morning star with the evening star (ever since Frege a much used example in logical analyses) is based on the recognition that one and the same trunk of world-lines (the four-dimensional representation of the planet Venus) is the object of reference of the two designations, referring to alternative segments of that trunk. Only within the system of Kepler's kinematics and of ordinary geometrical optics can this identity be explicated and warranted. This and the preceding example concerned the identity of things (continuants), or more precisely speaking, the identity of the designation of a name with the designatum (descriptum) of a description or else the identity of the descripta of two descriptions of thing-like entities. But quite similar considerations hold for concepts (predicates of various levels). The identity of the concept of 'electric current' defined by various definite descriptions such as those based on the magnetic, chemical or thermal indications can be defended against empiricist or operationist doubts only after a full-fledged system of electrodynamics enables us to *deduce* those various effects from a unitary theory of electricity, magnetism, electrolysis, and heat. Those doubts could of course never be removed with finality. We not only admit but would even emphasize the empirical or inductive basis which underlies all such identifications in the realm of factual knowledge. The only kind of identification that can be proved with finality is found in the purely formal sciences. Despite the fundamental difference between the situation in empirical knowledge and that in pure mathematics, there is an instructive structural analogy here. Two different infinite series, for example, may be used for the definition (unique description) of one and the same number, as, e.g., in the case of π. But such mathematical proofs to identity also presuppose a frame of concepts and postulates. Only within such a frame can we assert meaningfully and demonstrate validly the identity of the object of two descriptions. (A perfectly obvious illustration is the arithmetical identity of 2^3 with $\sqrt{64}$). The frame of arithmetic, i.e., the postulate system of Peano in the Frege-Russell interpretation, is logically or analytically valid. The situation is radically different in empirical geometry. For example, the identity of two points or line-segments characterized in different ways depends upon the factual adequacy of the geometrical postulates. The same holds, *a fortiori*, for the identifications in the natural sciences. Returning to an illustration previously introduced, the identification of the temperature of a gas with the mean kinetic energy of its molecules depends of course upon the truth of the molecular-statistical theory of heat. But if the truth of the theory is assumed, the strict identity of reference becomes a matter of logical deduction. Temperature as a macro-concept refers to the state of a body which is only more fully characterized by the theoretical description of its micro-structure. Once the theory is adopted it would make no sense to speak of the temperature as something distinct and different from that set of micro-conditions. Only the pictorial connotations of the world 'temperature' that remind us of thermometers or of the directly felt heat of a body seem to make the corresponding concept merely 'parallel' to that of molecular thermodynamics.

The logical principle that underlies our argument is, as indicated before, simply a variant of Leibniz' principle of identity. The meaning of a concept is determined, not by its pictorial connotations, but by the system of rules which implicitly defines that meaning. If two terms, no matter what words or symbols they are and no matter what pictorial appeals they may convey, are mutually substitutable for each other because they fulfill precisely the same functions in a system of rules, then they have the same meaning, they are the same concepts.

The application of these considerations to the mind-body problem must by now be fairly obvious. Relative to the 'molar' (or macro-) account given by behavioristic psychology, the neurophysiological account is a micro-description of the very same events and processes. The pictorial connotation of the two accounts are of course different, since the images attaching to the behavioristic terms represent stimulus-response situations, while the images connected with the neurophysiological language are apt to represent observations of nervous tissues. The notoriously greatest difficulty however arises here from the pictorial connotations of the mentalistic terms that owe their introduction to a third avenue of approach to the same processes – *introspection*. The qualities of direct awareness, the facts of stimuli and responses, the directly observable data of the neurophysiologist are of course not to be identified with one another. We have already warned against the fallacies involved here. But we contend that the designata of the mentalistic language are identical with the descripta of the behavioristic language, and that both are identical with the designata of the neurophysiological language. Utilizing the distinction suggested before, we may say that the factual reference of some of the terms in each of these different languages (or vocabularies) may be the same while only their evidential bases differ. A state of mind, conceived as an event in the spatio-temporal-causal structure of the world, may thus be characterized by concepts that are evidentially anchored in quite heterogeneous areas. It is this anchoring that gives the concepts their particular place in one or the other vocabulary. But if we are sure not to confuse their factual reference with their evidential base we may rightly say that they have the same meaning. This holds unless we countenance in principle unconfirmable assertions or unless the facts of psychology themselves force upon us an interactionistic dualism. The last proviso indicates the systemic nature of the proposed identifications. On the whole, I should think, the available evidence points with remarkable consistency in the direction of a system of psychology, psycho-physics and psychophysiology which provides for the monistic solution here outlined. But this is the empirical, the factual issue which philosophical analysis cannot decide and should not prejudge. We can do no more than clarify the logical structure of the problem and remove unfounded objections to the identity theory which perhaps owing to a 'failure of nerve' seems to have been temporarily eclipsed by a return to parallelism – if not even interactionism. The view we are proposing here should not be construed as a metaphysical doctrine. It again has merely some historical affinities with certain forms of epiphenomenalistic materialism, panpsychism, or the double-aspect or double-knowledge theories. If a label is wanted, then perhaps 'double-language-theory' is still the least misleading I can suggest. Within the conceptual system which fulfills the intersubjective confirmability condition and is at the same time the simplest account compatible with the accumulated facts of psychology, the terms of the behavioral-psychological and of the introspective language are (systemically) synonymous. If further factual discoveries should force upon us a radical revision of the conceptual system, then conceivably, this claim of synonymy may have to be modified or even abandoned. In the meantime it is well to remember that the tentative identifications which generally underlie synonymies of this type are among the most fruitful devices in the search for unifying explanations in the progress of science. The identification of light with a special kind of electromagnetic oscillating field; ferro-magnetism wih the spin of electrons; of heat with molecular motion; of chemical valences with certain dynamical features of the atoms; of the medium of inheritable traits with the gene-structure of the chromosomes, etc., are only some of the more noteworthy cases in point.

One last critical question requires discussion. The entire preceding argument, it may be argued, depends upon the presupposition that the vocabulary of introspection is part of an intersubjective language and thus really interpreted behavioristically. Introspective terms are then introduced on the evidential basis of linguistic responses and are therefore in any case logically on a par with those terms that have their basis in non-linguistic responses of the organisms. Thus, it may be urged, that the real difficulty of the mind-body problem has been avoided rather than resolved. This objection obviously implies that the language of introspection is to be taken as phenomenal, purely experiential and thus strictly subjective. My reply, very briefly, is this: The problem thus proposed is the epistemological question of the relation between the 'private' (if not solipsistic) language of data (phenomena) to the language of 'public', intersubjective 'constructs' (thing-concepts). It is highly questionable as to whether the idea of a phenomenal language in this sense can even be consistently maintained, let alone fully elaborated. But to those who cling to this '*Aufbau*'-phase of positivism I would offer the suggestion that there can be only a correspondence, but never a translation between the phenomenal language and the thing-language. If introspective descriptions are not to be taken as referring to events which are at least in principle confirmable by the much more indirect route of behavioral (or physiological) evidence then they are indeed severed from the language of intersubjective communication and doomed to solipsistic privacy. There is no bridge between such a private language and the language of science except one of isomorphic correspondence. Structurally the situation bears a certain resemblance to the one in the reconstruction of the rational numbers on the basis of the natural numbers. Certain ordered pairs of natural numbers are introduced, they define rational numbers. But the rational numbers (like $\frac{3}{1}$, for example), which represent integers (3 in this example), merely correspond to them, but are not identical with them. This isomorphism here consists not only in the one-to-one correspondence of certain elements of one realm to all elements in another, but in the one-to-one correspondence of the results of all arithmetical operations with corresponding elements. The analogy with the (however much more complex) field of epistemology lies in the isomorphism between certain statements in the phenomenal language and those in the intersubjective scientific language. As Carnap pointed out long ago [1932b] epistemological reconstruction may be attempted in either of two ways. The protocol-propositions may be part of the system of the scientific language or they are outside of it. In the latter case we must have some statements in the scientific language that correspond to the protocol propositions. The correspondence, however, must not be confused with what is traditionally called psycho-physical or psycho-physiological parallelism. Parallelism has always been a doctrine according to which two different types of processes or two aspects of one and the same process are related by laws of coexistence or contemporaneity. The correspondence of the protocol propositions with propositions of the intersubjective system is a purely formal relation which arises exclusively out of the constructive definitions, involving differences in Russellian type-levels, by means of which the terms of the physical language are supposedly constituted out of terms belonging to the language of data. This is the position a consistent phenomenalist must take. But the many difficulties of that position have impelled Carnap and other physicalists to replace it by the reconstruction on an intersubjective basis. The analogy of this procedure in mathematics is of course the axiomatic method by means of which the total system of numbers (real numbers) is introduced and no problems of the '*Aufbau*'-type are then encountered. If the protocol propositions, i.e., the names and predicates occurring in them, are part of the total symbolic system of the language of science, then we have here before us the sort of 'realistic' reconstruction which underlies the systemic identity view of mind and body.

Résumé: Logical Empiricism in its present phase possesses the logical tools for a reformulation of the identity or double-language view of the mental and the physical. As in so many other issues of philosophy, this solution represents an equilibrium that has been reached only after several oscillations toward untenable extreme positions. The identity proposed is neither the reductive definitional one of phenomenalism or of behaviorism, nor is it an identity that presupposes a metaphysical realism. It is rather the hypothetical identity of the referents of terms whose evidential bases are respectively: introspective, behavioral or physiological. It is granted that the relations between the evidential indicators (linguistic responses, overt behavior and the data of neuro-physiology) must be interpreted as empirical laws. But this does not in the least preclude the identity of the factual reference of the concepts which characterize the causal processes and events in terms of which the facts in each sphere of evidence may become explainable and predictable to an ever increasing extent. It is this hypothetical, systemic, referential identity that has been overlooked by those who retreated to a timid parallelism. The alleged difficulties of the identity view are mainly due to a confusion of pictorial appeals with cognitive meanings. A more adequate discussion of the points touched upon as well as of the many related questions and difficulties would of course require much more space than is available here.

NOTES

[1] This widely-held position may be traced back to Spinoza, and is represented in various metaphysical versions also by Leibniz (in a certain sense also by Kant), Schopenhauer, Fechner, Clifford, Riehl, Paulsen, the American monistic critical realists, especially R. W. Sellars, D. Drake, C. A. Strong; by one phase of B. Russell's thought; by R. Ruyer in France; by the psychologists Ebbinghaus, M. Prince, Warren; the Gestalt psychologists, especially Köhler and Koffka; by L. T. Troland, E. G. Boring, C. K. Ogden, and others.

[2] The realism of pure semantics and pragmatics is outlined in H. Feigl [1950b]; W. S. Sellars [1948b] and [1949]. An analysis of the problem of justification may be found in H. Feigl [1950a].

[3] Even these first two examples could be analyzed in terms of individual-concepts (unit-classes) instead of things, whose identity is under examination.

17. PHYSICALISM, UNITY OF SCIENCE AND THE FOUNDATIONS OF PSYCHOLOGY

[1963d]

The present essay attempts to analyze the meaning and to appraise the validity of the various theses of physicalism. Since I have had the privilege of discussing these issues with Carnap intensively and extensively on many occasions ever since 1926, I shall only rather briefly deal with some of his earlier views of this matter, and dwell more fully on recent modifications in his outlook – most of which do not exist in published form but are known to me from personal conversations. One of the purposes of the following observations then is to invite Carnap to react critically to my own suggestions and formulations on several basic points.

I shall begin by stating informally and relatively independently of Carnap's contributions what I consider to be the commonsense background of the doctrines of physicalism. After this introduction I shall go on to scrutinize some of the more strenuous and rigorous formulations of the theses of physicalism and of the unity of science. The first thesis of physicalism or the thesis of the unity of the language of science is essentially the proposal of a criterion of scientific meaningfulness in terms of intersubjective confirmability. 'Unity of science' in this first thesis means essentially a unity of the confirmation basis of all factually cognitive (i.e., non-analytic) statements of the natural and the social sciences. A corollary to this thesis is the assertion of the unity of scientific method. Despite the tremendous variety of special scientific techniques in the various disciplines, there are basic common features of the inductive and the hypothetico-deductive methods of establishing knowledge claims in all sciences. Contrasted with this first thesis which Carnap always regarded as well established by logical analysis, is the second thesis of 'unitary science' (as I shall call it for short) which Carnap considers only as a fruitful research program of the sciences, but by no means as sufficiently established by the progress of research to date. This second thesis of physicalism claims that the facts and laws of the natural and the social sciences can all be derived – at least in principle – from the theoretical assumptions of physics. We may formulate this second thesis as the belief in the possibility of a unitary explanatory system.

The first thesis of physicalism may in a preliminary and informal manner be construed as the principle of the primacy of sensory observation in the validation of the statements of empirical knowledge. Statements about the objects and events of the world – be they classified as physical, chemical, biological, psychological, sociological or historical – are generally confirmed (or disconfirmed) by means of sense perception. The data of observation can serve as confirming or disconfirming evidence, of course, only if certain principles of interpretation and of confirmation are presupposed. These principles, if explicitly stated, would tell us which data are evidence for which 'facts' and how strongly the data support the assertions of 'fact.' Issues of inductive logic aside, the important point here is simply this: Knowledge claims in common life, and certainly in science, are disregarded if they are so conceived as to be absolutely incapable of intersubjective check. No matter how strong our own subjective conviction or the force of 'self-evidence,' we would not consider a judgment justified if it could not conceivably be tested by others. Suppose, by way of an extreme example, someone claimed telepathic or chairvoyant intuition of distant events which are inaccessible to him through the normal channels of sense perception. He might be subjectively convinced that, e.g., at this very moment his old friend N. in Vienna (from whom he has not heard for 25 years) is writing a letter to him. The 'target' (object) of his telepathic or clairvoyant act is clearly something distinct from the act itself. The fact that the act itself occurred, he could report on the basis of introspection; and others could presumably confirm the occurrence of this act on the basis of their observations of his behavior (including, of course, his verbal utterances). But the very meaning of the target proposition, and not only its validation, involves reference to something beyond his direct experience and can be understood only within the frame of the customary commonsense conception of the spatio-temporal world. 'The actions of the Viennese friend N.' is a phrase whose meaning could never be explicated exclusively in terms of anybody's 'telegnostic' insight. I am not here stressing the obvious need of a check on the reliability of telegnostic acts. I am rather concerned to point out the even more obvious fact that any tests of such reliabilities ineluctably require independent checks of the truth of the proposition about the target. And in the commonly accepted frame the target proposition requires confirmation by the usual evidence of sensory perception. This commonly accepted frame is precisely that of intersubjective confirmability. I doubt very strongly that we could even coherently imagine a *reversal* of this situation, i.e., that statements about 'external' or 'distant' events could be established on the basis of telegnostic insight alone, and that the validity of ordinary sense perception be checked by comparison with the 'more basic' extrasensory perception. I do not wish to deny that among the logically conceivable universes there might be some in which this situation prevails. That is to say, I don't think that an outright logical inconsistency is involved in this however utterly fantastic conception. Nevertheless – and this is all our little thought experiment was to demonstrate – the primacy of sense perception for the interpretation and the establishment of intersubjectively meaningful and valid knowledge claims is an extremely fundamental feature of *our*-world-as-we-are-accustomed-to-conceive-it. The old empiricism of Locke and the new empiricism of Carnap ultimately rest on the conviction that sensory experiences are much more reliable indicators of 'external' states of affairs than are thoughts, images, wishes, sentiments or other 'non-sensory' data. Just as the reliability of intuition (normal 'hunches' or alleged paranormal gnostic acts) would have to be ascertained by the *normal* inductive methods, so the very meaning of statements, even if they were paranormally *arrived at*, can be understood only within the *normal* frame of a spatio-temporal world in which the to-be-known objects can be causally related to the sense organs of the knowing subjects.

These remarks on the case of paranormal knowledge claims were made only in order to illuminate the idea of the intersubjective frame. The controversial issues of extrasensory perception are not part of our theme here. But the point of our remarks applies *mutatis mutandis* to the claims of normal introspective knowledge. It is now fairly generally admitted by psychologists even of predominantly behavioristic orientation that introspection or self-observation is not to be discarded or disregarded but to be used with caution, i.e., with the proper safeguards with respect to its reliability. But behaviorism conceived as a 'psychology of the other one' has long been able to provide an account in intersubjective terms of 'subjectivity', 'privacy', 'the phenomenally given', and its observation by 'introspection',[1] while the precise logical form of such an account is still in dispute – and will be discussed in greater detail a little later – its main emphasis may again be construed in terms of the primacy of sensory perception as the confirmation basis of all intersubjective knowledge claims. Perhaps the best way to get clear about just what this emphasis implies, is to ask what this thesis excludes or denies. The answer seems very plain and simple to me: Physicalism thus understood excludes as scientifically meaningless sentences which could be confirmed *only* subjectively: Analytic philosophers, especially those practicing the methods of G. E. Moore and Wittgenstein,[2] have in various ways rather convincingly argued that the *absolute* privacy or subjectivity which for some philosophers constitutes *the* criterion of the mental is an idea begotten by confusions and pregnant with unresolvable perplexities. There are important passages in

Carnap's formulations of 1932 which anticipate in very compact form much of what has been dialectically (and partly independently) elaborated by the British analytic philosophers. As an illustration of this point consider the problem of the silent thinker. We are inclined to say: What goes on in his mind is 'private' to him, that is to say that *only he* knows what he is thinking about, and that the practical situation is such that no other person, no matter how closely he observes the behavior of the thinker could possibly find out. But practical impossibility of finding out was soon distinguished from absolute impossibility. It is generally admitted that the present state of scientific techniques (including kymographic registration of subvocal speech responses, electro-encephalograms, lie detectors, etc.) does not enable us to obtain highly reliable information about the thought contents even of non-silent thinkers. By 'transcending the limit', or illegitimately extending the ordinary usage of terms some philosophers have concluded that there are intersubjectively absolutely unknowable mental contents, qualia or 'raw feels.' These private states may be 'had', 'experienced', 'enjoyed' (or 'suffered'), 'lived through' by the individual subject, but are distinct from and something over and above the intersubjectively observable or discoverable behavioral or physiological processes and could for this reason not themselves be the objects of intersubjective knowledge. This way of conceiving the problem of 'other minds' leads notoriously to such unanswerable questions as: Are other persons' experienced raw feels (colors, sounds, smells, itches, tickles, etc.) quite similar to those with which I am familiar by direct acquaintance or could they be utterly different, i.e., systematically interchanged, such that the other person 'privately' experiences *green* when looking at ripe cherries and *red* when looking at grass? (The puzzle of the 'inverted spectrum.') Do other persons experience anything at all even if they behave in every respect as if they did? (One form of the solipsism puzzle.) Analytic philosophers have been alert to point out that these puzzles are quite similar to those that have been posed in connection with our knowledge of the past, or of physical objects. A historian might say: The present evidence is in every respect as if such and such had really happened in Egypt four thousand years ago. To which the skeptical philosopher responds with the query: Can you ever be sure that it really happened that way? Could not the laws of nature themselves have changed in the meantime, so that your inference might really be invalid? Or, might it not be that the world with all 'traces', 'remnants', 'memories', etc., sprang into existence only five minutes ago, and that therefore all 'history' is nothing but an illusion? Admittedly, no philosopher can raise such questions without shamefaced blushing. But the special hygiene or therapy for the prevention or elimination of these anomalies and perplexities needs to be made quite explicit by analytic philosophy. Since the resolution of these curious quandaries is by now fairly familiar, I shall state it here very succinctly and in my own way. Philosophers afflicted by extreme doubts (of the sort described) insist on direct verification as a means for the removal of such doubts. But in the nature of the case, as they themselves conceive it, direct verification is *ex hypothesi* excluded as either logically or physically (i.e. not just practically) impossible. The continued discussions of the 'other minds' puzzle have shown, I think, that 'having another person's experience' is a self-inconsistent phrase. Direct verification of knowledge-claims whose very conception allows only for indirect verification (confirmation) is thus to be recognized as a *logical* impossibility. But even logical impossibilities of this sort are inconsistencies within a special pre-supposed conceptual frame. Just because the very statement of the philosophical doubts inescapably (if even only implicitly or unwittingly) requires adoption of that frame, the puzzles posed can only be surreptitiously arrived at and thus reveal themselves as the gratuitous pseudoproblems as which they have always been diagnosed by positivists and analytic philosophers. To make this more specific, let us first consider the conceptual frame of historical knowledge. Common sense conceives of a sequence of events in temporal order which by and large (e.g., especially in the astronomical, geological, paleontological, and partly even in the human phases of history) is what it is, independently of whether it is or is not known. Embedded in this sequence of world events are cognizing human beings who make it their business to interpret evidence, i.e., to reconstruct past events or predict future ones, or to infer contemporaneous but not directly observed states of affairs. The philosophically uncorrupted historian does not deplore the impossibility of literally 'going back to the past' (as with a 'time machine' à la H. G. Wells). He knows implicitly that if he, the historian, was born in the twentieth century, he could not conceivably also have been born in the first century B. C., and thus might have been able to witness Caesar's assassination. What I am trying to point out is simply the fact that implicit in the ordinary conceptual frame of our cognitive activities is the distinction between direct and indirect verification. Although this distinction can be formulated more or less restrictively (down to an extremely limited notion of direct verification), and perhaps never quite sharply, it remains nevertheless a very clear and indispensable distinction.

Now, just as in the case of historical knowledge, we must distinguish between the evidence and that which is evidenced, so in the case of our knowledge of other minds it is imperative to distinguish between the behavioral symptoms and the mental states they symptomatize or indicate. Anxious to avoid pseudoproblems of the type mentioned before, Carnap, and with him many others in the early radical phase of logical positivism, maintained that psychological statements describe *nothing but* actual (or possible) behavior. Psychological concepts expressed by mentalistic terms were conceived as logical constructions erected on a basis of purely behavioral concepts. This was plausible enough in all cases except that of human introspection. If, for example, I report that I feel elated, grieved, or that I hear a ringing sound, there are – to be sure – a great number of behavioral symptoms which may indicate with varying degrees of reliability that I am actually in the mental state described by myself in introspective terms. The very utterance of sentences containing those introspective terms is an important symptom and may, depending on further circumstances, be taken as fairly reliable evidence for the presence of the corresponding mental state. But in experiencing or having (living through, '*erleben*') the mental state I am in the privileged position of being able directly to confront a statement (no matter whether uttered by myself or by some other person) with the pertinent mental state.[3] The asymmetry that this privilege involves is clearly borne out by the fact that while for everybody else it would be possible to have the usual (empirical) doubts as to whether I really hear a ringing in my left ear, or as to whether I merely behave as if I did, I myself could not without a special kind of absurdity say: 'I am not sure whether I experienced a ringing sound.' I might of course be in doubt as to whether it is just a buzzing in my ear or whether I hear a distant squad car siren or a telephone bell. I might also doubt as to whether 'ringing' rather than 'whistling', 'hissing', or 'tinkling' would best describe the sound. But I would not doubt that it is a sound rather than a smell that I am experiencing. I could not possibly doubt the occurrence of the experience itself while it lasts. The questions 'how do I know?' or 'on the basis of what evidence do I believe' that I have that experience seem utterly inappropriate. But the question how some other person knows, i.e., on the grounds of what evidence he could infer that I hear a ringing sound, is perfectly appropriate.

So much then by way of a characterization of the difference between direct and indirect verification of knowledge claims. Directly verifiable introspective reports about immediately experienced states utilize phenomenal predicates such as 'hot', 'cold', 'loud', 'soft', 'red', 'green', etc., i.e., without any attempt at interpretation as in the physical mode (did I hear a telephone bell?) or in the psychological mode (did I feel cold because I was 'chilled' by fears or anxieties?)

What then is the position of physicalism regarding the directly verifiable knowledge claims of self-observation? The discussions of the last few decades should certainly tend to moderate and modify the originally rather 'crass,' 'materialistic' pronouncements of behaviorists and physicalists. The epistemological arguments against classical materialism have been directed also against the more sophisticated versions of modern physicalism. It is urged that the very knowledge of the 'physical' behavior of organisms rests on a basis of evidence which when analyzed sufficiently far down to the immediately certifiable must be expressed in phenomenal terms. A more defensible form of physicalism must therefore render account of the epistemic primacy of immediate experience and of the difference between direct verification (confrontation with immediate experience) and indirect verification (inference with the help of laws or statistical rules).

Before sketching some possible forms of such a more liberal physicalism, let us first ask once more just what the first thesis of physicalism opposes or excludes. Perhaps a good way to get to the heart of the matter is to consider Carnap's critique of the analogical inference of mental states in other persons (in 'Psychologie in physikalischer Sprache', [1932a, pp. 118ff]). I shall put the illustration in my own way: Suppose we compare the entirely unproblematic inference of the presence of brains in as yet unopened skulls, with the (allegedly) quite different and philosophically problematic inference of mental states associated with the behavior and/or brain states of other persons. It is of course admitted that inductive or analogical inference is essential and indispensable in the establishment of empirical knowledge claims. Thus we might formulate the inference in simple symbols (S_1 = skull of a first person, S_2 = skull of a second person, B_1 and B_2 the corresponding brains):

$$S_1 : B_1 = S_2 : \mathrm{B}_2$$

The italicized symbols stand for observed facts, i.e., both the skull and the brain inside it have been observed in the case of the first person, but only the skull of the second person has been observed, it has not yet been examined as to its internal contents. The inference of B_2, i.e., the presence of a brain in S_2, is the more probable the more similarities are noted between S_1 and S_2, or the more cases of such similarities we have observed for a large number of skulls containing brains (as revealed by opening them). This is clearly the familiar case of empirical inference.[4] Now consider by way of contrast the inference of mental states on the basis of observed behavior or brain states. Let B_I stand for *my* molar behavior and/or brain states; M_I for my mental states as I can describe them in phenomenal terms on the basis of introspection; B_{II}, and M_{II}, corresponding states of a second person. The analogical inference might again be symbolized by

$$B_I : M_I = B_{II} : M_{II}$$

Early physicalism as represented by Carnap in 1932 declared this inference as illegitimate for the reason that M_{II} (the mental state of the other person) is not independently certifiable as is B_2 (the brain of the other person) in the previous example. In other words, Carnap maintained that inductive inferences are legitimate, in the sense of meaningful or at all permissible (not necessarily in the sense of reliable) if and only if the inferred conclusion is capable of independent test. Obviously, much in this argument will hinge on just what one will admit as 'independent test.' If one insists on *direct* independent test, then a very large class of inferences would be ruled out. Carnap saw this clearly in a later phase of his physicalism (in 'Logical Foundations of the Unity of Science', [1938], reprinted in [Feigl and Sellars, 1949, p. 419f]). The example of the electric charge on a raindrop which far away from any observer falls into the ocean demonstrates clearly that with a liberalized formulation of the meaning criterion, statements about such only *practically* unconfirmable but theoretically confirmable statements must be admitted as perfectly meaningful. Or, consider as another example the impossibility for *me* to verify directly the state of my own body when in total anaesthesia or after my death.[5]

Returning to the parallel of the impossibility of direct verification in history and in the psychology of 'other minds' we might say: It is impossible directly to verify that the Grand Canyon was formed by erosion; but nobody, unless afflicted by philosophical doubt, would question the legitimacy (i.e., the meaningfulness, not necessarily the reliability) of the inference that at a time long before there were human beings present to observe the formation of the canyon, erosion was the main factor in the process.

Similarly, we might say that, while *I cannot* verify directly the presence of a feeling of elation in my friend, I can legitimately infer it on the basis of his 'radiant' expression, lively behavior, speech, etc. or – more reliably – on the basis of various psychological tests; and this inference *is* legitimate precisely because there are a number of independent avenues for its confirmation. But in the case of the inference of the mental state, i.e., the feeling of elation, as something distinct from actual or possible behavior or brain processes, we could say that my friend himself is in a privileged position and can verify its presence directly, independently of any behavioral or neurophysiological evidence that *others* could marshal. The reason why physicalism in its early form did not pay any attention to direct verification by the person concerned was of course the supposedly 'purely subjective' character of such verifications. Behavioral and physiological tests could presumably be carried out by any observer properly equipped with the instruments and techniques of observation and experimentation. But if there were a domain of immediate experience radically private and secluded, i.e., absolutely isolated and insulated, hence completely inaccessible even through the most indirect routes to test by other individuals, then by this very character such immediate experiences could never be or become a subject matter for *science*. This is of course merely an obvious analytic consequence of the definition of scientific knowledge which insists on intersubjective testability. Before we examine some of the philosophical implications of this definition of science, it may be well to remember that mental states which are absolutely private in the sense just indicated would also be precluded from behavioral manifestation of any sort. Neither facial expression, nor verbal report, nor even the intonation of verbal utterances could in any lawful way be connected with these private states; for if they were, these behavioral symptoms could be used as a confirmation base for statements about those (in this case not '*absolutely*') private experiences.[6] Philosophers – some as early as the sophist Gorgias and the Cyrenaics, others as recent as C. I. Lewis [1929] – who raised the issue of the inverted spectrum and pursued its consequences to the bitter end, must however, have had absolute, unmitigated privacy in mind. That is to say, their assertion that person B might see the grass 'really' red while person A sees it 'really' green, is understood in such a way, that neither color vision tests, nor any other behavioral, neuro-anatomical or neurophysiological evidence would reveal the discrepancy which is assumed to exist exclusively in the pure 'qualia' of the direct experience of the two persons.

I would urge that these assertions, while extremely fanciful, if not absolutely groundless in the light of the normal principles of commonsense and scientific inference, are nevertheless, *not* absolutely meaningless. I am also inclined to think that the assertion of the survival of a totally isolated stream of experience after bodily death[7] makes perfectly good sense – in a sense of 'sense' which must then of course be classified as purely subjective, and which *ex hypothesi* could not be the sort of intersubjective meaning which must be attributed to typically occultist hypotheses according to which the 'surviving mind' ('soul', 'psychoid', etc.) can manifest itself in alleged mediumistic physical phenomena – such as, e.g., giving messages by plucking piano strings or speaking through a living human 'medium.' Now, while I am personally utterly skeptical about 'survival' in either form, I have used these excursions into the domain of scientifically 'taboo' ideas, simply to point out the difference between two proposals for the delimitation of factual meaningfulness. Subjective confirmability is clearly the wider and more tolerant proposal; intersubjective confirmability is more restrictive in that it excludes all those assertions which could be checked by only one subject and are 'in principle' unconfirmable to others. When scientists repudiate what they call (often rather loosely) 'mysticism' or 'supernaturalism', I think they have primarily reference to assertions which are not open to public test. The positivist scientists and naturalistic philosophers of various types suspect that knowledge claims of this sort are illegitimate because (a) they may be no more than expressions of emotions, and thus only because of the grammatical form of the sentences confused with genuinely cognitive assertions, and/or (b) while they may have the modest cognitive content of autobiographical, introspective reports, they pretend to knowledge of something over and above the experience itself (religious, mystical, etc.); but this 'something more', by its very conception is in principle removed from independent intersubjective check, and thus the suspicion remains that the 'apprehension of an Absolute' in mystical experience – even if this mystical experience be similar for many individuals – may well be an illusion of the sort that can be produced by hypnosis or autosuggestion. Physicalism is the explicit formal expression of this scientific attitude.

In the interest of the very clarity advocated by analytic philosophers and logical empiricists we must now ask two searching questions: (1) What is the logical status and the justification for the physicalistic criterion of factual meaningfulness? (2) Is, as its critics often maintain, logical empiricism (and physicalism) merely one form of metaphysics – namely a rather negativistic one?

It is today generally agreed among logical empiricists that the criterion of factual meaningfulness is to be construed as a norm proposed for the purpose of avoiding unanswerable questions. Just as certain purely syntactical rules, such as Russell's rule of types, are designed to eliminate logical antinomies, so the *additional* requirement of confirmability-in-principle eliminates pseudo-problems, i.e., problems which by their very construction can be recognized as absolutely insoluble. By regarding the meaning criterion as a proposal rather than as a proposition it becomes impossible to subject it to its own jurisdiction or to ask whether it is true or false. What *is* true is the tautology that in a language conforming to the meaning criterion, unanswerable questions (of pretended *factual intent*) can not even be asked, let alone answered responsibly. For example, if absolute space is so conceived as to permit not even incompletely or indirectly confirmable answers to questions regarding the positions or motions of observable bodies with respect to that absolute space, then sentences which embody attempted answers to such questions are absolutely meaningless on the proposed criterion. A conservative way of putting all this might be: The rules of logical syntax together with the requirements of confirmability-in-principle form at least a necessary (but possibly not sufficient) condition for the factual meaningfulness of linguistic sign combinations. I am inclined to think that even metaphysicans or theologians cannot pursue their own purposes without some such delimitation of sense from non-sense.[8]

The justification for the adoption of such criteria of meaning can of course be only a practical one. If we wish to avoid the agonies and perplexities of problems which through our own making are unresolvable, then a criterion in terms of confirmability will have the desired salutary effect.[9] This is to say that the vindication of the adoption of and adherence to, a meaning criterion must refer to the purposes one aims at in using the language of cognition. Now, since the aim of the *scientific* enterprise is generally so conceived as to provide knowledge which is susceptible to *inter-subjective* test, it is clear that *purely* private, only subjectively confirmable knowledge claims are to be ruled out, i.e., declared as scientifically meaningless. As we have tried to indicate by our illustration above, subjective confirmability may be fulfilled even where intersubjective confirmability is absent. I may be able to confirm a strange continuation of my own stream of thoughts and emotions in the complete absence of sense data concerning the extradermal or intradermal world. This might lead me to think, once this sort of thing had happened, that this part of my self (constituted by thoughts, images, remembrances and sentiments) had survived my 'bodily' death. And, *ex hypothesi*, no other person could possibly confirm this. Here then is a fork in the philosophical road: Which of the two meaning criteria – subjective or intersubjective – are we to adopt? Why is it that scientifically oriented thinkers strongly oppose adoption of a criterion of meaning on a purely subjective basis? Why is it that they insist on the *public* character of knowledge as a defining (necessary) condition of the scientific enterprise? This insistence of scientifically-minded thinkers seems to rest on the belief that there is nothing in heaven or on earth (or even beyond both) that could not possibly be known, i.e., there are no assertions about reality which could not conceivably be confirmed or disconfirmed on the basis of sense perception. Translating from the epistemological into the cosmological idiom, this amounts to the thesis that whatever there is in any shape or form, things, events, states, anywhere, at any time, 'inorganic', 'organic', 'mental', 'social', etc., can be causally related – even if only very indirectly by complicated chains – to the sense organs of human organisms. In other words, there is nothing absolutely isolated, causally completely unrelated to those parts of the world which form the stimuli of sense perception. Put in this way, this appears like a very bold belief about the nature of the universe. It is this belief which metaphysicians, by turning the tables on the logical empiricists, have called the 'metaphysics of positivism.' Again and again have we heard the criticism that the positivists rule out extra-scientific knowledge by declaring the scientific method the *only* method by which knowledge claims can be established, and that by 'arbitrary' decree any other sources or methods of knowledge are ruled out of court.

And yet, metaphysicians or theologians often defend the 'rationality' of their beliefs in the existence of extra-natural or non-physical entities by arguments of a typically inductive or hypothetico-deductive flavor. Many have come to realize that appeal to logic (purely deductive or dialectical) or to self-evidence simply will not do. But if they base their arguments on empirical evidence – even if 'empirical' covers for them a much wider range than sensory experience – they will have to face the question as to whether they can justify the positing of transcendent entities in the manner in which, for example, the assumptions of atomic physics can be justified by the hypothetico-deductive method as applied to the data of experimental physics and chemistry. In this day and age it is obvious that we have extremely 'good reasons' for the assumptions of the atomic theory, and that these assumptions cannot possibly be interpreted as merely 'shorthand expressions' for the regularities on the macrolevel of observed phenomena [cf. Feigl, 1950b; Kneale, 1953; Beck, 1950; Braithwaite, 1953]. The case of the mystic (theologian or metaphysician) is plausible only as long as naturalistic explanations of religious and mystical experience are not available. 'Naturalistic explanations' here refers to the type of account given in various psychologies of religion – Jamesian, Freudian, etc. Even if many of the striking features of religious and mystical experience have not been explained in detail, the majority of psychologists are quite confident that the available evidence on the whole points in the direction of explanations within the present frame of psychological (and cultural-social) regularities and will not require the introduction of fundamentally different categories. Whether one formulates the principles underlying this scientific confidence as an aspect of the policy of induction and theory construction, or as an aspect of the rules of simplicity or parsimony, or simply as norms of giving 'good reasons', they are in any case characteristic of the sort of conservatism without which scientific research would be unprotected against the dangers of groundless and limitless speculation. There is, one may hope, less reason to fear the opposite danger, viz., that scientific conservatism may degenerate into a rigidly dogmatic retention of a given frame of explanation. The tremendous and often revolutionary advances of science since the Renaissance, and especially in our century, bear ample testimony to the flexibility and the highly imaginative and ingenious character of scientific theorizing. The notorious difficulties of an exact delimitation of the concepts of the 'natural' or the 'physical' reflect the often surprising expansions of scientific concept formation and theory construction.

Returning to the case of mystical experience, we may say that the present prevalent scientific attitude acknowledges the occurrence of these unusual experiences, but doubts the interpretation in terms of transcendent entities that the mystics themselves (or some theologians or metaphysicians) impose upon them. I am inclined to think that the scientific attitude should be very different (and perhaps will be very different in the near future) with respect to the phenomena of parapsychology. If it were fully established that the phenomena of extrasensory perception, i.e., clairvoyance and telepathy, and perhaps even precognition and psychokinesis, do not result from experimental or statistical errors (not to mention self-deception or outright fraud), then our conception of the basic laws of nature may well have to be revised at least in some essential aspects. Curious 'actions at a distance' – spatial as well as temporal, and – conceivably though by no means necessarily – some alterations in our basic psychophysiological assumptions might have to be introduced. The only alternative would seem to be the assumption of some cosmically pre-established sets of 'spurious' coincidences and correlations – an assumption which in any other field governed by statistical evidence would seem objectionably ad hoc and thus bound to obstruct scientific progess.

The foregoing considerations are to call attention to (1) the flexibility or 'openness' of the concept of the 'physical', and consequently (2) the need to re-examine the two theses of physicalism in their relations to one another. If 'physical' means the sort of entities, no matter how inferential or hypothetical, whose assumption can be justified on the basis of *sensory* confirmation, then the first thesis of physicalism does imply the assertion of a certain generic feature of the universe and is thus clearly not a truth of pure logic (pure syntax or pure semantics). Carnap himself pointed out [10] the factual nature of intersubjectivity and its analogy to the intersensory character of ordinary perceptual objects. In the latter case it is a matter of empirical fact that certain objects are accessible tactually as well as visually, and that – epistemologically speaking – the existence of single objects is predicated upon the regular concomitance[11] of sensory data, or 'appearances' in the various modalities. Similarly, the assertion that everything there is in our world is in principle susceptible to at least indirect confirmation by sensory experience of *any* human observer, not only amounts to an assumption about the universe, but also specifies at least very sketchily certain features of the laws of the universe. These general features consist in the assumption of a spatio-temporal-causal network in which the knowing subjects are embedded as genuine parts. This is a thesis common to most forms of philosophical naturalism – a thesis, which despite its vagueness, has certain implications for the second thesis of physicalism. This second thesis, it will be recalled asserts that scientific theories attain progressively more and more unifying syntheses of their subject matter, and that they tend toward a unitary set of explanatory principles. The thesis furthermore asserts that these explanatory principles will be (note the unavoidable vagueness!) somewhat like the most comprehensive postulates of present-day theoretical physics. The progress of physics in the last few centuries, the great syntheses achieved successively by classical mechanics, classical electromagnetics, the atomic theory, the theory of relativity and quantum mechanics, the prospect of incorporating the bio-psychological sciences (possibly with the help of cybernetics) into an ever more adequate grand scheme – these have been some of the encouraging factors in the various stages of monistic philosophies. The unitary-science thesis may be regarded as a twentieth century sequel and incisive revision of eighteenth century materialism. Some of the culturally understandable motivations may be similar, but present-day physicalism displays a much greater logical and epistemological sophistication.

The second thesis of physicalism, in asserting that the facts and laws of mental life can be given a 'physical' explanation, while not strictly implied by the first thesis, is at least rendered rather plausible. If there is nothing in the realm of mental phenomena that is in principle excluded from sensory confirmation, then all mental phenomena must in some way be part of the nomological network (the causal, or at least statistical order) which alone makes indirect confirmation possible. The notions of '$physical_c$', i.e., an object in principle connectible with the sensory confirmation basis, and '$physical_e$', i.e., object of explanation in terms of the basic laws of nature, are thus seen to be much more closely related than Carnap's original sharp distinction of the two theses of physicalism suggested.

Before we return to the epistemological analysis of physicalism let us try to assess its cosmological aspects. If the term 'physical' designates the objects of the laws and theoretical assumptions of physics, then obviously the first question to be asked is: of *which* physics? It should scarcely be necessary here to review the drastic and pervasive changes wrought by the successive revolutions in theory construction mentioned above. The concept '*physical*$_e$' has expanded tremendously beyond the original identification with the 'mechanical' so characteristic of the natural philosophy of the seventeenth and eighteenth centuries. The field theories of the nineteenth century, the revisions due to the relativity and quantum theories in our century have affected profoundly our concepts of space, time, substance and causality. These alterations were required precisely because a larger range of observable phenomena was to be encompassed by increasingly comprehensive and increasingly unified systems of explanation. There are in present-day physics principles of continuity (fields), of discontinuity (quanta of energy, as in the interaction of fields with particles, and of particles with one another, etc.); important new relations of spatial and temporal magnitudes (involving an upper limit for the propagation of causal influences – according to the theory of relativity); the mutual transformability of radiation-energy and basic particles; and principles of organization, fundamental for the formation and structure of atoms and molecules (as formulated in W. Pauli's exclusion principle).[12] It would seem hazardous to assume that the concept of 'physical$_e$' will in the future undergo no further radical alterations or expansions. The most cautions definition one might suggest would be a dated one ('physical' in terms of the respective scientific theories of, e.g., 1687, 1900, 1905, 1925, 1958, etc.). The only alternative to this sort of definition would be the much vaguer, but nevertheless more fruitful one of defining 'physical$_e$' simply as the object of any more or less comprehensive explanatory system whose concepts are defined implicitly by a set of postulates, partially interpreted in terms of a sensory confirmation basis. This definition of 'physical$_e$' in terms of the hypothetico-deductive procedure with a basis in intersubjectively testable observation propositions recommends itself in that it reflects (a) the elasticity and openness of the explanatory concepts of advancing science, and (b) the 'objectivity' which has always been a prime desideratum (and often an achieved virtue) of the natural sciences.

The openness of the concept 'physical$_e$' which frees the second thesis of physicalism from dogmatic dependence upon a given stage of physical theories allows for a non-metaphysical interpretation of emergent novelty and emergent evolution. The important point in the notion of emergence is not so much that there are in the course of the history of the universe completely new entities – qualities, relations, structures, events or processes – but rather that a certain set of concepts and laws sufficient for the explanation and prediction of a given range of phenomena may not be sufficient for the explanation of a wider range of phenomena. The triumphant achievements of the mechanistic world view – until about the middle of the nineteenth century – are responsible for the confidence (characteristic also of the much more speculative views of the ancient atomists, as well as of modern atomic theory at least since the days of Rutherford and Bohr) that the striking novelties connected with high complexity of structures are derivable with the help of mathematical-geometrical devices only. That is to say, that no special *physical* composition laws are required for the explanation of the behavior of structures of higher degrees of complexity. Geometry plays the role of a 'silent partner,' very much like logic and arithmetic do, in these derivations. This is of course not to deny the empirical character of applied geometry; it is merely to emphasize its subsidiary role of being 'presupposed.' Modern atomic theory has its own peculiar *physical* composition rules, especially the Pauli principle. The behavior of electrons in the context of atomic structure cannot be derived from the laws of motion of free electrons (as in cathode or β-rays); additional *physical* principles are needed. But again, once the basic laws of atomic structure and dynamics have been ascertained in the simpler cases of atoms like hydrogen, helium and lithium, the rest of the periodic table as well as the structure of molecules are found to be derivable from those laws *more geometrico*. Whitehead's suggestion that the behavior of electrons within living organisms may be fundamentally different from that in inorganic compounds can of course not be refuted *a priori*. It is entirely a matter of empirical research to find out how broad or complex a basis of evidence is needed in order to permit us to glean those laws which then applied by purely mathematical-geometrical computations, will also be sufficient for a range of phenomena of greater breadth or complexity. It must be admitted that it is conceivable that as we advance in the study of structures of higher and higher complexity, there might never be an end to the emendation of the laws of nature. In view of the difficulties of current theory perhaps something of this sort may even be expected as research penetrates to deeper levels of the structure of matter. Nevertheless, according to the prevalent – and perhaps somewhat sanguine – view of many scientists, nature while extremely complex is not hopelessly difficult to unravel into basic regularities. This is no doubt what Einstein means by the famous epigram (exhibited at Princeton): "God is sophisticated, but he is not malicious." Metaphysicians will insist that this optimism of the scientists is really an 'act of faith', and as such as unjustifiable on empirical grounds as are transcendent theological beliefs. But empiricists need not be disturbed over this. 'Belief in the ultimate simplicity of nature' is not a logical presupposition of science, except in the trivial and tautological sense that – to the extent to which science achieves, on some level of analysis (i.e., in terms of a certain set of variables), an adequate and unitary explanatory system, nature has – on that level – as much simplicity as is reflected in the given explanatory system. Empiricists are therefore perfectly justified in viewing the 'principle of simplicity' as a guiding maxim of research, as part and parcel of the policy of the inductive and hypothetico-deductive procedures of science, rather than as a metaphysical postulate [Pap, 1949, 1953; Feigl, 1954]. Any speculation regarding the 'ultimate', 'rock bottom' structure of nature is bound to be an utterly irresponsible piece of dogmatism. Elevating the best-established laws of a given stage of science to the rank of a rigid philosophical *a priori* – as, e.g., Kant did with the principles of Newtonian physics – is not only unjustifiable, it is also pernicious in that it is apt to impede the progress of research.

The implications of the preceding excursion for the two theses of physicalism will now be summarized and discussed:

The decision to rule out as scientifically meaningless statements which are not even indirectly confirmable intersubjectively on a sensory basis, and the confidence that this decision will not exclude anything in existence from the realm of science, reflects the conviction that whatever is subjectively verifiable is in principle also intersubjectively confirmable. This conviction – if it is not to be a 'metaphysical presupposition' – must therefore be so construed as to fall under the jurisdiction of the principles of induction. In plain and ordinary words this means that naturalists claim to have good empirical reasons for their belief that nothing is inaccessible to study by the scientific method. If the first thesis of physicalism is to formulate more than a tautological consequence of a definition of 'scientific method', it would indeed seem to be the expression of an inductively grounded assumption, and would thus be in principle subject to refutation.

Inductive validation of the first thesis of physicalism is, however, not the simple straightforward sort of thing with which we are so familiar in common life and in the empirical sciences. Ordinary inductive justification occurs within a frame of spatio-temporal-nomological structures which are usually unquestioningly assumed although they are not in principle unquestionable. The adoption of this frame can be practically justified (vindicated) by its entailed consequences. This vindication is in part deductive, in that the demonstration of the entailed consequences is a matter of purely logical derivation. But the adoption and retention of a certain frame contains an irreducibly inductive element, the 'by their fruits ye shall know them' maxim. This maxim clearly refers to continued appraisal in the light of expected and forthcoming 'fruits.' This may be illustrated by a reflection upon the adoption of the traditional and customary spatio-temporal (4-dimensional) frame of physical description. In the usual concerns of everyday life and of the experimental sciences the employment of this frame is a matter of unquestioned practice. Innumerable factual questions of various degrees of specificity or generality are formulated as well settled within this frame. But there have been occasions when the frame itself was subjected to questions – as, e.g., in the topological modifications required for the Einsteinian cosmology, or in the suggestions regarding a genuinely spatial fifth dimension as in a now largely forgotten theory of Kaluza's. – I am concerned to point out that the 'empirical method' is a matter of various levels. Simple questions on the level of observation are so convincingly decidable just because the frame [cf. Carnap, 1950a] within which they are asked is accepted and not at all questioned in this context. When it comes to the acceptance of a certain set of natural laws or hypotheses, simple observations or measurements alone will not be decisive, because scientific laws and theoretical assumptions constitute confirmation rules and thus furnish the frame for the decision of more specific descriptive questions. But the acceptance or rejection of theoretical assumptions themselves is regulated by frame principles of still more fundamental and generic significance – such as the norms of factual meaningfulness and validation. While these norms are best construed as prescriptive proposals rather than as descriptive propositions, they differ from the rules of deductive logic in that they are not matters of merely formal or intralinguistic relevance, but do reflect certain basic and pervasive features of our world. It is, however, impossible to describe these features directly as one would describe specific facts or regularities within the world. Wittgenstein (at least in the mood of his *Tractatus*) would have said that these features 'show forth' in the successful application of a language, in the adequacy of a certain type of conceptual system. To say of space that it is three-dimensional does not make sense in the same way in which it makes sense to say of a specific scrap of paper that it is triangular. Similarly, to say that whatever there is in the world is in principle (no matter how indirectly) causally connectible with human sense organs does not make sense in the same way in which it makes sense to say, e.g., that lights of a specific minimum luminosity and a specified maximum distance are visible to persons with normal vision. This becomes clear by reflection upon the conditions of confirmation or disconfirmation of the respective assertions. Specific empirical assertions (such as those about triangularity, visibility, or the like) may be confirmed or disconfirmed by clearly circumscribed types of evidence; whereas the categorical or generic statements about the tridimensionality of space or about the causal accessibility of everything in the world, while not *a priori* in the analytic manner, might nevertheless be regarded as *a priori* pragmatically or functionally.[13] That is to say, the adoption of certain norms of meaningfulness and of validation is not a matter of arbitrary decision, but is guided by the consequences of their adoption as these consequences are disclosed in the management of cognition in a world 'we never made.'

The first thesis of physicalism may then be regarded as a new formulation of the principles of empiricism: (1) Statements are to be regarded as *scientifically meaningful* only if they are in principle intersubjectively confirmable or disconfirmable. If a statement, by the very interpretation imposed upon it, is in principle incapable even of the most indirect sort of intersubjective test, then though it may have meaning of a purely logical sort, or may be significant in that it carries pictorial, emotional or motivative appeals, or may even be testable in an *exclusively* subjective manner, it cannot be accepted as an answer to a *scientific* question. The phrase 'in principle intersubjectively confirmable or disconfirmable' should be understood in the most liberal manner. The sort of indirect testing of assertions here allowed for includes of course the testing of only partially interpreted postulate systems. It countenances as scientifically meaningful, statements about the most remote, the most intricately concealed or difficult to disentangle states of affairs. It includes statements about unique and unrepeatable occurrences, if only they are of a type that places them within the spatio-temporal-nomological net which itself has an intersubjective confirmation base. (2) Statements are to be accepted as scientifically valid only if they are sufficiently highly confirmed by in principle intersubjectively available evidence. The precise meaning of 'sufficiently highly confirmed', as well as the exact explication of 'degree of confirmation', 'inductive probability', or 'evidential support' need not be discussed in the present context.

The preceding formulations render briefly but, I trust, sufficiently adequately Carnap's present liberal empiricist views. The early forms of rational reconstruction – both the phenomenalist reduction of the *Logischer Aufbau der Welt*, and the radical physicalist reduction of the 'Unity of Science' phase of the early thirties have been completely abandoned. Epistemological analysis no longer consists in the retracing of logical constructions to elementary concepts on the phenomenal or on the macro-behavioral ground level. Statements concerning 'physical events' are *not translatable* into statements about sense data (actual or possible). Statements about mental events are *not translatable* into statements about (actual or possible) overt behavior.[14] In both cases epistemological analysis consists in making explicit the conceptual structure of the *nomological* relations between the data of observation and posited events or processes for which the data serve as evidence. The meaning of statements (at least in one very important sense of 'meaning') is to be identified with their factual reference, and not with their evidential basis. The slogans of early logical positivism and of ultra-operationism about meaning and verification – while helpful in the repudiation of transcendent metaphysics – despite their imprecision were far too restrictive to do justice to the actual conceptual structure of knowledge. According to Carnap's present view the essential requirements of empiricism are fulfilled if the nomological net which implicitly defines the concepts of science is tied in a sufficient number of points to concepts of the observation base. While it is useful, and perhaps indispensable for a logical analysis of unfinished and developing science to distinguish between observation language and theoretical language, it is equally instructive to reflect the more stabilized parts of science or the tentatively anticipated parts of highly unified[15] science in a reconstruction in which all terms belong to the same language. In his important essay 'Über Protokollsätze' [1932b, pp. 215–228], Carnap discussed the relative advantages and disadvantages of each form of reconstruction. Even in a unitary-language reconstruction there will of course still be a distinction between observation terms (terms designating immediately experienceable qualities or relations) and other terms designating unobservables. It is fairly generally agreed [cf. Carnap, 1939; Hempel, 1952; Braithwaite, 1953] now that the unobservables (i.e., concepts designating unobservables) are implicitly defined by the partially interpreted postulate system which – together with its explicit definitions and derived theorems – formulates the nomological net. The observables usually occur only as highly derived terms of the system (although there is no *a priori* reason why some of them might not figure as primitives). The meaning of the unobservables (theoretical concepts of physics, for example) is thus specified through their place in the network; or – if a metaphor be permitted – the meaning of the unobservables is fixed by 'triangulation in logical space' from points on the observation base. The richer the various connections in the net, the more fully and definitely can the meaning of each concept be specified with respect to other concepts of the net.

If the unitary language is strictly intersubjective, the characterization of the observables themselves can be achieved internally, i.e., by reference to their peculiar locations in the nomological net. The lessons of behaviorism point the way here. (We shall criticize certain behaviorist reductive fallacies a little later.) Subjectivistic and phenomenalistic epistemologies have traditionally employed some principle of acquaintance or the notion of ostensive definition in this connection. The terms for unobservables (universals or particulars) were to be understood through reference to aspects or items of direct experience or intuition. In the intersubjective reconstruction this analysis is supplanted (or, if you will, paralleled) by an account of the habitual regularities (acquired through learning processes) of the use of certain words in connection with various exteroceptive or interoceptive stimuli or stimulus configurations. Ostensive definitions had been something of a vexing anomaly anyway. They could not be written down – as any normal definition can be. They had therefore better be reinterpreted as rules for the use of symbols to be incorporated by drill in our linguistic habit system. The intersubjective description of the behavior of human organisms gives an account of the use of direct observation terms including the so-called 'subjective' terms referring to 'private' experience [cf. esp. Skinner, 1945, 1953; W. Sellars 1954. (Carnap anticipated the basic idea of Skinner's analysis in his [1932a].)]. We acquire the use of such phrases as 'I feel happy' (or 'tired', 'indignant', 'elated', 'depressed', etc., etc.) in a way not fundamentally different from the way we acquire the use of color, sound, taste or smell terms. In the process of education we learn to associate certain words with certain situations, things, feelings, etc. through the familiar processes of learning (conditioning, imitation), i.e., through reinforcement by our social environment. Once our linguistic abilities have matured more fully, we can also make up our own words for experienced qualities which are 'private' in the two senses of: 'not shared by anyone else' and 'resulting from intradermal stimuli.' Thus it would be possible for someone experiencing utterly strange qualities (as under the influence of drugs or Yoga practices) to label them with 'α', 'β', 'γ', etc. and thus to develop a partly 'private' language. But it is obvious that a perfectly intersubjective account of this private language can be given, as soon as the causal relations between the eliciting internal states and the associated verbal responses are ascertained. Returning to the more usual private experiences, such as headaches, memory images, dreams or the like, it is generally plausible that their introspections may be viewed intersubjectively as processes caused by co-present or immediately preceding central states, sometimes – but not necessarily – issuing in overt verbal responses. Given a fuller development of neurophysiology the details of such a causal analysis of 'private' experience and of introspection could presumably be filled in quite satisfactorily.

The controversial problem of the certainty of direct observation statements can perhaps be resolved by distinguishing between the subjectively felt *certitude* of statements made during their actual confrontation with the data they describe, and the *objective degree of certainty* (degree of confirmation) that can be ascribed to them on the basis of the best intersubjective evidence that can be marshalled for their support. Furthermore, it seems possible to account for the high degree of subjective certitude of direct observation statements within the intersubjective frame. The relatively short and usually smoothly functioning set of processes that connect a cerebral state with a learned verbal response assure on the whole a high objective probability for statements involving a minimum of inductive extrapolations.

At this point we have to deal with the notorious and perennial objections which will be raised by all those philosophers who maintain that behavioristic as well as neurophysiological accounts *necessarily leave out something essential*, namely a description of direct experience as we 'have' it, 'live it through', 'enjoy' or 'suffer' it; in short what they miss is an account of precisely the subjective awareness of, or acquaintance with, the 'raw feels' of direct experience. Another equally notorious and insistent objection concerns the irreducibility of 'meaning', 'reference', 'intentionality', 'norms', etc. to physicalistic categories. Since I agree in a certain respect with this last line of objections and since there is no space for their detailed discussion here, I shall restrict myself to some extremely brief suggestions. [For a fuller analysis of these issues cf. W. Sellars, 1952, 1953c.] I do agree that physicalistic categories do not and could not possibly provide a basis for an adequate analysis of the normative aspect of meaning. But logical empiricists have admitted this throughout. What is under discussion here is the psycho-logical, not the psycho-physical problem. Ever since Frege's and Husserl's devastating critiques of psychologism, philosophers should know better than to attempt to reduce normative to factual categories. It is one thing to describe the actual regularities of thought or language; it is an entirely different sort of thing to state the rules to which thinking or speaking *ought* to conform. Whether we deal with rules of inference or rules of designation, the only aspect that a causal-descriptive, behavioristic account can possibly cover are those dealt with in *descriptive* syntax, semantics or pragmatics. The meaning-relation (reference, intentionality) of *pure* semantics – Carnap uses the term 'designation' – is neither phenomenal nor behavioral. It is a concept in a purely formal discipline developed for the express purpose of reflecting the norms of designation and of inference by the construction of an ideal language model. But in the *descriptive* disciplines of semiotic we deal with linguistic behavior, and with the various aspects of the actual use of words – in connection with other words, with events inside or outside the communicator, etc. If 'meaning', 'reference', 'intentionality', are understood psychologically, then these concepts belong to *descriptive* semiotic and no insuperable difficulties arise. But it is, and will always remain, a category mistake, to attempt to reduce *pure* to *descriptive* semiotic.

I turn now to the other problem which is much more germane to our general topic: Is the physicalistic reconstruction of mental life necessarily incomplete in that it cannot include an account of direct experience of the 'raw feels'? This question constitutes perhaps the most perplexing central issue of the modern mind-body problem and is basic for an appraisal of the status of psychology in the system of the sciences. The contention of behaviorism that psychology is a natural science (continuous with biology) must now be more carefully scrutinized. Certain naive forms of behaviorism are easily repudiated. In the early phase of 'logical behaviorism' Carnap[16] tended to regard statements in the mentalistic, subjectivistic language as logically equivalent with statements about overt behavior. But even in that early phrase Carnap qualified this radical view, on the one hand, by pointing out the dispositional form of many psychological concepts, and on the other by reference to the possibility of neurophysiological explanations. The dispositional form of psychological concepts was especially emphasized later in 'Testability and Meaning'. To ascribe a psychological predicate to a certain organism was declared equivalent to (or shorthand for) a test condition → test result conditional, or to an open set of such conditionals (or equivalences) as formalized by means of unilateral or bilateral reduction sentences.[17] If psychological statements are to be intersubjectively confirmable, they must be established on a sensory confirmation basis. In the terminology previously suggested this would make psychological concepts and statements physical_e, while their status as regards physical_c is thereby not prejudged. The empirical regularities of behavior (which are compressed into dispositional statements) may however eventually become derivable from neurophysiological premises. Encouraging precedents of such derivations or explanatory reductions are plentiful in the history of physics, chemistry and biology. Countless macro-laws have been explained, i.e., logico-mathematically derived from assumptions about micro-structures and micro-processes in those disciplines. Even if psychological research at present may more fruitfully concentrate on macro-behavior, the progress of neurophysiology, especially since the advent of cybernetics, should be a sufficient warning to the protagonists of exclusively molar forms of behaviorism. They should not repeat in the field of psychology the sort of error committed in physics by the opponents of atomic theory (Mach, Ostwald, and other positivists some fifty years ago). [Cf. Feigl, 1951b; see references at the end of essay.]

Given this general outlook it becomes obvious that the naive peripheralistic forms of behaviorism must be repudiated and their shortcomings remedied by the admission of central states and processes as the genuine referents of psychological terms. Although at the current stage of neurophysiological research specific identifications of these states and processes are mostly quite problematic, this does not make the idea in principle objectionable. Many terms in physics, chemistry and biology were construed as having a surplus meaning (beyond the empirical dispositionals which 'anchored' them in the confirmation basis) long before that surplus meaning could be more precisely specified in terms of micro-structures confirmed by new and independent sorts of evidence. This 'promissory note' feature of many scientific concepts must therefore be regarded as an essential part of their meaning. Concepts such as memory trace, habit strength, unconscious wish, etc. may plausibly be taken to refer to (as yet very incompletely specified) central conditions. The philosophical puzzles of the mind-body problem become poignant only in the case of psychological terms which in the introspective situation have a direct reference to items of the phenomenal field, i.e., to 'raw feels.' Once the naive behavioristic identification of the referents of phenomenal terms with overt peripheral behavior is abandoned, the new identification with central states may be questioned on similar grounds. The liberal meaning criterion of physicalism$_c$ does not necessarily rule out dualistic interpretations of either the interactionistic or parallelist type. One would of course want to be very careful in explicating the precise meaning of such dualistic doctrines. But as long as mental events are at all located in the intersubjectively anchored nomological net, they are thereby physical$_c$, even if there be some doubt as to their precise status in regard to physical$_e$. But if the application of the label 'physical$_e$' is not limited to some given stage or style of explanatory theory, then whatever the place of mental events in the nomological net, no matter what type of laws are characteristic of them, they would be physical$_e$. This is surely too easy and too cheap a demonstration of a monistic physicalism. Contemporary naturalistic monists have often espoused so omnivorous a concept of 'nature' that their thesis became unassailable – at the price of triviality. If 'physical$_e$' is understood in the extremely tolerant way, it becomes practically indistinguishable from the meaning of 'physical$_c$.' It is clear, however, that monistic physicalism, past and present, is conceived somewhat more narrowly. It excludes, at least and especially, the ascription of typically theological features to 'purely' mental factors or events in that it insists on explanations of teleological and purposive behavior on the basis of organic structures and processes. It rejects teleological explanations of the vitalistic type as anthropomorphic and mythological, and insists on the explanation even of conscious purposive behavior in terms of such neurophysiological mechanisms as, e.g., negative feedback. 'Teleological explanation' is rejected, not necessarily as meaningless (it *could* be physical$_c$), but as presumably superfluous, i.e., because a scientific account of 'teleological mechanisms' (no longer a contradiction in terms) by means of non-teleological concepts is becoming increasingly successful.[18]

Limitations of the meaning of 'physical$_e$' of the sort just suggested would seem to exclude typically interactionistic theories of the relation of the mental to the physical. The essential core of such theories which always seemed extremely objectionable to physical monists was the assumption that an 'immaterial' agent could in some way organize and direct the course of 'material' processes toward certain ends or outcomes. The terms 'material' and 'immaterial' are, however, typical of the earlier, mechanistic phase of natural science. If we disregard the philosophically irrelevant pictorial connotations of the term 'material' (e.g., hard little balls in perpetual motion), it means the sort of structures and processes describable and explainable by the principles of Newtonian mechanics. The fields and particles of modern physics are then clearly not material in *this* sense. They are, however, physical$_e$ (as well as, of course, physical$_c$). Now, the doctrine of immaterial agents directing the behavior of organisms was evidently suggested by the introspectively founded interpretation of voluntary action. Action carried out with an end in view is one of the most familiar features of our experience. Many forms of pre-scientific thought adopted this sort of will-directed action as a paradigm of explanation. But with the advance of modern science we have learned that explanation does not necessarily or even usually coincide with familiarization. The vast majority of present-day psychologists still oppose in principle animistic explanations and insist on physical$_e$ explanations. With the qualifications made above this is no longer the trivial all-embracing thesis of a vague naturalism but a testable hypothesis, or at least a research program whose success can be appraised in the light of empirical evidence. The introspectively impressive 'efficacy' of intentions and volitions must then be explained in a manner compatible with physical$_e$ principles.

Epiphenomenalistic parallelism was never a plausible doctrine. Voluntary action, the role of attention, as well as psychosomatic phenomena, such as hysterical symptoms, appeared as strong evidence against a doctrine which would make mental events a dispensable luxury, a causally superfluous and inefficacious by-product of neurophysiological processes. For this reason various double aspect, double knowledge, double language or identity theories strongly recommended themselves to scientifically oriented thinkers. The sort of solution of the old puzzle has been very plausible also because it harmonized well with the familiar analytic clarification of the free-will problem [cf. Riehl, 1894; Schlick, 1918; Drake, 1925, 1933; R. W. Sellars, 1932; Hobart, 1934]. Once the notorious confusions of the free-will perplexity were removed, it became clear that it makes perfectly good sense to say that our volitions are free to the extent that they are determined by our basic personality, i.e., to the extent our interests, knowledge and deliberations are causally effective in the actions we perform. A general account of mental phenomena in physical_e terms thus seems to face no overwhelming 'meta-physical' objections. The available scientific evidence points on the whole strongly in the direction of a monistic solution. I shall now try to explicate it more precisely, and to test its strength against various criticisms. We thus return to the question: Is physicalism defective in that it omits or excludes from consideration anything essential to the science of psychology? Our answer, briefly is this: If by 'physical' we mean, as specified, both physical_c and physical_e then a physicalism in the form of an identity theory of the mental and the physical can be formulated which, though contingent in its validity upon certain pervasive features of empirical evidence, is nevertheless not necessarily defective; and can be defended against a number of charges that have been traditionally levelled against psychophysical monism.

First then a more precise statement of the physicalistic identity theory:[19]

It claims that there is a synthetic (basically empirical) relation of systemic identity between the designata for the phenomenal predicates and the designata of certain neurophysiological terms. This sort of identity differs *in its mode of ascertainment* from accidental identities as well as from ordinary nomological identities. An accidental identity would be formulated, for example, by the statement: 'The woman named Ann E. Hodges (32 years old) of Sylacauga, Alabama, is the person who was hit by a meteorite weighing nine pounds in December 1954.' A nomological identity: 'The metal which has a specific heat of 0.24 and a specific gravity of 2.7 has an electric resistivity of 2.8 microhms per cc.' Systemic identity differs from nomological identity in that it requires a background of scientific theory and of semantical analysis. Systemic identity might also be called 'theoretical identity' because it is only in the light of a theory that this sort of identity can be recognized. In a complete theory of the atomic structure of matter the nomological identities (as well as other relations formulated by empirical laws) which our previous example illustrates would be deducible from the postulates of the theory. The systemic identity would then hold between the referent of the description of a certain metal in terms of its atomic (or molecular) structure and the referent of a description of a certain metal (in our example aluminum) in terms of its directly observable or measurable physical and chemical properties. Other examples: The macro-concept of temperature designates the same state of matter that is designated by (a disjunction of) micro-descriptions in terms of molecular motions; the macro-concept of electric current (as used in experimental physics) designates the same process, which in great detail is described by the micro-account (again in disjunctive form) of the motion of electrons through a lattice of atoms. It should be noticed that in all the identities illustrated so far we deal with cases of empirically ascertainable synonymy (or perhaps I had better call it co-reference?). This is as it should be. Logically it is improper, if not inconsistent, to say that two *things* are identical. If they were, they (?) would be *one*. Identity can legitimately be ascribed to the referent of two different names, of a name and an (individual) description, of a predicate and a (generic) description, two different descriptions (individual or generic), etc. Empirically valid synonymies[20] of the accidental, nomological or systemic types must of course be distinguishable from purely logical synonymies, as for example,'$2^3 = + \sqrt[2]{64}$'; or 'x is *earlier than* y = y is *later than* x.'

The systemic identity of the referent of phenomenal terms with neurophysiological ones (such as, perhaps, 'mounting anxiety = increasing hypothalamic activity') can consistently be maintained only if the phenomenal terms are used intersubjectively, i.e., if their referents are physical_c events. The application of the neurophysiological term (systemically synonymous with the phenomenal term) provides the setting for incorporation in the nomological net of physical_e concepts and laws. *Absolutely* private phenomenal terms would according to our earlier discussion not even be physical_c and would thus not fulfill the necessary condition for being physical_e.

A familiar objection[21] to the identity theory maintains that the 'raw feels' of direct experience could not conceivably be the referents of neurophysiological terms. Neurophysiology deals with the processes in the sensory-neural-glandular-muscular structures, it has reference to the electrochemical aspects of the 'firing' of neurons, etc. – and so it is argued, how could directly experienced qualities such as colors, sounds, smells, pains, emotions, or the like, be identical with neural processes whose properties are so fundamentally different? It is usually granted that these two types of processes may be lawfully related, so that to a given quality of experience there corresponds a certain neural state or process (or a disjunction thereof) either by way of simple concomitance or as a consequence of causal relations of interaction between 'mind' and 'brain.' Since what is regarded as the decisive point in this objection depends on various emphases, we shall have to consider each of them.

First of all it must be pointed out that according to our epistemological point of view the designata of the concepts of physical science are by and large totally unfamiliar, i.e., unknown by acquaintance. Only phenomenal terms are directly associated with certain qualities and relations in the field of immediate experience. A Martian super-scientist who did not share any of our human repertory of immediate data could nevertheless (conceivably) attain a perfect behavioral and neurophysiological account of human life. He might not 'know by acquaintance' what colors look like, what pains feel like, what it 'means' to experience 'pity', 'reverence', 'regret', etc. As has often been pointed out, a congenitally blind (human) scientist, equipped with the necessary instruments and intelligence, could achieve not only an adequate knowledge of the *physics* of colors and radiations, he could also arrive at a (behavioristic and neurophysiological) account of color perception and imagination. Similarly a clinical psychologist completely deprived of certain sectors in the area of emotional experience would in principle be able to introduce the behavioral or neurophysiological equivalents of such (to him completely unfamiliar) emotions in his 'psychology of the other one.' Of course, it must be admitted, that (a) without *some* basis of immediate experience neither the Martian superscientist nor the emotionally 'blind' clinical psychologist could ever get started in his cognition of anything in the world; and (b) that possession of a repertory of experience of a certain breadth will be immensely helpful in a *heuristic* way of the projection of tentative hypotheses or laws concerning the regularities of human experience. In taking himself as an instance or sample of the type of object ('person' in this case) to be investigated, the psychologist will have a certain advantage if he finds in himself the kind of processes which he studies in others. On the other hand there are of course also certain dangers of error involved in overestimating the interpersonal similarities. But it is clear that direct acquaintance with, e.g., melancholia, or megalomania, is not an indispensable prerequisite for the psychiatric diagnosis or etiological explanation of these mental conditions. The Martian may be completely lacking experiences of the sort of human piety and solemnity, and hence unable to 'understand' (empathize) what goes on in the commemoration of, e.g. the armistice – but this would not in principle make it impossible for him to give a perfectly adequate causal account of the behavior of certain human groups on a November 11th at 11 a.m. [example taken from Eddington, 1929]. Quite generally, the significance of intuition, insight, empathetic understanding consists in the power of these processes to *suggest* hypotheses or assumptions, which, however, could *not* be established, i.e. confirmed as scientific statements except by intersubjective methods.

Returning to the central issue, the distinction between 'knowledge by acquaintance' and 'knowledge by scientific description' can be drawn in such a way that the first reduces strictly to familiarity in the sense of ability to recognize a quality immediately when experienced, i.e., the ability to affix the proper phenomenal label. Knowledge by acquaintance also involves in some areas, but not generally or necessarily, the ability to imagine certain qualities or configurations. One may rightly wonder whether the word 'knowledge' should at all be applied to acquaintance or familiarity in the sense just explained. If it is the mere having ('*erleben*') of certain contents of experience, no truth-claim is connected with it. If it is the ability of correct labeling, then it is perhaps 'knowing how', but again not 'knowing that' which alone makes a truth-claim.

The electrochemical concepts of neurophysiology, like all concepts of the natural sciences, have their epistemic roots in the area of sensory evidence. If one confuses evidence with reference, as positivists and phenomenalists stubbornly do, then of course it would seem that the meaning of physical concepts had to be indentified with the sensory data that serve as a confirmation basis. Very naturally when we hear of 'cerebral processes' we think of a brain-as-seen-when-opening-the-skull, or of nervous-tissue-as-seen-under-the-microscope. It is this 'root-flavor' which is so often mistaken for the factual meaning of our statements or concepts. More precisely, it is the pictorial appeals (usually the visual imagery) which masquerade as the 'true meaning' of our concepts. But while as empiricists, we insist on 'rooting' our concepts in a sensory confirmation base, this does not imply that our concepts *refer* to it. The concept of the electromagnetic field, for example, must of course be introduced in such a manner that it is not completely disconnected from the data of sensory experience, but its referent is not visualizable at all. 'Thou shalt not make graven images unto thyself' is a warning to be heeded in the philosophical interpretation of the concepts of physics; this notwithstanding the admittedly often great but always limited heuristic (or didactic) value of images and models. The *prima facie* implausibility of the identity thesis arises, I believe, mainly from the psychological incompatibility of images such as of nervous tissue or of molecular structures (as pictured by didactic tinker-toy models) with the qualities of some data of consciousness, such as sounds, smells or emotions. More fundamentally, perhaps the most perplexing difficulty of the mind-body problem can be avoided by distinguishing between phenomenal and physical space.[22] Visual, tactual, and kinaesthetic data contribute the 'intuitive' character of phenomenal space (or spaces). The geometry employed in the description of physical space is a conceptual system which, though based upon the evidence of the sensory kind of spatiality, is itself not adequately intuitable (visualizable, etc.). This implies that the neurophysiological concepts which are used in the description of cerebral processes are not to be 'visualized' in terms of the phenomenal data on whose basis they are confirmable. Some parts of direct experience (the visual, tactual, etc.) have *phenomenal* spatial extension, others (emotions, volitions, etc.) have at best a very vague and diffuse phenomenal localization. In opposition to Descartes I feel tempted to say that it is only the mental, i.e., the phenomenal data, which have (intuitable) spatial extension, whereas physical objects as conceived in physical science have only abstract conceptual (non-intuitable) topological and metrical relationships. Hence there is no conflict and no incompatibility in regard to the 'location' of, e.g., a directly experienced patch of color. It is where we 'see' it in phenomenal space. The systemically identical cerebral process is assigned a place in the abstract 3-dimensional manifold of physical space; and a detailed analysis of the central process in its relations to afferent and efferent impulses should be able to account for the behavior relevant in place learning, spatial orientation, optical illusions, etc.

The psychophysiological isomorphism assumed by the Gestalt psychologists may well be interpreted as the identity of certain items or aspects of the phenomenal field with certain global or configurational aspects of the (in dualistic terms: 'correlated') neurophysiological processes. The criticism that the physical language necessarily omits reference to the experienced aspect may then be rejected because reference is here confused with the evocative appeal of certain terms of our language. Many psychological terms of the intersubjective language of ordinary communication carry such an evocative appeal. This comes simply from the way their use has been learned. If, in the utopian future of a complete neurophysiology, children could be taught to use the appropriate neurophysiological terms on the basis of introspection, these terms would then have the same sort of emotive (pictorial, emotional, motivative) appeals that psychological words have in common language; and there would be the additional advantage of getting rid of the spurious dualism that is essentially linguistic. The incorporation of words which fulfill a phenomenal-introspective function into the total terminology of scientific explanatory terms could thus be achieved.[23]

Philosophically more interesting is the closely related question whether the physical language is not bound to disregard the experienced *uniqueness* of particulars as well as of universals. There is, to be sure, a certain sense of 'uniqueness' which escapes all efforts of conceptual characterization. The 'absolute' uniqueness of the 'now' (*this* present moment), of the 'I' (*this* present person), as it is experienced directly, and reflected in the expressions of poets and the anguished stammerings of existentialists, is indeed a matter of *acquaintance* and not of knowledge at all. The only way in which the uniqueness of particulars and of universals can be cognitively represented deprives it of its absolute character and assigns to individuals and to qualities the sort of singularity that they have in a total relational structure. In the scientific description of the world the 'now', the 'here', and the 'I' – no matter how poignantly and uniquely experienced subjectively – are supplanted, respectively, by a moment in time (among a continuum of other moments), a point in space (in a continuum of other points) and a person (among other persons and things). Whatever can be conceptually formulated about the *empirical* singularity of certain moments, places or persons is so formulated with the help of definite descriptions (unambiguous characterizations) which owe their uniquences to either a sufficiently plausible (but always problematic) singleness of the qualitative or relational setting, or to a unique association of proper names with their designata. The attachment of proper names (or of coordinates) to their designata is a pragmatic affair, presupposed in the semantical analysis of language.[24]

The question 'Why am I, I?' not only looks queer in print, but is apt to provoke needless perplexity. If it is not an empirical question as to the causal factors which contributed to the formation of my personality, it is probably a manifestation of deep bewilderment with 'existence', the sort of emotional expression which Heidegger mistook for profound philosophy. The thought experiments regarding the inverted spectrum show clearly that the empirically assertible uniqueness of the qualities is inseparable from the nomological net in which they have their cognitive, conceptual place. In this way scientific knowledge does symbolize the qualities, they are not excluded, omitted or disregarded. The rules according to which the symbols are used reflect the total relational structure. Even the correctness of the affixing of a symbolic label to an experienced quality can be *checked* – subjectively or intersubjectively – only with reference to the conceptual structure. But the *use* of phenomenal terms, being a matter of *doing* rather than of a *knowledge claim* can be considered as a result of training or conditioning, and as such is not a conceptual affair. But as soon as we wish to give even the sketchiest *account* of such training and its results (habits of verbalization) then, of course, we are making a knowledge claim, and this can occur only within the frame of a conceptual structure, i.e., a nomological net.

Moritz Schlick, in his London lectures on 'Form and Content'[25] had developed an analysis of cognition in terms of directly experienced content and conceptual structure. He was aware of the dangers of a metaphysics of 'ineffable' contents, although in his somewhat metaphorical manner of writing he could perhaps not completely avoid gratuitous perplexities. But his doctrine jibes well with his original critical-realistic solution of the mind-body problem.[26] A purified doctrine of form and content would merely insist on distinguishing between (a) *having* a datum and its *description*; (b) the *evoking of data* (images, feelings) through the pictorial or emotional uses of language, and their symbolization by the representative, descriptive function of language; (c) the *activities* and *abilities* of affixing names (*or predicates*) to data or aspects of data), the *empirical description* of these activities and abilities (in descriptive pragmatics), and their metalinguistic formalization (in pure pragmatics); (d) descriptions of *data* and descriptions of *non-data* (inferred or posited on the basis of presupposed nomological relations). Once these distinctions are recognized, there is no occasion for a metaphysics of the inexpressible, ineffable or unknowable.[27] The qualitative features of the raw feels directly designated by phenomenal terms may also be indirectly but (empirically) uniquely characterized by their place in the nomological net of neurophysiological descriptions.

At this point it may be well to consider briefly another critical question. If raw feels are designated by certain neurophysiological concepts, how do we decide as to whether butterflies, earthworms – or amoebas, for that matter – are 'sentient' beings? In keeping with the two theses of physicalism as interpreted thus far, this comes down to the question of the degree of similarity between the processes in these various organisms. Systemic identity presupposed there is no additional cognitive significance in the ascription of raw feels over and above the identifiable physiological processes. Certain philosophers have adopted a panpsychistic position, but this doctrine is to be rejected not because it is unconfirmable but because it does not really make its predications, as it pretends, on the basis of considerations of analogy. The differences of organization between organisms are so tremendous that the sort of structure which is characteristic of human mental life is either extremely impoverished or completely absent in lower organisms. But wherever the similarities warrant it, the application of (human) phenomenal terms, preferably with proper qualifications and cautions will add the emotive appeals which give ethical and quasi-ethical questions their peculiar poignancy. Does my fellow man really suffer agonies when subjected to torture? Does the wriggling of the worm on the fishhook manifest genuine pain? If these questions, though admittedly sometimes extremely difficult to answer, are not to be made absolutely unanswerable (by our own perverse intellectual devices), then proper attention to analogy is the only responsible way to settle them. Lately the same sort of puzzle has been posed in regard to man-made robots. Electronic computers ('thinking machines') might be equipped with additional devices which mimic responses to various 'perceptual', 'pain', etc., stimuli and which even display 'manifestations of emotion' [cf. esp. Turing, 1950; Scriven, 1953].

We do not *seriously* apply mentalistic terminology to such machines because their internal structure is so radically different. All response mechanisms are deliberately built into them, and it seems extremely unlikely that a machine consisting of inorganic parts could in *every* respect duplicate the complex, adaptable, docile, purposive behavior of human beings. If, on the other hand, a *homunculus* could be produced by synthesis and combination of essentially the sort of materials (proteins, etc.) of which human organisms consist, then only a theologian might be reluctant to ascribe mental life to it.

Two final crucial questions: How does the here proposed systemic identity theory differ from parallelistic dualism or from a physicalistic emergentism? Dualistic parallelism has for a long time been the implicit if not the explicit preference of cautious philosophers and psychologists. 'The facts of psychophysiology indicate a *correlation* of the mental and the physical', this is the customary formulation. 'Mental' means here at least the phenomenal, but usually also the much larger rest of the subject matter of psychology, including the unconscious processes and dispositions of psychoanalysis. 'Physical' is usually understood as a rather unclarified combination of (picturized) materiality and subject matter of physical science (intersubjective, nonteleological, etc.). A little reflection shows that 'correlations' of the 'mental' and the 'physical' taken in these senses cannot be interpreted consistently. There are empirical relations between the processes in the central nervous system and the peripheral sensory and motor processes. These are *causal* relations of a complex structure, scientifically analyzable into relatively more 'micro' linkages, of neural and ultimately of molecular-atomic-electromagnetic nature. They do *not* have the character of an ultimate 'parallel' concomitance. This sort of parallelism is indeed more usually asserted to hold between the phenomenal (directly given) states of consciousness and the 'corresponding' neural processes. The experimental situation in which this kind of parallelism could presumably be ascertained, would be the observation of one's own brain processes along with some other directly given mental states. Let us assume a device, the 'autocerebroscope', were available for this sort of observation. While one experiences, e.g., a sequence of musical tones (or of odors, tastes, feelings, emotions, etc.) one would simultaneously also be confronted in one's visual field with certain shape and color patterns, which according to plausible (realistic) interpretations, would indicate in great detail the configurations of neural processes in one's own brain. Now, it is easy to see that precisely on the basis of the parallelism hypothesis the relation between the visual patterns and the brain processes correlated with (parallel to) the sequence of tones would again be of a *causal* nature. That is, the visual patterns would be *effects* of immediately preceding brain states, made visible by the cerebroscope. In no conceivable experimental setup could one 'observe' both a brain state *and* its so-called mental correlate *simultaneously*. The parallel correlation would therefore always be a matter of *interpretation* of the autocerebroscopic data. Now, it must be admitted, the liberal empiricist meaning criterion would not rule out as meaningless the assumption of two parallel series – the brain states on the one hand and the mental states on the other. But the positive assertion of their duality could be justified only by the sort of evidence (autocerebroscopic, parapsychological, occultist or otherwise) which, assuming the parallelistic hypothesis, could not be forthcoming. It is therefore much more in keeping with the usual procedures of inductive and theoretical science to identify the designata of certain phenomenal descriptions with the designata of certain neurophysiological descriptions. This avoids the introduction of superfluous hypotheses. This empirical core combined with the epistemological considerations concerning acquaintance and description, and the semantical analysis of systemic synonymy, constitutes the difference between the physicalistic identity (or double language) view of mind and body, and dualistic parallelism, i.e. the theory according to which the mind-body relation is analyzed in terms of a general empirical equivalence of mentalistic and physicalistic propositions. It must be emphasized again that the identity theory stands or falls with the empirical evidence, and can therefore never be regarded as justified by purely logical considerations alone. Evidence of the independent existence of 'mind-like' agents as conceived by some vitalists or parapsychologists would indeed have to be scrutinized very seriously. As long as the scope of naturalistic (physical_e) explanations is uncertain, nothing more compelling can be offered than confidence in certain inductive generalizations on the one side and a miscellaneous group of puzzling and recalcitrant phenomena on the other. While the analytic philosopher might well be completely neutral in such issues, philosophers of science are apt to express certain predilections. Physicalists (like myself), in any case, are so impressed with the triumphs of scientific explanation, that they would, if necessary, admit all sorts of revisions in the (physical_e) laws of the universe, rather than to abandon the identity view and thus to open the door for typically animistic doctrines.

The sort of revisions which physicalists should not too strenuously oppose are those (briefly touched upon before) of the doctrines of evolutionary emergence. Since emergence need not conflict with scientific determinism, or with the degree of determinism countenanced by modern physics, it is conceivable that the concepts and laws required for the explanation of biological or psychological processes will not be reducible to those sufficient for inorganic processes. Emergence understood in this way [cf. Meehl and Sellars, 1956; Bergmann, 1944] would be entirely compatible with the first thesis of physicalism. But it would qualify the second thesis to the effect that certain concepts of biology and psychology would have to be implicitly defined by the nomological net, i.e., that they could not all be defined explicitly on the basis of the primitives sufficient for the physics and chemistry of inorganic phenomena. Or, even if the biological phenomena were completely reducible to physicochemical ones, it is conceivable, though (to me) not plausible, that the explanation and prediction of the 'raw feels' of psychology may require genuinely irreducible (i.e., primitive) concepts connected by nomological rather than by merely logical relations with the primitives of (inorganic) physics.[28]

I conclude by a succinct statement of the philosophically most challenging points of our critical review of the theses of physicalism.

Both theses of physicalism reflect certain assumed basic features of our world. The first thesis, far from being a purely syntactical criterion of meaningfulness, asserts that subjective and intersubjective confirmability coincide in their extensions. This expresses one essential aspect of scientific optimism, namely the belief that there is nothing in the realm of existence which is in principle inaccessible to examination and exploration by the scientific method. Metaphysicians must however be cautioned not to feel triumphant about this. Given the methods – the only ones we can justifiably call 'objective'

– for the certification of knowledge claims, any speculations concerning realities beyond the reach of those methods remain as boundless and hence as irresponsible as they have always been regarded by empiricists. For all we know, and possibly may ever know, confirmability on a sensory basis is a necessary condition for the intersubjective meaningfulness of factual knowledge claims. As long as this is not regarded as an absolutely unquestionable *a priori* presupposition, but as a frame principle of human knowledge, dovetailing with what commonsense and science tell us about the embeddedness of knowing organisms in the known or to-be-known world, there is no danger of relapsing into the rationalistic metaphysics which logical empiricists have consistently opposed. The 'logical behaviorism' of some twenty years ago tended to assert that mentalistic terms are attached to behavior by sheer convention. This would make the first thesis of physicalism analytic. A more adequate appraisal of the total empirical and logical situation however forces us to formulate the basic psychophysical relation as one of synthetic, empirically grounded systemic (or theoretical) synonymy. But in being a relation of synonymy, rather than of mere empirical equivalence, the 'identity' of the referents of introspective and certain physical terms, reflects – if it holds – a very fundamental trait of our world. As such it is much more fascinating, but of course also much more problematic than a purely analytic thesis could ever be.

The second thesis of physicalism expresses a related but even bolder belief characteristic of the optimistic attitude of scientists. In endorsing a program of theory construction which attempts to subsume a maximum of facts under a minimum of unitary basic postulates, and in conceiving these basic postulates according to the paradigm of modern physical theory, physicalism amounts to a monistic view of scientific explanation, and therefore – in a sense – also of the universe. Despite a certain unavoidable and even desirable vagueness in the conception of 'physical explanation', the thesis is definite enough in what it rules out. Animistic or irreducibly teleological explanations are quite clearly excluded – if not as scientifically meaningless, than certainly as superfluous. Apart from the impressive success of the physicalistic mode of explanation, and apart from ideological (anti-obscurantist) motivations, there is perhaps a more trenchant and philosophically more significant argument in favor of the second thesis. Just as *testability* is a necessary condition for the *possibility* of scientific knowledge (first thesis of physicalism), so is *predictability* a necessary condition for the *success* of the scientific endeavor. Even if vitalistic or animistic hypotheses could be formulated in conformity with the liberalized meaning criterion, they would, if true, restrict predictability (and explainability) quite severely. Whether our universe will accommodate the scientific quest for more and more complete physicalistic reduction, will in principle always remain an open question.

The general philosophical lesson to be drawn from these conclusions would seem to be: A properly reformulated physicalism contributes greatly to our analysis of scientific method. But since both these involve assertions about the world, physicalism cannot and should not claim to settle by logical analysis any issues in the strife of the '*Weltanschauungen.*' It will be helpful, if in the future, purely analytic philosophy, i.e. clarification without cosmological presuppositions or commitments, were even more sharply than heretofore separated from the advocacy and justification of such presuppositions.[29]

NOTES

[1] Cf. E. A. Singer [1924]; A. P. Weiss [1925]; and with greater philosophical and scientific sublety respectively, Gilbert Ryle [1949]; and B. F. Skinner [1953]. It is to be noted that Carnap's first formulations of the unity of science thesis and of physicalism (or logical behaviorism) were made quite independently of E. A. Singer's and A. P. Weiss's contributions, and that he anticipated in the main points much that is essential in the basic outlook of Ryle and of Skinner. Carnap's own views had developed in this area as much as elsewhere under the predominant influence of Bertrand Russell. But the abandonment of the earlier Mach-Russell type phenomenalism in favor of physicalism was largely due to Karl Popper's critique of observation propositions (later published in Popper's *Logik der Forschung* [1935]) and by Otto Neurath's enthusiastic, though logically often defective, advocacy of the unity of science thesis. Carnap's most important pronouncements on the subject are: [1928]; [1931, pp. 432–465]; [1934]; [1932a, pp. 107–142]; [1933, pp. 215–228]; [1937]; [1935b, pp. 43–53]; [1936b, pp. 129–135]; [1936a, pp. 419–471]; and [1938, pp. 42–46]. Cf. also C. G. Hempel [1949].

[2] A. J. Ayer [1946]; [1953, pp. 1–20]; and [1954]; Max Black [1948]; B. A. Farrell [1950]; Stuart Hampshire [1952]; Gilbert Ryle [1949]; Michael Scriven [1953]; John Wisdom [1952b]; Ludwig Wittgenstein [1953].

[3] I am referring to such cases of introspection as, for example, one would undergo when asking oneself (or when asked by a psychotherapist) whether one feels anxiety when contemplating one's insufficiencies, whether one feels a glow of proud satisfaction when remembering a great achievement, etc., . . . Of course, even in the so-called 'physical' examination of one's eyes by the oculist, or of one's ears by the otologist, some of the questions asked are answered on the basis of introspection: 'I still see the last row of letters a bit too blurred to be able to read them', 'the ringing is in my left ear, not in my right ear', etc.

[4] Doubts about it could be raised only by those who on philosophical grounds are perplexed with the legitimacy of induction. We are not concerned with these perplexities here. I think they have been satisfactorily resolved, by the analytic philosophers as well as by Reichenbach and Carnap. For a general summary of these results and my own

analysis of the problem of induction, see: 'Scientific Method without Metaphysical Presuppositions' [Feigl, 1954a], and the references at the end of that article.

[5] This illustration was suggested by my friend, P. E. Meehl.

[6] For a suggestive discussion of a closely related point, see the brief article by P. E. Meehl, [1950b].

[7] Suggestively discussed by V. C. Aldrich [1938], and by C. Lewy [1969].

[8] In fact, I believe that the continuing controversy could be considerably clarified if metaphysicians and theologians came forth with at least an outline of their own criteria of meaningfulness.

[9] I have dealt in some detail with the problem of the meaning and the limits of justification, not only of the meaning criterion, but also of the principles of deductive and inductive logic, and the moral principles, in my essay, 'De Principiis non Disputandum ...?' [1950a].

[10] It may be noticed however that these facts, though of empirical nature, are of far wider range than single empirical facts or even specific natural laws. We are concerned here with a perfectly general structural property (*ordnungshafter Zug*) of experience which is the basis of the possibility of [an intersensory as well as intersubjective] science. (From: Carnap, [1934].)

[11] Or as we might say nowadays, the truth of certain lawful connections as expressed in subjunctive conditionals.

[12] We might mention also some even more drastic, and hence more problematic, departures from classical conceptions, such as the Wheeler-Feynman theory of advanced potentials with its apparent time reversals); present theories of the role of mesons in nuclear structure (with its completely unvisualizable duality of particle and wave aspects, already introduced in earlier phases of quantum mechanics); and the Bondi-Gold theory of the continuous accretion of matter on a cosmic scale.

[13] Cf. C. I. Lewis [1929]; Victor Lenzen [1939a]; Arthur Pap [1946]; and Wilfrid Sellars [1953b].

Corresponding to Carnap's notion of *P*-rules of inference ('*P*-transformation rules' in his *Logical Syntax of Language*) it may be suggested that the 3-dimensionality of space (or the 3 + 1-dimensionality of space-time) could be formulated as *P-formation* rules of the language of science. In field physics certain variables (electric, magnetic, gravitational, etc. magnitudes) are ascribed to space-time points. In a well-formed formula (representing a singular proposition) a functor is assigned to a quadruple of numbers.

[14] In the *Aufbau* ideology, 'translatability' meant mutual logical deductibility. That this relation does not apply here is shown in some detail in my article 'Existential Hypotheses' [1950b].

[15] 'Unified' in the sense of the unitary explanatory system to which the second thesis of physicalism refers.

[16] Cf. especially [Carnap, 1935b, pp. 43–53]. Also the closely related (now equally superseded) presentation by C. G. Hempel [1949].

[17] It must be remembered that the term 'reduction sentence' is used by Carnap for a formula which provides a partial and conditional definition of a dispositional concept. Dispositional concepts, Carnap emphasized, are not explicitly definable on the basis of observation predicates, they can only be introduced just through this sort of partial specification of meaning, on the basis of empirical regularities among observables. Reduction in this sense must therefore be sharply distinguished from the other, more

customary methodological concept of reduction in the sense of explanatory derivation. 'Reduction' in this latter sense is illustrated, for example, by the derivation of the laws of classical thermodynamics from the assumptions of the kinetic theory of heat (statistical mechanics).

[18] The only reservation here to be made concerns again the questionable and puzzling phenomena of extrasensory perception – especially precognition. The physical monist will have to await further developments in this field with an open mind.

[19] In the development of my own formulation of the identity theory I was stimulated by the work of Schlick and by continued discussions and correspondence with Carnap. Readers will be interested in the following crucial passages (fairly literally translated by me from the German) contained in a letter Carnap wrote me from Prague, June 21, 1933, in response to some critical queries I had then submitted to him:

	"Example:	B_1	'The organism of N. is in the state of house-imaging'
A	'N. has a visual image of a house.'	B_2	'In the organism of N. there is an electrochemical condition of such a kind' (described in terms of electrochemistry).

Both B_1 and B_2 are translations of A. According to my recently adopted terminology I assert: A is equivalent ('*gehaltgleich*') with both statements on the right side: viz., L-equivalent (*logically* equivalent) with B_1: but P-equivalent (*physically* equivalent) with B_2, i.e., mutually translatable (derivable) using besides the logical laws also natural laws as rules of inference, incorporated as transformation rules in the scientific language. You are therefore right in saying that B_2 is only synthetically equivalent with A. This holds also of your example [quoted from Whitehead] about the relation of the tactual and the visual breakfast; and for the agreement between direct and indirect measurements of distance . . . "

"The difference between natural laws and logical laws is admitted, but it is not as enormous as we (with Wittgenstein) supposed it to be. Even the logical principles of language may be modified if this appears as expedient; and this not just on the basis of purely speculative considerations, but possibly prompted by the facts of experience."

. . . "The whole 'riddle of the universe' [Schopenhauer's '*Weltknoten*', i.e., the mind-body problem] seems finally to come to this: one will have to make clear to oneself in an appropriate manner that brain processes are, on the one hand, *objects* of scientific sentences, and on the other hand *causes* of the emission of sentences. This, in itself by no means mysterious, situation should then be so formulated that people with emotional (not to use the offensive word 'metaphysical') headaches can accept it more easily. As to whether these aches can be completely eliminated is a psychological question, or perhaps a practical task of psychoanalysis."

[20] In order to forestall a possible misunderstanding, it should be noted that the phrase 'empirically valid synonymy' as understood in the cases of accidental, nomological or systemic identities refers to something different from the sort of synonymies that a mere descriptive-semantical examination of a natural language can disclose. That 'tepid' and 'lukewarm' are used synonymously in English can be ascertained by observation of linguistic customs without a special study of the materials to which these predicates may be applied. That 'temperature' refers to the state of material objects which is also the

referent of micro-physical descriptions can be established only by special experimental research and theoretical interpretation.

[21] For incisive (but in my opinion inconclusive) arguments against various forms of physicalistic monism, cf. especially C. J. Ducasse [1951]; N. Jacobs [1937]; C. I. Lewis [1941]; Arthur Pap [1951] and [1952].

[22] Cf. especially the chapters on qualitative and quantitative knowledge, and on the psychophysical problem in M. Schlick's *Allgemeine Erkenntnislehre* [1918] which contains a superb and undeservedly neglected clarification of these issues.

[23] Perhaps I should at this point reassure emotionally tender persons that I am using this fantasy merely as a thought experiment, and that I am not seriously proposing this sort of language reform. I too happen to have a certain romantic attachment to the homey, christmassy, or poetic appeals of many words of ordinary introspective language.

[24] For an important recent analysis of the logic of egocentric particulars (or token-reflexive words) cf. Y. Bar-Hillel [1954].

[25] Cf. *Gesammelte Aufsätze* [1938]. (The lectures 'Form and Content' are printed in the English language.)

[26] Schlick, in his *Allgemeine Erkenntnislehre* [1918] had expounded, perhaps with greater clarity than other monistic critical realists of that period, a physicalistic identity theory which – due to the predominantly phenomenalist tendencies of the Vienna Circle – had later been suppressed (even by Schlick himself). If I am not altogether mistaken it is precisely this theory which, properly formulated in modern semantical terms, deserves resuscitation and which impresses me as the sort of solution that should be especially acceptable to Carnap's present way of thinking.

For other largely independent older or more recent similar doctrines of various degrees of epistemological sophistication, cf. Alois Riehl [1894]; F. Gätschenberger [1920]; Durant Drake [1925]; R. W. Sellars [1932]; Curt Weinschenk [1936]; and Hans Reichenbach [1938, esp. Chap. IV].

[27] Cf. Carnap's pertinent and trenchant replies to the criticisms of E. Zilsel and K. Duncker in [Carnap, 1933]. These were written, however, in his syntactical phase, i.e., before the indispensability of pure semantics and pure pragmatics were realized.

[28] In the terminology of Meehl and Sellars this would make all psychological concepts physical_1 (part of the intersubjectively confirmable nomological net) but not physical_2 (part of the intersubjective nomological net sufficient for the explanation of all inorganic and [most?] biological phenomena).

[29] The present paper was completed and submitted to the editor, Professor Paul A. Schilpp, in December 1954. I have subsequently written a further and more elaborate essay on 'The "Mental" and the "Physical"' [1958b]. – H. F.

18. MIND-BODY, *NOT* A PSEUDOPROBLEM

[1960]

Any serious effort toward a consistent, coherent, and synoptic account of the place of mind in nature is fraught with embarrassing perplexities. Philosophical temperaments notoriously differ in how they react to these perplexities. Some thinkers apparently like to wallow in them and finally declare the mind-body problem unsolvable: '*Ignoramus et ignorabimus.*' Perhaps this is an expression of intellectual masochism, or a rationalization of intellectual impotence. It may of course also be an expression of genuine humility. Others, imbued with greater confidence in the powers of philosophical insight or in the promises of scientific progress, offer dogmatic solutions of the old puzzle. And still others, recognizing the speculative and precarious character of metaphysical solutions, and deeply irritated by the many bafflements, try to undercut the whole issue and declare it an imaginary problem. But the perplexities persist and provoke further efforts – often only minor variants of older ones – toward removing this perennial bone of contention from the disputes of philosophers and scientists. Wittgenstein, who tried to 'dissolve' the problem, admitted candidly [1953, Section 412]: "The feeling of an unbridgeable gulf between consciousness and brain-process . . . . This idea of a difference in kind is accompanied by slight giddiness," but he added quickly "which occurs when we are performing a piece of logical sleight-of-hand."

As I see it, Wittgenstein's casuistic treatment of the problem is merely one of the more recent in a long line of positivistic (ametaphysical, if not anti-metaphysical) attempts to show that the mind-body problem arises out of conceptual confusions, and that proper attention to the way in which we use mental and physical terms in ordinary language will relieve us of the vexatious problem. Gilbert Ryle, B. F. Skinner, and, anticipating all of them, R. Carnap, have tried to obviate the problem in a similar way: The use of mental or 'subjective' terms is acquired by learning the language we all speak in everyday life; this language, serving as a medium of communication among human beings, is by its very nature *intersubjective*; it is on the basis of publicly accessible cues that, for example, the mother tells the child 'you feel tired', 'now you are glad', 'you have a headache', etc., and that the child learns to use such phrases as 'feeling tired', 'being glad', 'having a headache' as applied not only to others, but also to himself when he is in the sort of condition which originally manifested itself in the cues (symptoms, behavior situations and sequences, test conditions and results, etc.) observable by others. But there is the rub. Even if we *learn* the use of subjective terms in the way indicated, once we have them in our vocabulary we *apply* them to states or conditions to which we, as individual subjects, have a 'privileged access.' If I report moods, feelings, emotions, sentiments, thoughts, images, dreams, etc., that I experience, I am *not referring* to my *behavior*, be it actually occurring or likely to occur under specified conditions. I am referring to those states or processes of my direct experience which I live through (enjoy or suffer), to the 'raw feels' of my awareness. These 'raw feels' are accessible to other persons only indirectly by inference – but it is *myself* who *has* them.

I do not wish to deny that ordinary language serves many purposes quite adequately. As I see it, ordinary language unhesitatingly combines mental (phenomenal) and physical (behavioral) terms in many descriptions and explanations of human and animal conduct or behavior. 'Eagerness was written all over his face'; 'He was trembling with anxiety'; 'No doubt his gastric ulcer is due to his suppressed hostility'; 'An attack of the flu left him in a discouraged and depressed mood for several days'; 'A resolute decision finally enabled him to overcome his addiction.' As these few illustrations indicate, ordinary language clearly reflects an interactionistic view of the relations of the mental and the physical. As long as we are not too particular about squaring our accounts with the facts established, or at least strongly suggested, by the advances of psychophysiology, we can manage to keep out of logical troubles. Some philosophers, such as Ryle, Strawson, Hampshire, and other practitioners of the ordinary-language approach, have most persuasively shown that we can talk about the mental life of '*persons*,' i.e., about episodes, dispositions, actions, intentions, motives, purposes, skills, and traits, without getting bogged down in the mind-body puzzles. But, notoriously, there is in this approach scarcely any reference to the facts and regularities of neurophysiology. Moreover, not all is well logically with these neobehavioristic analyses. 'Persons' remains a term insufficiently explicated, and what I could glean from Strawson's analysis [1958] is that he defines 'person' as a sort of synthetically glued-together unity of a living body and its mental states. Strawson accounts for introspection in terms of "self-ascription." While this is helpful, it cannot be the whole story about mental states: infants, idiots, and at least some of the higher animals undoubtedly have 'raw feels', but are not 'self-ascribers.' If highly learned men nowadays express (philosophical) doubts about *other minds*, and debate seriously as to whether or not very complex robots have direct experiences, then obviously a better philosophical clarification of the relations of the mental to the physical is urgently needed.

The crucial and central puzzle of the mind-body problem, at least since Descartes, has consisted in the challenge to render an adequate account of the relation of the 'raw feels', as well as of other mental facts (intentions, thoughts, violations, desires, etc.) to the corresponding neurophysiological processes. The problems may fairly clearly be divided into scientific and philosophical components. The scientific task is pursued by psychophysiology, i.e., an exploration of the empirically ascertainable correlations of 'raw feels', phenomenal patterns, etc., with the events and processes in the organism, especially in its central nervous system (if not in the cerebral cortex alone). The philosophical task consists in a logical and epistemological clarification of the concepts by means of which we may formulate and/or interpret those correlations.

Scientifically, the most plausible view to date is that of a one-one (or at least a one-many) correspondence of mental states to neurophysiological process-patterns. The investigations of Wolfgang Köhler, E. D. Adrian, W. Penfield, D. O. Hebb, W. S. McCulloch, *et al.*, strongly confirm such a correspondence in the form of an isomorphism of the patterns in the phenomenal fields with the simultaneous patterns of neural processes in various areas of the brain. The philosopher must of course regard this isomorphism as empirically establishable or refutable, and hence as *logically* contingent. It is conceivable that further empirical evidence may lead the psychophysiologists to abandon or to modify this view which on the whole has served so well at least as a fruitful working hypothesis. It is conceivable that some of the as yet more obscure psychosomatic phenomena or possibly the still extremely problematic and controversial 'facts' of parapsychology will require emergentist or even interactionistic explanations. (As an empiricist I must at least go through the motions of an 'open mind' in these regards!) But tentatively assuming isomorphism of some sort, a hypothesis which is favored by many 'naturalistic' philosophers, are we then to interpret it philosophically along the lines of traditional epiphenomenalism? Although Professor Köhler [1938] does not commit himself explicitly to this view, I am practically certain that this is the general outlook within which he operates. If the basic physical laws of the universe should be sufficient for the derivation of biological and neurophysiological regularities, if the occurrence of neural patterns (physical *Gestalten*) is not a case of genuine emergent novelty but a matter of the combination of more elementary physical configurations, and if, finally, the experiential patterns correspond in some way isomorphically to neural process patterns, then this *is* epiphenomenalism in modern dress.

It will be best here not to use the somewhat ambiguous label 'parallelism.' Psychophysiological parallelism, as held by some thinkers in an earlier period, allowed for a 'mental causality' to correspond to 'physical (i.e., neurophysiological) causality.' Sometimes it even connoted an all-pervasive correspondence of mental and physical attributes (in the manner of Spinoza), and thus amounted to a form of panpsychism. But the favored outlook of modern psychophysiology amounts to postulating causal relations, i.e., dynamic functional dependencies only on the physical side, and then to connect the neural process patterns merely by laws of (simultaneous) coexistence or co-occurrence with the corresponding mental states. Only a small subset of neural processes is thus accompanied by mental processes.

Traditionally the most prominent objection to epiphenomenalism has been the argument from the 'efficacy of consciousness.' We seem to know from our direct experience that moods, pleasure, displeasure, pain, attention, vigilance, intention, deliberation, choice, etc., make a difference in the ensuing behavior. But, of course, this subjective impression of the causal relevance and efficacy of mental states can easily be explained by the epiphenomenalist: Since, *ex hypothesi*, some dynamically relevant physical conditions are invariably accompanied by mental states, there is, then, also a regular occurrence of certain types of behavior (or of intra-organismic events) consequent upon mental states. For empiricists holding an essentially Humean conception of causality, it is then quite permissible in this sense to speak of the causal efficacy of mental states. There are, it should be noted, countless highly 'teleological' processes that occur in our organism evidently without the benefit of any mental influence, guidance, or instigation. For example, the kinds of regenerations and restitutions that are involved in recoveries from many types of physical injury or disease appear as if they were most cleverly 'designed', yet for many of these phenomena purely physiological (and perhaps ultimately physicochemical) explanations are available. Yet according to the epiphenomenalistic doctrine such explanations are sufficient also for behavior which we ordinarily consider instigated, regulated, or modulated by mental factors. If an effort of concentration facilitates learning algebra, piano playing, or the like, then consciousness cannot be regarded as a causally irrelevant or superfluous 'luxury.' I don't think we need to apologize for arguments of this sort. It is true, radical Materialists and Behaviorists reject such arguments as 'tender-minded', but then radical Materialism or Behaviorism typically *repress* or *evade* the mind-body problem. They do not offer a genuine solution. Epiphenomenalism, while not evading the problem, offers a very queer solution. It accepts two fundamentally different sorts of laws –

the usual causal laws and laws of psychophysiological correspondence. The physical (causal) laws connect the events in the physical world in the manner of a complex network, while the correspondence laws involve relations of physical events with purely mental 'danglers.' These correspondence laws are peculiar in that they may be said to postulate 'effects' (mental states as dependent variables) which by themselves do not function, or at least do not seem to be needed, as 'causes' (independent variables) for any observable behavior.

Laws of concomitance in the physical world could usually be accounted for in terms of underlying *identical* structures. Thus, for example, the correspondence of certain optical, electrical, and magnetic properties of various substances, as expressed in simple functional relations between the refraction index, the dielectric constant, and the magnetic permeability, is explainable on the basis of the atomic structure of those substances. Or, to take a slightly different example, it is in terms of a theory of *one* (unitary) electric current that we explain the thermal, chemical, magnetic, and optical effects which may severally or jointly be used in an 'operational definition' of the intensity of the current. Similarly, it is at least a partially successful working program of psychophysiology to reduce certain correlated macrobehavioral features to underlying identical neurophysiological structures and processes. It should be emphasized, however, that a further step is needed if we are to overcome the dualism in the epiphenomenalist interpretation of the correlation of subjective mental states with brain states.

The classical attempts in the direction of such unification or of a monistic solution are well known: double-aspect, double-knowledge, twofold-access, or double-language doctrine have been proposed in various forms. The trouble with most of these is that they rely on vague metaphors or analogies and that it is extremely difficult to translate them into straightforward language. I can here only briefly indicate the lines along which I think the 'world knot' – to use Schopenhauer's striking designation for the mind-body puzzles – may be disentangled. The indispensable step consists in a critical reflection upon the meanings of the terms 'mental' and 'physical', and along with this a thorough clarification of such traditional philosophical terms as 'private' and 'public', 'subjective' and 'objective', 'psychological space(s)' and 'physical space', 'intentionality', 'purposiveness', etc. The solution that appears most plausible to me, and that is entirely consistent with a thoroughgoing naturalism, is an *identity theory* of the mental and the physical, as follows: Certain neurophysiological terms denote (refer to) the very same events that are also denoted (referred to) by certain phenomenal terms. The identification of the objects of this twofold reference is of course logically contingent, although it constitutes a very fundamental feature of our world as we have come to conceive it in the modern scientific outlook. Utilizing Frege's distinction between *Sinn* ('meaning', 'sense', 'intension') and *Bedeutung* ('referent', '*denotatum*', 'extension'), we may say that neurophysiological terms and the corresponding phenomenal terms, though widely differing in *sense*, and hence in the modes of confirmation of statements containing them, do have identical *referents*. I take these referents to be the immediately experienced qualities, or their configurations in the various phenomenal fields.

Well-intentioned critics have tried to tell me that this is essentially the metaphysics of panpsychism. To this I can only reply: (1) If this be metaphysics, make the least of it! (2) It is not panpsychism at all – either the 'pan' or the 'psyche' has to be deleted in the formulation. By way of very brief and unavoidably crude and sketchy comments let me explain my view a little further. The transition from the Logical Positivism of the Vienna Circle to the currently prevalent form of Logical Empiricism, as I interpret it, involved a complete emancipation from radical phenomenalism, behaviorism, operationism and their all-too-restrictive criteria of factual meaningfulness. Parallel with the critique of philosophical doubt by the Neo-Wittgensteinians, Logical Empiricists nowadays have no patience with skeptical questions regarding the existence of physical objects or of other minds. 'Skeptical doubts' of these sorts are illegitimate not because the beliefs in question are incapable of confirmation or disconfirmation, but because doubts of this pervasive character would call into question the very principles of confirmation and disconfirmation that underlie all empirical inquiry – both on the level of commonsense and on that of science. There can be no question that assertions of the existence of stars and atoms, or of the occurrence of conscious and unconscious mental proccesses, are subject to the normal procedures of inductive, analogical, or hypothetico-deductive confirmation or disconfirmation. It is preposterous (not to say philosophically perverse or naughty) to deny that we have well-confirmed knowledge concerning imperceptible physical objects or concerning the mental states of other human beings. A mature epistemology can make explicit the principles of such, often highly indirect, confirmations or disconfirmations. And along with this a liberalized meaning-criterion can be formulated, broad enough to include whatever is needed by way of commonsense or scientific hypotheses, and yet sufficiently restrictive to exclude transcendent metaphysical (pseudo-) beliefs. Freed from the torments of philosophical doubt and from the associated reductive tendencies and fallacies of phenomenalism as well as of radical behaviorism, we can now with a good intellectual conscience embrace a genuinely critical and empirical realism.

Once this position is attained, a mind-body-identity theory of the kind sketched above appears as the most adequate interpretation of all the relevant facts and considerations. This is not panpsychism for the simple reason that nothing in the least like a psyche is ascribed to lifeless matter, and certainly at most something very much less than a psyche is ascribed to plants or lower animals. The panpsychists claimed to reason by analogy, but this is precisely what they did not do in fact. The difference between the nervous system of, say, an earthworm and of a human being is so tremendous that we should in all consistency assume a correspondingly large difference in their respective mental states. And even on the human level there is no need whatever for the assumption of a psyche in the traditional sense of a soul that could act upon the brain, let alone be separable from it. One may, of course, doubt whether a purely Humean conception of the *self* (as a bundle and succession of direct data) will be sufficient for an adequate psychology. Nevertheless no substantial entity is required. Events, processes, and their properly defined organization and integration, should be perfectly sufficient. Professor Stephen C. Pepper suggested to me in conversation that my view might be labeled 'pan-quality-ism.' While this locution is not pleasant to the ear, it does come much closer to a correct characterization than 'panpsychism.' But since Paul E. Meehl [1958] who understands my view at least as thoroughly as does Professor Pepper, has designated me a 'materialist', perhaps one last word of elucidation may be in order.

I am indeed in agreement with one main line of traditional materialism in that I assume, as does Professor Köhler, that the basic *laws* of the universe are the *physical* ones. But (and this is so brief and crude a formulation that I fear I shall be misunderstood again) this does not commit me in the least as to the nature of the *reality* whose regularities are formulated in the physical laws. This reality is known to us by acquaintance only in the case of our direct experience which, according to my view, is the referent also of certain neurophysiological concepts. And if we are realists in regard to the physical world, we must assume that the concepts of theoretical physics, to the extent that they are instantialized in particulars, are not merely calculational devices for the prediction of observational data, but that they denote realities which are unknown by acquaintance, but which may in some way nevertheless be not entirely discontinuous with the qualities of direct experience. But – "whereof we cannot speak, thereof we must be silent." If this is metaphysics, it seems to me entirely innocuous. I have little sympathy with the mysticism of Eddington or the psychovitalism of Bergson. I reject the former because there is literally nothing that can be responsibly said in a phenomenal language about qualities that do not fall within the scope of acquaintance. Extrapolation will carry us at most to the concepts of unconscious wishes, urges or conflicts as postulated by such 'depth-psychologies' as psychoanalysis. And even here, future scientific developments may be expected to couch these concepts much more fruitfully in the language of neurophysiology and endocrinology. And I reject psychovitalism because it involves dualistic interaction. At the very best 'intuition' (empathetic imagination) may be heuristically helpful in that it can suggest scientific hypotheses in psychology (possibly even in biology), but these suggestions are extremely precarious, and hence must always be relentlessly scrutinized in the light of objective evidence [Feigl, 1958c].

Does the identity theory simplify our conception of the world? I think it does. Instead of conceiving of two realms or two concomitant types of events, we have only one reality which is represented in two different conceptual systems – on the one hand, that of physics and, on the other hand, where applicable (in my opinion only to an extremely small part of the world) that of phenomenological psychology. I realize fully that the simplification thus achieved is a matter of *philosophical* interpretation. For a synoptic, coherent account of the relevant facts of perception, introspection, and psychosomatics, and of the logic of theory-construction in the physical sciences, I think that the identity view is preferable to any other proposed solution of the mind-body problem. Call my view metaphysical if you must; I would rather call it *metascientific*, in the sense that it is the result of a comprehensive reflection on the *results* of science as well as on the logic and epistemology of scientific *method*. But I admit that for the ordinary purposes of psychology, psychophysiology, and psychiatry an epiphenomenalist position is entirely adequate, if only the traditional, picturesque but highly misleading locutions (e.g., 'substantial material reality and its shadowy mental accompaniments') are carefully avoided.

I conclude that the mind-body problem is not a pseudoproblem. There are, first, a great many genuine but unanswered questions in psychophysiology. And, secondly, there is plenty of work left for philosophers in the logical analysis of the intricate relations between phenomenal and physical terms. Problems of this complexity cannot be relegated to the limbo of nonsensical questions. I doubt quite generally whether many issues in modern epistemology can be simply 'dissolved' in the manner in which some artificially concocted pseudoproblems can be disposed of by a minimum of reflection on the proper use of terms. Questions like 'How fast does Time flow?', 'Do we really see physical objects?', 'Why is there anything at all rather than nothing?', 'Why is the world the way it is?', etc., can indeed be very quickly shown to rest on elementary conceptual confusions. But the issues of perception, of reality, and of the mental and the physical require circumspect, perspicacious and painstaking analyses.

NOTE

In a long essay [Feigl, 1958b] written nearly three years ago I have attempted to do fuller justice to the complexities and the unresolved issues of the mind-body problem than I possibly could in the preceding brief comments. A very ample bibliography is appended to that essay. Since its publication I have found a welcome ally in J. J. C. Smart [1959]. Carnap's early article, 'Psychologie in physikalischer Sprache', which anticipated much of the neobehavioristic arguments of Ryle, Skinner, and Wittgenstein is at last available in English translation [Carnap, 1959b]. A brief but perhaps not sufficiently elaborate critical reply to the Wittgensteinian position on the problem of other minds is contained in my symposium article (a response to Norman Malcolm) [Feigl, 1958c]. An exposition and critical analysis of Carnap's physicalism is presented in [1958e]. For a forthright but philosophically unsophisticated physicalistic solution of the mind-body problem, see [Smith, 1958]. The brilliant psychologist and methodologist Paul E. Meehl has dealt with the mind-body problem and related issues in several chapters of a book [Meehl, 1958] which despite its primarily theological and religious intent contains large parts of scientifically and logically important and incisive discussions.

19. SOME CRUCIAL ISSUES OF MIND-BODY MONISM

[1971a]

The following considerations concern exclusively one of the (at least) three components of the traditionally as well as currently discussed mind-body problems. It is the *sentience* aspect rather than the aspects of sapience or selfhood that I wish to review in brief compass. Although I admit that the obvious interconnections of these three strands make it hazardous to separate them and thus to concentrate only on sentience (traditionally: consciousness, awareness, direct acquaintance, the phenomenally given, etc.), it nevertheless seems timely, even urgent, to attempt once again to clarify and reappraise what is right and what is wrong with physicalism (or the new materialism) as well as with some antiphysicalistic points of view.

My remarks are made within a framework of assumptions or presuppositions that may be best characterized as those of a scientifically oriented critical realism, and a tentative (physicalistic) reductionism. Since this position has been amply argued for in numerous publications (e.g., by R. Carnap; P. Oppenheim and H. Putnam; J. J. C. Smart; D. M. Armstrong; myself; and many others) I shall review now very succinctly only the points relevant for the important qualifications which differentiate my own views from those of the others.

Having originally taken my cues from the (by now 'classical') critical realism of the early Moritz Schlick and the later Bertrand Russell, as well as having favored the somewhat similar views of the American monistic 'naturalists' (especially R. W. Sellars, but also C. A. Strong and Durant Drake), I assume that there is – *epistemologically* – an important distinction to be made between the data of immediate experience and the world of 'things-in-themselves.' But in radical disagreement with the agnostic doctrine of Kant (and in complete agreement with Schlick and Russell) I consider the things-in-themselves knowable. And I hasten to add that such scientific knowledge as we possess of them, and keep expanding, is 'structural', i.e., it is knowledge by description (in B. Russell's sense, recently explicated more formally by R. B. Braithwaite, R. Carnap, and G. Maxwell with the help of the Ramsey-sentence approach). While very few of the deep-rooted beliefs of common sense – as for example those philosophically formulated in 'direct realism' – survive epistemological and scientific criticism, some of the basic tenets of common sense remain (relatively) unscathed. If the well-known disastrous slide into phenomenalism (or instrumentalism) and ultimately into a solipsism of the present moment is to be avoided, Berkeley's *esse est percipi* must be rejected from the start, and the *existence* (independently of perception) of the 'external world' assumed.

The much-discussed quantum-mechanical, partial dependence of the observed or measured situation upon the act of observation or measurement can be safely disregarded in the present context. The influence of observation on observed objects maintained by a majority of present-day physicists is in any case negligible in regard to macroobjects. ('It does not hurt the moon to look at it', even if electrons do get a 'kick' out of being looked at.) As to whether the finer details of neural processes (e.g., in the synapses of the central nervous system) require considerations of complementarity and indeterminacy is, however, a relevant question in regard to the identification (individuation) of neural events. Moreover, it seems quite questionable as to whether the 'subjectivistic' interpretation of quantum mechanics is really called for.

For our purposes, and with some simplification, we can regard the assumptions of critical realism as among the premises of a hypothetico-deductive system. But just as the theories of the factual sciences generally, so also even these frame assumptions must be testable (in principle, and no matter how indirectly) by the data of immediate observation. This minimum of empiricism is in any case indispensable if the obvious fundamental difference of pure logic and mathematics from factual knowledge is to be properly understood.

One other assumption of common sense survives: It is the identity of mental states, events or processes (sensations, thoughts, intentions, desires, volitions, moods, sentiments, etc.) referred to by persons who 'have' them and other persons who come to know about them on the basis of behavioral (including, of course, linguistic) or neurophysiological evidence. To illustrate: My doctor knows that he is causing me the experience of a pain when he lances an abscess on my arm, and I *feel* that pain and can report (or 'avow') it. This much seems clear, and is indeed insisted upon by the philosophers of ordinary language (Wittgenstein, Austin, Strawson, Malcolm, *et al.*). Despite their opposition to the idea of a private language (or perhaps because of that opposition!) they do not countenance the notion that the word 'pain' has different meanings in first person experimental reports and in ascriptions to other persons (or by other persons to me).

This point, however, requires a crucially important reexamination. I maintain (again in agreement with Russell) that without analogical conception and reasoning by analogy the identity (in the sense of synonymy) of first-person (self-ascribed, reported, avowed, or otherwise indicated) experience with experience ascribed to that (first) person by other persons can be understood only in the very different sense made widely familiar by the (logical) behaviorists, physicalists, or the Australian materialists. This other sense is quite clear: it retains essentially the point of view of the 'psychology of the other one.' Anxious to make psychology an honest natural science, the behaviorists – from Watson to Skinner – have consistently dealt with mental phenomena within the frame of an intersubjective account of the world. This trend of thought began with the crude identification of mind with behavior (J. B. Watson, E. A. Singer), and became more sophisticated in the views of R. Carnap, G. Ryle, and B. F. Skinner. These two philosophers and that brilliant psychologist attempted (in different ways) to show that assumptions of a publicly (intersubjectively) inaccessible mental life are either outright meaningless (though pictorially and/or emotionally significant) or, in any case, redundant for the aims of scientific description and explanation of human behavior.

There are two reasons – and I trust not merely *argumenta ad hominem* – which have for a long time convinced me of the indispensability of a subjectivistically understood conception of immediate (first person) experience. There is, first, the epistemic primacy of the data of immediate experience. Lest I be accused of attempting to resurrect the doctrine of 'incorrigible' sense-data-statements, let me say right now that the empiricism to which I subscribe requires neither infallible 'protocol' or 'basic' statements; nor does it assume an atomistic ('pointillistic') structure of the phenomenally given. Moreover I would not only admit, but insist that sense-data as traditionally understood are the products (really 'destructs') of a very special sort of analysis. Ordinary, normal adult direct experience is perhaps best described as the life-world (*Lebenswelt*) of the phenomenologists and existentialists. That is to say that ordinary immediate experience is suffused with interpretation, remnants of memory, expectations, associations, etc. But even if by a wrench of abstraction we succeed in stripping off this overlay, what remains are hardly ever elementary ('atomic') sense data but *Gestalten* (i.e., patterns, configurations, etc.). Experimental psychophysiology has for some time suggested that the much discussed isomorphism holds between these *Gestalten* and certain global features of cerebral processes. (I shall return to this important point later.)

In the pursuit of a logico-epistemological reconstruction of our knowledge-claims regarding the external or physical world we are – in the last analysis – driven back to the phenomenally given as the ultimate testing ground. Consider such confirmations as, for example, of astrophysical hypotheses by means of telescopic, spectroscopic or other observations. Feyerabend's contentions to the contrary notwithstanding, the astronomer ultimately has got to *see* (or if he is blind, to hear or touch) something. Of course I admit that what is thus sensed would make no sense without some presupposed theory (as e.g., of optics, photography or the like). But those theories in turn have been (let us assume successfully) tested by previous observations. 'Incorrigibility' is not required. Even basic statements may have to be revised in the light of further observations and theories.

In this age of electronic computers, robots, etc., it has been argued that the human observer and his immediate experience could be replaced by machines that do the 'observing.' The reply to this objection is quite simple: How would we human beings ascertain the reliability of observation machines; and is it not epistemology for human beings (and neither for gods nor machines) that we are trying to work out?

The second point (that again I do not consider an *argumentum ad hominem* – let alone a sentimental and fallacious piece of reasoning) is that the very understanding of moral imperatives requires indispensably references – literally – to direct experience. 'Thou shalt not wantonly inflict pains on humans or animals' will serve as a simple example. The word *pain* here has a surplus (factual, not purely emotive!) meaning over and above the one that radical behaviorists, physicalists or materialists countenance. I trust that I am not taken to be moralizing in this context. What I am saying is that there is a cognitive presupposition in ethical imperatives. Even someone who holds a purely noncognitive (emotivist) position in moral philosophy can, I hope, understand and agree.

Having been a member of the Vienna Circle (ca. 1924–1930), I realize that logical positivists (empiricists) or radical physicalists will shake their collective heads and accuse me of apostasy. But I have been a renegade from that movement and became a critical (or hypercritical) realist at least thirty years ago. Along with others I have tried to liberalize the empiricist meaning criterion in a manner that still excludes the 'pernicious' transcendence of certain types of metaphysics and theology and yet allows for analogical conception and inference. If a theologian wishes to construe his beliefs in this manner, I will not tell him that what he is asserting is meaningless. I shall merely ask him by what sort of reasons he can justify his beliefs. Far-out scientific hypotheses, such as those of current nuclear theory or of cosmology, are merely precariously transcendent. Hence I think that empiricists should consider theological (or metaphysical) assertions as meaningful provided they are at least incompletely and indirectly testable.

Considered superficially, analogical conception and inference of other minds appear to share the features of perniciously transcendent metaphysics and theology. To use the well-known Peirce-James formulation, there seems to be "no difference that makes a difference" in regard to all conceivable evidence. But I submit that our 'private', 'privileged' access (each must speak for himself in this matter) to our own immediate experience *is* a cognitive matter. The arguments of the Wittgensteinians (especially of Norman Malcolm) are utterly implausible to me. Introspective reports or avowals are either true or false. (I have already admitted that they are not incorrigible. They may conflict with other basic statements, with background knowledge or well-established theories.) Reference to one's own immediate experience is the (epistemological!) prototype of all designations of objects, properties or relations by the words of our language. Never mind how we come to use language. Very likely, reinforcement along with the existent (innate) set of our (brain-mind) capacities accounts for that. Carnap was historically the first (anticipating both Ryle and Skinner) to point out how the use of introspective phrases (e.g., 'I am glad', 'I am tired', 'I am afraid', 'I hope . . . , ' 'I am thinking of . . . , ' etc.) is acquired; i.e., how the child learns, in the context of environmental circumstances, 'taught' by his elders, the uses of those subjective expressions. Of course the physicalists are right in regarding avowals as caused by central states. But to the extent that whatever human beings do or say is in principle open to causal analysis (within the possible limits implied by quantum mechanical indeterminacy), this is entirely compatible with a semiotic account (i.e., by syntax, semantics and pragmatics). This renders it possible to apply the metaconcepts of designation (representation) and of truth and falsity. In opposition to many of the Oxford linguistic philosophers (and also to Malcolm) I think it makes perfectly good sense to say 'I know that I have a pain', 'I know that I am glad', etc. But if the finesses of lexicography should speak against such use of 'I know that . . . ', I shall not be adamant on this point, and simply choose some other locution.

As I see it, the neo-Wittgensteinian approach to the notorious vexations of the 'other minds' problem is still positivistic. The 'beetle in the box' is just as ascertainable as the brain in the skull. But, I admit, the ascertainment of the 'raw feels' (E. C. Tolman's term for immediate awareness or the phenomenal qualities) cannot be achieved in the simple manner of independently verifying the conclusion of an analogical inference. Obviously this has been the bugbear of the positivists. For similar reasons statements about the historical past should have caused them the same sort of misgivings but, strangely enough, hardly anything of the sort appears in the positivistic literature.

From the point of view of the liberalized criterion of factual meaningfulness we don't have to restrict the meaning of mental state ascriptions to the behavioral 'criteria.' The criteria of which the neo-Wittgensteinians speak are allegedly quite different from symptoms. Nor are they to be understood as logically equivalent to, or entailing, the ascriptions in question. They can serve as (empirically?) necessary and sufficient conditions of those ascriptions only under 'normal' circumstances. Perhaps this can be accepted as a fairly adequate analysis of the way we actually ascribe – in ordinary situations and in terms of the commonly used language – mental states to other persons. But is there then really that essential difference between (fairly reliable) *symptoms* and *criteria*?

For the consistent physicalists, as well as those behaviorists who at least admit that there are central states (i.e., who do not insist on the 'black box' or 'empty organism' outlook) peripheral behavior, including linguistic utterances, facial expressions, etc., serves as a probabilistic indicator of central states induced by stimuli and/or apt to (causally) produce overt responses.

Quite generally, and especially ever since the developments in pure semantics (Tarski, Carnap, *et al.*) it is now quite legitimate and simple to distinguish the truth conditions from the confirming evidence of a given knowledge-claim. The ascription of mental states, no matter (for the present purpose) whether understood phenomenally or in terms of brain states, is clearly and radically different from statements about the behavioral evidence. Only in the case of first-person direct-observation (phenomenal, experiential) statements is it plausible to identify their truth conditions with their confirming evidence. If the term 'self-evident' had not suffered traditionally so much from misuses and ambiguities, I would not hesitate to characterize statements about the phenomenally given as self-evident.

Along with other proponents of the mind-brain identity thesis, I have never asserted an identity of mental states with actual or possible peripheral behavior. It should be understood without elaboration that I also repudiate as fallacious the identification of mental qualities with aspects of the stimuli. Obviously a color sensation, for example, is not identical with the radiation (of a certain intensity and frequency pattern) that elicits that sensation. And, as indicated above, I also reject the phenomenalistic identification of physical objects with complexes of 'elements' (Mach) of direct perception or configurations of sensation. What in more modern parlance is termed the translatability thesis of physical into phenomenal language is untenable. If the errors of these types of reductionism are to be eliminated, then all the relationships mentioned, far from being identities, had better be viewed as lawful (causal) relationships between distinguishable states or events.

There is, however, one type of identification that recommends itself to our favorable attention. This is the reduction of one sort of entity to another as it occurs in the context of scientific explanation. This sort of reductive identification is often (but not always or necessarily) a macro-micro reduction. Much used examples are the kinetic theory of heat; the electron-gas theory of electric currents; the identification of table salt with sodium chloride; of genes with DNA helical molecules; of (short term) memory traces with reverberating neural circuits; etc., etc. The identification of gravitational fields with a Riemannian (i.e., non-Euclidean) structure of space-time in the general theory of relativity provides an example in which no microreduction is utilized. As I have emphatically pointed out in previous publications, these identifications (i.e., the ascertainment of such reductive identities) are empirical in nature. It requires empirical evidence to substantiate them. They are thus fundamentally different from identities in logic or pure mathematics.

As a first approximation in the logical reconstruction of scientific explanation this way of analyzing theoretical reductions still seems to me plausible and fruitful. But taking into account the forceful arguments of P. K. Feyerabend, a more adequate and precise way of explicating those reductions would be in terms of replacement or supplantation. This is required wherever the reducing theory is logically inconsistent with the reduced theories or empirical laws. In that case we do not have identities but replacement by "successor concepts." But since, against Feyerabend, I maintain that the successor concepts coincide in many cases at least for a certain range of the relevant variables (and approximately) with the replaced original concepts, I shall allow myself for our purposes the simpler locutions of "identity" and "identification."

Physical theories that have attained a fairly high degree of completeness characteristically provide the premises for the derivation of empirical laws. Most importantly, these empirical laws include statements about the functioning of pertinent measuring instruments, such as e.g., thermometers, ammeters, photometers, etc. Thus the expansion of the thermometric substance in a thermometer is derivable (at least with very high probability) from the kinetic theory of heat (in statistical-molecular mechanics). This still does not render the equivalence of thermometrically measured temperature with the average kinetic energy a matter of analytic synonymy. First of all the postulates of the microtheory that furnishes the premises are synthetic and empirical. Secondly, the simple relation between thermometric indications and the corresponding set of microconditions is part of the confirming evidence of the theory. Redefining "temperature" in terms of the microtheory can and has been done, but this sort of conventionalistic device merely conceals the empirical character of the "bridge law" which certainly cannot be denied, especially if classical (phenomenological) thermodynamics is compared with statistical mechanics.

Bridge laws or correspondence rules are in any case indispensable if a scientific theory is to be understood as an empirically testable set of postulates. Just where we place the bridgeheads, i.e., the concepts in the theoretical network and the observables, is to some extent a matter of decision, and depends mainly on the aims of clarification and logical reconstruction. Thus there is, for example, some leeway as to what should be selected as the observables in the Ramsey-sentence approach.

I admit that if a given physical theory has achieved the identifications, reductions, or replacements, then one is tempted to think that there is no need for bridge laws (or nomological danglers) because the theory is then – in a sense (!) – complete. This, however, is an illusion. The 'anchoring' of a theory in data of experience is precisely what distinguishes physics from pure mathematics.

I shall now try to show that the 'nomological danglers' can be understood in a way that is entirely unobjectionable. They do not violate any principle of parsimony (often referred to as Occam's razor). Of course I agree that parsimony in the sense of factual simplicity is one of the guiding principles of scientific theorizing. According to Newton's first *regula philosophandi* we should not assume more causes than are necessary for the explanation of given phenomena. But in the case of the mind-body problem whatever parsimony or simplicity can be achieved should result from a proper epistemological analysis of the differences and the relations between physical and psychological concepts. In other words, parsimony should be achieved as a byproduct of a clarification rather than from a wilful application of Occam's razor.

The first thing to do then is to reflect on the differences between the concepts of the physical sciences and the concepts that designate immediately the phenomenally given events and their qualities and relations. To put it very briefly: the concepts of the physical sciences are invariant with respect to the different sense modalities in which they may be (ostensively) 'anchored.' To illustrate: although it would be more difficult for a congenitally blind person to arrive at physical or astronomical knowledge, it is not impossible in principle. For the blind person could be equipped with photoelectric cells which react to incoming light rays from spectroscopes, telescopes or microscopes, etc. The photoelectric cells might then be connected through amplifiers with radio sets or other devices which through their emitted sounds, etc., would furnish discernible kinds of information in the auditory or tactual modalities. If this seems at first glance a bit fantastic, it should be remembered that our knowledge of stars that emit only either ultraviolet or else radio waves is just as indirect. And so is our knowledge of the structure of atomic nuclei, of the spin of subatomic particles, etc.

I contend, consequently, that a being with something like human intelligence (i.e., capacities for inference, theory construction, critical reasoning, etc.) might have a repertoire of sense modalities, and of immediate experience generally, that is utterly different from that of us earthlings; and that such a being (say, a Martian) might well arrive at the same concepts and theories as we do in the physical sciences. And if mind-body monism holds, the Martian could (in principle) also achieve complete knowledge (by description) of the psychology of human beings, including, of course, their sensations, perceptions, thoughts, emotions, moods, volitions, intentions, etc. – all that on the basis of such behavioral and neurophysiological evidence as is accessible through his (the Martian's) sense modalities.

The piece of science fiction naturally leads to the question: If that is the nature of physical knowledge, can the physical sciences ever include the private, purely subjective aspects of mental events? I shall try to answer yes, if the question is understood in one way, and no, if understood in another way. The answer customarily given in traditional philosophy is negative because it is felt that the qualities of immediate experience are 'homeless' in the physical account of the universe. Some thinkers (e.g., Bergson, Poincaré, Eddington, Schlick, et al.) have in their various ways maintained a doctrine of the ineffability, inexpressibility or incommunicability of the 'content' of direct experience. Only the structure, so they say, is intersubjectively knowable and communicable. But on closer analysis this contention boils down to the obvious truth contained in the distinction between *having* (or living through, enjoying or suffering) an immediate experience (and thus being able to achieve 'knowledge by acquaintance' of experienced qualities) and knowing *about* it ('by description').

The – at present – utopian kind of physical knowledge can give an account of introspective, self-knowledge, avowals, etc. We may (somewhat speculatively) say that some part of the cerebral cortex 'scans' the processes in some other part of the brain. And since the scanning part would be the one connected to the motoric nerves of the speech organs, we can thus sketch – at least by way of a 'promissory note' – what the scientific account of private mental states (and their avowels) would be like. It is this sort of speculation that makes (especially the Australian) materialists so confident that a physicalistic central state theory of mind is possible, and that it need not be incomplete. Indeed, in the frame of intersubjective science nothing need be left out – except the 'feel' of the raw-feels.

This is why we must admit something *is* omitted in the intersubjective, scientific account. But what are omitted are not ineffable qualia or the like, for even in their introspective description we deal with their *structural* features. Whatever genuine knowledge we can attain is *propositional*. It reflects, for example, the similarities, dissimilarities (and degrees thereof) of the immediately experienced qualities. Propositional characterizations of these qualities would then isomorphically correspond to some structural features of cerebral processes – all this, of course, only if physicalism is assumed.

I think what is omitted or left out in the physicalistic (intersubjective) account of the world is not any event, process or feature. It is – rather obviously – the *egocentric* perspective which the intersubjective world view quite deliberately displaces and replaces. It is in the egocentric perspective (prominently stressed by Descartes, Berkeley, Hume and the positivists) that we label the qualia of immediate experience directly. (The 'successor' concept in the physicalist account is the scanning process mentioned above.) Now, the customary (and often ambiguous) uses of ordinary language to the contrary notwithstanding, it is surely a 'category mistake' (of a special kind) to combine egocentric-subjective language with the intersubjective-physicalistic one. Most of the philosophical puzzles of the mind-body problem can be shown to originate from this sort of mixing of terms belonging to two categorially different conceptual systems. The *phenomenal* (visual, tactual, kinesthetic, etc.) spatialities of the egocentric account are to be emphatically distinguished from the nonpictorial, unvisualizable concept of *physical* space. Similar distinctions have already been suggested above for the categorial differences between experienced qualities and the properties of physical objects.

In short, concepts which directly designate qualities of immediate experience and concepts whose meaning is largely independent of the specific (ostensive) anchoring in one or another of the sense modalities, and which only by probabilistic indication refer to entities, events or processes in what we (somewhat misleadingly) call the physical world, are of an entirely different semantical type. Hence, if we wish to formulate a mind-brain identity thesis that involves more than the reduction of behavioral to neurophysiological concepts, a further step needs to be made. The identification of mentalistic (subjective, egocentric, private) phenomena with neurophysiological events or processes is then (as James Cornman put it) a 'cross-categorial' one. In fact, I should think this is the only cross-categorial identification required if we wish to relate the egocentric to the physicalistic accounts. All the other reductive identifications (as in physics, chemistry, and biology) are not really cross-categorial as long as we replace the naive realism (or direct realism) of common sense with a clarified critical realism. But even from the point of view of common sense, as Keith Gunderson has convincingly shown, we can clear up some of the perplexities of the mind-body puzzles by reflecting upon the epistemic asymmetries that obtain between one's knowledge of one's own mental states and one's knowledge of the external world as well as the mental states of others. Just as one cannot see one's own eyes (without the help of mirrors, etc.) so one cannot in the purely egocentric perspective perceive oneself entirely as a part of the physical world. Just as one's eyes are not a part of one's visual field (Wittgenstein, *Tractatus*), so is one's brain (without the aid of an auto-cerebroscope) not part of the world perceived. Thus even within the "manifest image" (W. Sellars) of the world it is well-nigh impossible to escape the egocentric orientation. But this is remedied in the intersubjective scientific conception. Here the individual self becomes an organism among other organisms and all the rest of things and events that constitute the universe. Linguistically this 'great transformation' manifests itself by the disappearance of the egocentric particulars 'now', 'here', and 'I', along with all their cognate expressions. They are replaced by proper names, coordinates, definite descriptions, etc. The 'existentially poignant uniqueness' of the self of which we have heard so much in recent philosophy can be understood only within the frame of the egocentric perspective. Once the 'democratization' so characteristic of the intersubjective frame is achieved, whatever uniqueness may remain is the usual empirical one of differentiating and individuating properties and relations. The directly experienced uniqueness is indeed inexpressible within the categorial frame of science.

NOTES ON CONTROVERSIAL QUESTIONS

1. The arguments for the identity theory here suggested are primarily based on logical and epistemological considerations. While I think that the empirical evidence of psychophysiology makes it plausible that most forms of dualism

and interactionism (along with the older forms of animism and vitalism) can be disregarded, the question of evolutionary emergence is still open. Equally unsettled is the question as to whether emergentism (if it could only be clearly formulated; the essay by Meehl and Sellars is a notable attempt in that direction) is compatible with some form of monism. Personally I believe that the accumulating evidence points increasingly toward a (qualified) reductionism in biology and behavioral psychology.

2. The identity theory that I favor does of course not require and could not accommodate nomological danglers in the intersubjective (scientific) theory of the world. But in the cross-categorial identification there remains the brute fact of the isomorphism of phenomenal with physical *Gestalten*. There does not seem any ready explanation of the difference in 'grain' between the phenomenal continuity (for example, of a smooth color expanse, or the homogeneity of a musical tone) and the atomic structure of the corresponding brain processes. It should be clear that the identity formulation (whose ascertainment depends upon empirical confirmation) holds only in extensional contexts. Like any other logically contingent identity it cannot be expected to hold *salva veritate* in all intensional contexts. Thus, to *know* that you are experiencing a certain mental event does not entail that you *know* with which brain process it is identical. (This is analogous to the case of the child who knows that the milk is warm without knowing anything about the mean velocities of the molecules of which the milk is composed.)

3. The unconscious mental processes as assumed in psychoanalytic theories can be viewed as quite legitimately postulated by means of explanatory theoretical concepts. These concepts are to be understood as analogues to familiar concepts designating conscious phenomena and/or as behaviorally indicated central processes whose neurophysiological nature is still largely unknown. Methodologically the situation here is somewhat similar to that of thermodynamics before the development of the kinetic theory of heat, or of chemistry before the introduction of the atomic theory. Philosophers who maintain that the idea of unconscious mental processes is inconsistent, merely reveal that they stick to an (unfruitful) terminological decision according to which 'mental' is defined as 'phenomenal' or 'conscious.'

4. Does the identity theory suggest a panpsychistic metaphysics? No, if we conform to the rules of analogical inference, the differences between lifeless matter and the living organisms (and especially those equipped with central nervous systems) are too enormous for assuming similarities of their respective 'inner natures.' Nevertheless a pan-quality-ism (S. C. Pepper's term) is not unreasonable provided that the intrinsic qualities of inorganic things or

systems are conceived as incomparably more 'colorless' than the qualities of human experience. It should be evident without further discussion that any similarities of my identity theory with the metaphysical solutions of Spinoza's attribute theory or of Leibniz's monadology are only very superficial and largely coincidental.

5. If a metaphysician is dissatisfied with the purely structural characterization of independently existing physical entities, i.e., if he wishes for something more than definite descriptions or Ramsey sentences, perhaps he can be consoled by being permitted to 'introject' the intuitive notion of existence. This notion, of course, has no place (makes no sense) in the intersubjective scientific conception of the world. But since Descartes was surely not talking outright nonsense when (in the famous *cogito ergo sum*) he ascribed existence to himself, we may grant that this intuitive and subjective idea of existence is significant in the egocentric perspective. But, of course, it makes no difference that amounts to an intersubjectively testable difference in the scientific world perspective.

6. The cross-categorial identity thesis should also be helpful for a better understanding of the causal account of perception. It enables us to relate in greater detail the egocentric (phenomenal or phenomenological) account of knowledge to the scientific one. In the intersubjective (scientific) account of perception, we view (as we can and do already in everyday life) the knowing subject (person, organism) and the incoming stimuli as it were from the side. In this lateral perspective it is clear even to untutored common sense that and why a blindfolded person cannot see environmental objects, and why, once the blindfold is removed, he can perceive them. This triviality can serve as an excellent antidote to the exclusively egocentric epistemologies of Berkeley and his positivist followers. Once the identity thesis is adopted, the main (philosophical) puzzles of the causal theory of perception resolve themselves. The detailed psychophysiology of perception is, of course, still far from complete.

7. Now to answer the perennial question: why not parallelism or epiphenomenalism instead of identity? As stated above, I don't think Occam's razor alone furnishes the answer. It is the mistaken conception of the *physical* (much more than that of the mental) that is largely responsible for such dualistic theories as parallelism or epiphenomenalism. I propose we mean by 'physical' the (structural) type of conception and/or whatever physical concepts (in this sense) designate or denote. And by 'mental' (but only for the purpose of the present essay) we mean what had better be called phenomenal. To recapitulate: as was already seen quite clearly by Schlick and Russell,

the physical (in the sense just defined) may well designate (or codesignate) those small parts of the world that are phenomenal as well. If so, one of the great mysteries of modern philosophy has been (largely) dispelled. There *is* an identity of properties if we abandon picture thinking (naive and direct realism) about physical objects. The identity is that of the *structure* of the phenomenally given with the *structure* of certain global aspects (*Gestalten*) of the processes in the cerebral cortex. The identity theory thus understood takes care of the traditional puzzles regarding the efficacy of the mental (as in deliberation, attention, intention, volition, desires, pleasure, displeasure, etc.). Mental processes (being cross-categorially identical with brain processes) are of course among the most important causes of our behavior. And surely there is interaction – namely, between the brain (as well as other parts of the nervous system) and the rest of the organism.

8. It seems that analogies and homologies remain favorite devices not only in science but also in philosophical speculation and analysis. I found the application (made by N. Brody and P. Oppenheim) of Bohr's doctrine of complementarity to the mind-body problem interesting and suggestive. But on closer analysis all I am able to accept is the categorial difference, and therefore (syntactical) noncombinability (incompatibility or noncompatibility seem to me inappropriate here) of the egocentric and the intersubjective conceptual frames. My differences from the outlook of the brilliant and sophisticated Australian materialism (of Smart, Armstrong, Medlin, Kekes, *et al.*) should by now be evident. When radical physicalists argue for the completeness of their world view, I object – not on the usual (and to them acceptable) grounds, i.e., that science can never be known to be finished-in-principle. Fully aware of the introduction of new entities throughout the history of science, they, of course, allow for such additions and modifications. (We all agree that science may well be, and probably *is*, an endless quest.) No, I am objecting to the physicalists' deliberate blindness in regard to something that admittedly does not amount to a *surplus* in the scientific-intersubjective frame. Hence I consider their arguments against the ψ-ϕ nomological danglers as an *ignoratio elenchi*. It should be clear by now that the physicalists' assertion of the (potential) completeness of a physicalistic account of the world amounts to the truism (indeed, the tautology) that there can be nothing within the intersubjective-physicalistic account of the world that is not intersubjective-physicalistic. An enormous amount of confused and fruitless disputes could be avoided once we recognize the game of the radical physicalists for what it is! These admittedly keen and clear-headed philosophers consistently (and, alas, often unwittingly) apply

the 'HYLAS TOUCH'! No wonder then, that whatever they deal with turns out to be physical!

9. Just a few words on the fashionable topics of intentionality, and the action vs movement distinction. As in my previous publications I still think that (Brentano's notion of) intentionality is best explicated in terms of the semantical concept of designation. No matter as to whether mental imagery or imageless thoughts or words are on the subject side of the relation, the object (existent or nonexistent) is the referent (symbolically) designated. Hence, despite first impressions, this part of the problem of sapience is not part of the genuine mind-body problem, but can be resolved within the context of the relation of the *psychological* (or physiological, or computer-ological, or robotological) to the *logical*. The fallacies of psychologism are in any case to be avoided.

The colossal literature on intention (in the other sense, in which it is connected with action) seems to me largely fruitless and exhibits glaringly the futility of the ordinary language approach in philosophy. Surely, there is a perfectly good meaning to *causal* explanations of intentional, purposive, goal-directed behavior. Desires, ends-in-view, etc., furnish (like most explanation in terms of motivation) only the trivialities that even nursemaids and fishwives know very well, and that are still quite distant from genuine scientific explanation. To make a little clearer what was already known for centuries may be lexicographically useful, but it does not solve any philosophical problems.

10. What sort of bearing my view has on the methodology of theory construction in psychology is clearly indicated in my essay 'Philosophical Embarrassments of Psychology' [1959]. See also my major previous publications on the mind-body problems: [1950d]; [1963d]; [1967a]; and [1974b].

20. NATURALISM *AND* HUMANISM

An Essay on Some Issues of General Education and a Critique of Current Misconceptions Regarding Scientific Method and the Scientific Outlook in Philosophy

[1949a]

The main purpose of this essay is to dispel certain confusions and misunderstandings which still prevent the much-needed constructive synthesis and mutual supplementation of the scientific and humanistic elements in general education. It is my contention that the philosophical foundations of both science and the humanities are widely misconceived; and that the frequently held claim of their basic incompatibilities arises out of philosophical prejudices which, owing to cultural lag, have unfortunately not as yet been completely relegated to oblivion. Science is still identified with an absurd mechanistic reductionism, but this is the caricature of science drawn by representatives of the humanities who are largely ignorant of the nature of modern science and also of the more recent scientific outlook in philosophy. The defenders of the humanities often enough increase the existing tension by holding an equally distorted view of the philosophical basis of the humanities.

The errors committed may well be characterized respectively as *reductive* and *seductive* fallacies. It is claimed that science either ignores (perhaps by its very method cannot help ignoring) or else explains away the most essential human values. Science is here charged with the reductive fallacy. Usually the same group of thinkers maintains also that there are aspects of the human mind, manifest especially in the domains of morality, religion, and the arts, which contain an irreducible spiritual element and for that reason will never be capable of explanation by the scientific method, no matter how far it advances. I call this fallacy seductive because it is usually committed by those who indulge in what William James called "tender-minded", that is, wishful and sentimental, thinking.

This impasse between seductive thesis and reductive antithesis can be overcome only by a constructive synthesis that retains and develops whatever valid suggestions or emphases we may discover underneath the grandiose verbiage of the first and the harsh austerities of the second. Neither a philosophy of the 'Something More' nor a philosophy of the 'Nothing But' will do for our time. Only an approach that is resolutely guided by the question 'What is what?' will avoid reading mysteries into the facts as well as refrain from impoverishing them by reduction to something less than experience attests them to be. Such a philosophical outlook, if not yet fully achieved, is fortunately very much in the making.

Especially in the melting pot of American thought, we find that the valuable elements of naturalism and humanism are gradually united in a new integration: the pragmatism of Peirce, James, Dewey, Mead, Otto, Kallen, and Hook; the naturalistic realism of Perry, Holt, R. W. Sellars, Drake, and Santayana; the scientific empiricism of Bridgman, Hull, Tolman, Lundberg, N. Wiener, P. Frank, C. Morris, Northrop, Carnap, Reichenbach, Nagel, and others; the liberal ('American Humanist') wing of Unitarianism. All these trends of thought and many others converge in a broad movement that one may well be tempted to regard as the twentieth-century sequel to the Enlightenment of the eighteenth century.

The humanism held in common to a very large extent in these scientifically oriented philosophies is too well known to require elaborate restatement. Suffice it to say that such human values as freedom and responsibility, rights and obligations, creative and appreciative capacities, are here disengaged from the theological and metaphysical ideologies that have traditionally pervaded their conception. Increasingly adequate and nonreductive analyses have been propounded in the last five or six decades. This reconstruction in philosophy has been and still is in the making. Under the impact of modern science, philosophy is abandoning some of its earlier grandiose and overambitious claims in favor of a humbler and more useful function: the clarification of the foundations of knowledge and valuation. I shall now attempt to apply what seem to me to be the most important insights and suggestions of these currents of thought to the issues of general education.

Clearly nothing is more urgent for education today than a social philosophy that will be appropriate and workable in an age of science. Among the various prominent philosophies of education I mention first two currents which may be styled traditionalistic and which have on the whole either ignored the facts of the age of science or have tried (unsuccessfully, I think) to dispute and combat them: Neo-Thomism and Literary Humanism. Reliance on theological or metaphysical presuppositions makes these views incompatible with the modern scientific outlook. More definitely products of the modern scientific attitude are two other schools of thought: Dialectial Materialism, the official philosophy of Soviet Russia, but also fashionable in certain English scientific groups; and, scarcely worthy of being called a philosophy, the attitude of Vocationalism, quite prevalent in American education. The common element of these two views is their exclusive interest in the practical, technological applications of the natural and the social sciences. This, as well as other shortcomings, makes both points of view appear objectionable or at least badly in need of correction or supplementation. A much more acceptable position is that of Scientific Humanism. This view, at least in general outlook, is related to what is known as Progressivism or Reconstructionism in American education: a synthesis of the scientific attitude with an active interest in the whole scale of human values. Education in both the sciences *and* the humanities is the urgent need of our time. But how can these two aims properly be combined? The question reveals an uneasy feeling as to the compatibility of science and humanism.

Misunderstandings of the nature of science are primarily responsible for the appearance of incompatibility here. A proper historical and analytical perspective on the development of the scientific outlook and its distinctive traits as compared with prescientific and non-scientific attitudes helps to show that mankind achieves intellectual adulthood only with the scientific way of thinking.

Our age is still replete with remnants of and regressions to such prescientific thought patterns as magic, animism, mythology, theology, and metaphysics. The outstanding characteristics of modern scientific method are mostly absent or at best only adumbrated in those less mature phases of intellectual growth.

CRITERIA OF THE SCIENTIFIC METHOD

What, then, are the basic characteristics of the scientific method? The often alleged difficulties of an adequate definition of science seem to me mainly a matter of terminology. We must first distinguish between pure mathematics as an exclusively formal-conceptual discipline, and the factual (or empirical, that is, the natural and the social-cultural) sciences. The certainty, complete exactitude, and necessity of pure mathematics depends precisely on its detachment from empirical fact. Mathematics as applied in the factual sciences merely lends its forms and deductive structures to the contents furnished by experience. But no matter how predominant mathematics may be in the formulations and derivations of empirical facts, factual knowledge cannot attain either the absolute precision or necessity of pure mathematics. The knowledge claimed in the natural and the social sciences is a matter of successive approximations and of increasing degrees of confirmation. Warranted assertibility or probability is all that we can conceivably secure in the sciences that deal with the facts of experience. It is empirical science, thus conceived as an unending quest (its truths claim to be held only 'until further notice"), which is under consideration here. Science in this sense differs only in degree from the knowledge accumulated throughout the ages by sound and common sense.

The aims of science are description, explanation, and prediction. The first aim is basic and indispensable, the second and third (closely related to each other) arise as the most desirable fruits of scientific labors whenever inquiry rises beyond the mere fact-gathering stage. History, often and nowadays quite fashionably declared an art, is scientific to the extent that it ascertains its facts concerning past events by a meticulous scrutiny of present evidence. Causal interpretation of these facts (in history, but similarly also in psychology, sociology, cultural anthropology, and economics) is usually much more difficult than, but in principle not logically different from, causal interpretation (that is, explanation) in the natural sciences. The aims of the pure (empirical) sciences are then essentially the same throughout the whole field. What the scientists are seeking are descriptions, explanations, and predictions which are as adequate and accurate as possible in the given context of research.

The quest for scientific knowledge is therefore regulated by certain standards or criteria which may best be formulated in the form of ideals to be approximated, but perhaps never fully attained. The most important of these regulative ideals are:

1. Intersubjective Testability. This is only a more adequate formulation of what is generally meant by the 'objectivity' of science. What is here involved is not only the freedom from personal or cultural bias or partiality, but – even more fundamentally – the requirement that the knowledge claims of science be in principle capable of test (confirmation or disconfirmation, at the least indirectly and to some degree) on the part of any person properly equipped with intelligence and the technical devices of observation or experimentation. The term *intersubjective* stresses the social nature of the scientific enterprise. If there be any 'truths' that are accessible only to privileged individuals, such as mystics or visionaries – that is, knowledge-claims which by their very nature cannot independently be checked by anyone else – then such 'truths' are not of the kind that we seek in the sciences. The criterion of intersubjective testability thus delimits the scientific from the nonscientific activities of man.

Religious ecstasy, the elations of love, the inspiration of the artist, yes, even the flash of insight on the part of a scientific genius are not in themselves scientific activities. All these processes may eventually become subject matter for scientific study. But in themselves they do not validate knowledge-claims. They may, as in the case of the scientific intuition (or empathy in the psychological-cultural field) be instrumental in the generation of knowledge claims. But it is these knowledge-claims which have to be, first, formulated in an intersubjectively intelligible (or communicable) manner, and, second, subjected to the appropriate kind of tests in order to ascertain their validity. Beliefs transcending all possible tests by observation, self-observation, experiment, measurement, or statistical analysis are recognized as theological or metaphysical and therefore devoid of the type of meaning that we all associate with the knowledge-claims of common sense of factual science. From the point of view of the scientific outlook in philosophy it may be suggested that the sort of significance with which the in-principle-unconfirmable assertions of transcendent theology and metaphysics impress so many people is largely emotive. The pictorial, emotional, and motivational appeals of language, no matter how indispensable or valuable in the contexts of practical life, art, education, persuasion, and propaganda, must, however, not be confused with the cognitive meanings (purely formal- and/or factual-empirical) that are of the essence of science. Each type of significance has its function, and in most uses of language both are combined or even fused. The only point stressed here is that they must not be *con*fused, that is, mistaken for one another, if we wish to be clear as to what we are about.

2. Reliability, or a Sufficient Degree of Confirmation. This second criterion of scientific knowledge enables us to distinguish what is geneally called 'mere opinion' (or worse still, 'superstition') from knowledge (well-substantiated belief). It may be considered as the delimitation of the scientific from the unscientific knowledge-claims. Clearly, in contrast to the first criterion, we face here a distinction of degree. There is no sharp line of demarcation between the well-confirmed laws, theories, or hypotheses of science, and the only poorly substantiated hunches and ideas-on-trial which may ultimately either be included in the corpus of scientific knowledge or else rejected as unconfirmed. Truth-claims which we repudiate as 'superstition,' and, quite generally, as judgments based upon hasty generalization or weak analogy (if they fulfill the criterion of testability), differ from what we accept as 'scientific truth' in the extremely low degree of probability to which they are supported by the available evidence. Astrology or alchemy, for example, are not factually meaningless, but they are considered false to fact in that all available evidence speaks overwhelmingly against them. Modern techniques of experimentation and of statistical analysis are the most powerful tools we have in the discernment between chance and law and hence the best means of enhancing the reliability of knowledge.

3. Definiteness and Precision. This obvious standard of scientific method requires that the concepts used in the formulation of scientific knowledge-

claims be as definitely delimited as possible. On the level of the qualitative-classificatory sciences this amounts to the attempt to reduce all border-zone vagueness to a minimum. On the level of quantitative science the exactitude of the concepts is enormously enhanced through the application of the techniques of measurement. The mensurational devices usually also increase the degree of objectivity. This is especially clear when they are contrasted with purely impressionistic ways of estimating magnitudes. Of course, there is no point in sharpening precision to a higher degree than the problem in hand requires. (You need no razor to cut butter.)

4. Coherence or Systematic Structure. This is what T. H. Huxley had in mind when he defined science as 'organized common-sense.' Not a mere collection of miscellaneous items of information, but a well-connected account of the facts is what we seek in science. On the descriptive level this results, for example, in systems of classification or division, in diagrams, statistical charts, and the like. On the explantory levels of science sets of laws, or theoretical assumptions, are utilized. Explanation in science consists in the hypothetico-deductive procedure. The laws, theories, or hypotheses form the premises from which we derive logically, or logico-mathematically, the observed or observable facts. These facts, often belonging to heterogeneous domains, thus become integrated into a coherent, unifying structure. (Theological and metaphysical systems have, frequently enough, ambitiously tried to imitate this feature of science; but even if they succeeded in proceeding *more geometrico*, the important difference from science remains: they either lack testability or else reliability in the senses specified in our previous points.)

5. Comprehensiveness or Scope of Knowledge. This final point in our enumeration of criteria of science also characterizes scientific knowledge as different in degree (often enormously) from commonsense knowledge. Not only through bold and sweeping hypotheses, but especially through the ingenious devices by means of which they are tested, science acquires a reach far beyond the limits of our unaided senses. With telescopes, microscopes, spectroscopes, Geiger counters, lie detectors, and the thousands of other contrivances of modern science we manage to amplify our senses and thus open up avenues of at least indirect access to the worlds of the very distant, the very large, the extremely small, or the disguised and concealed. The resulting increase in the completeness of our knowledge is, of course, popularly the most impressive feature of science. It must be kept in mind, however, that the scope thus achieved is a product of hard labor, and not to be confused with the sham completeness metaphysicians procure for their world

pictures by verbal magic. Instead of presenting a finished account of the world, the genuine scientist keeps his unifying hypotheses open to revision and is always ready to modify or abandon them if evidence should render them doubtful. This self-corrective aspect of science has rightly been stressed as its most important characteristic and must always be kept in mind when we refer to the comprehensiveness or the unification achieved by the scientific account of the universe. It is a sign of one's maturity to be able to live with an unfinished world view.

The foregoing outline of the criteria of science has been set down in a somewhat dogmatic tone. But this was done only for the sake of brevity.[1] The spirit behind it is that of a humble account of what, I think, an impartial and elaborate study of the history of thought from magic to science would reveal. In any case, these criteria seem unquestionably the guiding ideals of present-day empirical science. They may therefore be used in a definition of science as we understand this term today. It seems rather useless to speculate about just what this term, by a change of meaning, might come to connote in the future.

It should be remembered that the criteria listed characterize the *pure* factual (empirical) sciences. The aims of the *applied* sciences – the technologies, medicine, social and economic planning, and others – are practical control, production, guidance, therapy, reform, and so forth. Responsible activity in the application of science clearly presupposes information which is fairly well substantiated by the methods of the pure sciences. (These remarks intend to draw merely a logically important distinction. The obvious practical interpenetration and important mutual fertilization of the pure and the applied disciplines is of course not denied here.)

CRITIQUE OF MISCONCEPTIONS

Having indicated at least in broad outline the nature of scientific method we may now turn to the critique of some of the misconceptions to which it is all too commonly exposed. In what follows, a dozen typical charges against science are stated and answered consecutively.[2]

Science arises exclusively out of practical and social needs and has its only value in serving them in turn. (Dialectical Materialism and Vocationalism)

While this is important it does not tell the whole story. Science has always also been the pursuit of knowledge, the satisfaction of a deep-rooted curiosity. It should be recognized as one of the cultural values along with art, literature, and music. Better teaching of the sciences and their history can redress the balance. Fuller utilization of results and suggestions from the history and the philosophy of science would give the student a deeper appreciation of the evolution of scientific knowledge and of the scientific point of view. Through proper instruction, the student could be led to rediscover some of the important results of science. The intellectual gratification that comes with a grasp of the order of nature, with the understanding of its processes by means of laws and theories, is one of the most powerful incentives in the pursuit of pure knowledge.

Science cannot furnish a secure basis for human affairs since it is unstable. It changes its views continually. (Traditionalism)

While there is constant evolution, and occasionally a revolution, in the scientific outlook, the charge is a superficial (usually journalistic) exaggeration. The typical progress of science reveals that later views often contain much of the earlier views (to the extent that these have stood the test of repeated examination). The more radical or revolutionary changes usually amount to a revision of the conceptual frame of a scientific discipline. The criticism often also presupposes other sources of certainty which will simply not bear critical scrutiny. The quest for absolute certainty is an immature, if not infantile, trait of thinking. The best knowledge we have can be established only by the method of trial and error. It is of the essence of science to make such knowledge as reliable as is humanly and technically possible.

Science rests on uncritical or uncriticized presuppositions. It validates its outlook by its own standards. It therefore begs the question as regards alternative approaches for settling problems of knowledge and action.

Science has been clarifying and revising its basic assumptions throughout its development. Particularly since the beginning of the modern age and still more intensively since the beginning of our century, an increasing awareness of, and critical attitude toward, the fundamental presuppositions has been most fruitfully applied in the repudiation of dogmatic prejudices and in the articulation of the conceptual frame of scientific method. It can be shown (through logical analysis) that the procedure of science is the only one we are *certain* will yield the results (reliable knowledge, that is, valid explanation and predictions) *if* such results can at all be achieved. Any alleged rival method – theology, metaphysics, mysticism, intuition, dialectics – if it made any contributions at all could not be examined and appraised on any basis other than the usual inductive criteria of science. Generally, it seems that these alleged alternatives do not even aim primarily at knowledge but, like the arts, at the enrichment of experience. They may therefore more properly be said to be *non*scientific, rather than *un*scientific.

Science distorts the facts of reality. In its Procrustean manner it introduces discontinuities where there is continuity (and vice versa). The abstractions and idealizations used in science can never do justice to the richness and complexities of experience.

Since the task of science is to discover as reliable and precise a knowledge of what happens under what conditions, it always tries to approximate the facts as closely as the problem on hand requires and permits. Both continuity and discontinuity can be formulated mathematically and be given an adequate formulation only with the help of modern mathematics.

Science can deal only with the measurable and therefore tends to 'explain away' that which it cannot measure.

While measurement is eminently desirable in order to enhance the precision and objectivity of knowledge, it is not indispensable in many branches of science, or, at least, on their more qualitative levels of analysis. Science does not explain away the qualities of experience. It aims at, and often succeeds in, making these qualities predictable.

Science never explains, it merely describes the phenomena of experience. The reality beyond the appearances is also beyond the reach of science.

This is partly a terminological issue and partly a result of the (traditional but most misleading and useless) metaphysical distinction between appearance and reality. In the sense in which the word *explaining* is used in common life, science *does* explain facts – it deduces them from laws or theoretical assumptions. Questions which are in principle incapable of being answered by the scientific method turn out, on closer analysis, not to be questions of knowledge. They are expressions of emotional tensions or of the wish for soothing (or exciting) experience.

Science and the scientific attitude are incompatible with religion and the religious attitude.

If by religion one refers to an explanation of the universe and a derivation of moral norms from theological premises, then indeed there is logical incompatibility with the results, methods, and general outlook of science. But if religion means an attitude of sincere devotion to human values, such as justice, peace, relief from suffering, there is not only no conflict between religion and science but rather a need for mutual supplementation.

Science is responsible for the evils and maladjustments of our civilization. It is creating ever more powerful weapons of destruction. The employment of scientific techniques in the machine age has contributed to the misery, physical and mental, of the multitudes. Moreover, the biological facts of evolution imply the negation of all morality: the law of the jungle.

These are particularly superficial charges. It is the social-political-economic structure of a society that is responsible for these various evils. Scientific knowledge itself is socially and morally neutral. But the manner in which it is applied, whether for the benefit or to the detriment of humanity, depends entirely on ourselves. Scientists are becoming increasingly aware that they, even more than the average citizen, have to work for enlightenment toward the proper use of knowledge. The facts and theories of evolution have been construed in many ways as regards their implications for ethics. Julian Huxley reads them very differently from the way his grandfather Thomas Henry did [cf. Julian Huxley, 1947; also C. D. Broad, 1949]. It should be easy to see that the forces active on the level of human civilization and intelligent communal life are not completely reducible to those involved in the ruthless struggle for survival.

The ethical neutrality of scientific truth and the ivory tower situation of the pure researcher is apt to generate an attitude of indifference toward the pressing problems of humanity.

Only maladjusted individuals are unable to combine the detachment necessary for the pursuit of truth with an ardent interest in the improvement of the condition of humanity.

Scientific method, while eminently successful in the explanation, prediction, and control of physical phenomena, is distinctly less successful in regard to the facts of organic life and almost altogether hopeless in the mental and social realm. The methods of the physical sciences are essentially mechanistic (if not materialistic) and therefore reductionistic; they cannot do justice to the complex organismic, teleological, and emergent features of life and mind.

'Scientism' as a slogan of criticism and reproach is very fashionable these days. It is true that some scientists and especially some of the popularizers of science have indulged in reductive fallacies of various sorts. But the true scientific spirit as exemplified in some of the foremost researchers is free from that impatience and simple-mindedness that tries to finish the unfinished business of science by hasty speculation. Admittedly, there are tremendous problems yet to be solved. On the other hand what method is there but the method of science to solve them? Explanations of the mechanistic type (in *one* sense of the term) have been abandoned even in physics. But mechanistic explanation in the wider sense of a search for law (deterministic or statistical) is still the indispensable procedure of all sciences that have gone beyond the purely classificatory level. Organic wholeness, teleology, and emergence can be understood, if at all, only by causal analysis on the usual empirical basis. Purposiveness and freedom of choice, far from being incompatible with causality, presuppose causal order.

The methods of science can never replace the intuitive insight or empathic understanding of the practical psychologist, psychiatrist, cultural anthropologist, or historian. This claim is made particularly wherever the object of knowledge is the individual, the unique and unrepeatable.

It is only through the scientific method that the validity and reliability of the intuitive approach can be gauged. There is, on this ground, some doubt as to its more exaggerated claims. However, there is nothing in the principles of scientific method that would deny the occasional, or even frequent, efficacy of intuitive judgments based, as they must be, on a rich (but often not articulated) background of experience in the given field. Aside from the mere artistic contemplation of the unique and individual, knowledge, in the proper sense of the word, always means the subsumption of the specific case under general concepts or laws. This holds in the social sciences just as much as in the natural sciences.

Science cannot determine values. Since scientific knowledge can (at best) find out only what is the case, it can, by its very nature, never tell what ought to be.

This final challenge often comes from theology or metaphysics. It usually maintains that questions of aims, goals, and ideals cannot be settled by the methods of science but rather require recourse either to divine revelation, the voice of conscience, or some metaphysical *a priori* truths. The answer to this in a scientific age would seem to be that a mature mankind should be able to determine its own value standards on the basis of its needs, wants, and the facts of the social condition of man. But it is true that science cannot dictate value standards. It can, as in social psychology, ascertain the actual evaluations of groups and individuals, study their compatibilities and incompatibilities, and recommend (that is *applied* science!) ways and means of harmonizing conflicting evaluations. True enough, in many of the urgent issues that confront us, we do not possess enough scientific knowledge to warrant a course of action. This means that we have to act, as so often in life, on the highest probabilities available even if these probabilities be low in themselves.

But such estimates of probabilities will still be made most reliable by the scientific method. Common life experience and wisdom, when freed from its adherence to prescientific thought patterns, is not fundamentally different from scientific knowledge. In both we find the procedure of self-correction, so essentially needed if knowledge is to be a guide for action. There is an important common element in mature thinking (as we find it in science) and mature social action (as we find it in democracy): progress arises out of the peaceful competition of ideas as they are put to intersubjective test. Cooperative planning on the basis of the best and fullest knowledge available is the only path left to an awakened humanity that has embarked on the adventure of science and civilization.

The scientific view of the world that we have characterized and defended against criticisms from various quarters may with historical and terminological justice be called Naturalism.[3] It differs from mechanistic materialism (too often a mere straw man put up by theologians or idealistic metaphysicians) in that it steers clear of reductive fallacies. If uninformed persons insist on viewing science as essentially materialistic and the humanities as essentially idealistic (not to say spiritualistic) the hopes of fruitful collaboration of both fields in education are slim indeed. But science, properly interpreted, is not dependent on any sort of metaphysics. It merely attempts to cover a maximum of facts by a minimum of laws. On the other side, a mature humanism requires no longer a theological or metaphysical frame either. Human nature and human history become progressively understood in the light of advancing science. It is therefore no longer justifiable to speak of science *versus* the humanities. Naturalism *and* humanism should be our maxim in philosophy and in education. A Scientific Humanism emerges as a philosophy holding considerable promise for mankind – *if* mankind will at all succeed in growing up.

NOTES

[1] A thorough discussion of the logical, epistemological, methodological, and historical issues connected with the criteria would require a whole book, not just another essay.

[2] These charges are not straw men. In more than twenty years of reading, listening, teaching, and argument I have encountered them again and again in Europe and just as frequently in this country. If space permitted and time were less valuable, I could quote many well-known writers in connection with each charge.

[3] It should scarcely need mentioning that this meaning of naturalism has only a distant and tenuous relation to the other meaning in which it is applied to a certain type of literature.

21. VALIDATION AND VINDICATION

*An Analysis of the Nature and the Limits of Ethical Arguments**

[1952]

The following schematic dialogue was constructed with the intention of illustrating some of the typical turns and twists which occur almost invariably when argument in moral issues is pursued through successive levels of critical reflection. A more systematic formulation of the philosophical conclusions that may be derived from a study of such justificatory arguments will be presented in the second part of this essay.

I. A DIALOGUE

A.: Under what conditions can war be morally justified?

B.: Under no conditions. I am a convinced pacifist and conscientious objector. There is no greater evil than war and deliberate killing.

A.: Would you rather be killed or enslaved than do any killing? Are there no circumstances, such as a need for self-defense that would justify killing?

B.: There are none.

A.: If you were saying that wanton killing and cruelty are to be condemned, I should heartily agree with you. But there are occasions in which killing is the only choice: a necessary evil, surely, but justifiable because it may be the lesser evil in the given circumstances.

The point of view of the radical pacifist is unreasonable. More lives might ultimately be saved, and greater happiness for a larger number of people might result if the innocent victims of aggression were to wage a victorious war upon the aggressor. This is essentially the same reasoning that I would apply to the situation in which, for example, a robber threatened my own life or that of a friend.

B.: I admit that all these are very unfortunate situations. My sincerest efforts would be devoted to prevent their very occurrence (by whatever suitable means: education, reform, arbitration, compromise, reconcilation, etc.). But once such a situation arises I still believe that one should not kill.

A.: How do you justify this position?

B.: How does one justify *any* moral judgment? Obviously by deriving it from the basic moral laws. Respect for the life, the rights, the happiness of others is surely such a basic norm, is it not?

A.: I shall be curious to find out how such basic moral laws are proved or established. But before we enter into this deep question, tell me how you defend such a rigid adherence to non-violence, even if you yourself may easily become the victim of aggression or war.

B.: I shall not invoke religious principles here. Perhaps I can convince you if I make you aware of the consequences of the pacifist attitude. Once practiced by many it would tend to spread by way of emulation and thus sooner or later eradicate the evil of killing altogether.

A.: This is an optimistic assertion concerning the probability of certain consequences. In any case it is a question of fact which is not easily decided. However, your disagreement with me seems to go beyond whatever we may think about the facts, namely the conditions and consequences of attitudes. True enough, in your last remark you have tried to establish a common basis of evaluation. You appealed to a humanitarian principle which I do share with you. Still, I think that to kill is morally better than to be enslaved. Since you disagree with me on this, it is obvious that we diverge in *some* of our basic norms. This divergence in attitude can apparently not be removed by considerations of fact.

B.: Are ethical principles then a matter of personal whim and caprice?

A.: I did not mean to imply this at all. As our own cases show, we tend to have very strong and serious convictions in these matters. Far from being chosen arbitrarily, our moral attitudes are a result of the culture and the subculture in which our personalities are formed.

B.: We are not necessarily conforming to the prevailing patterns. I for one, am certainly not. I arrived at my views by independent and serious reflection.

A.: I don't wish to dispute it. And yet your attitudes are a causal consequence of many factors: heredity, environment (physical, and especially social; the influence of parents, friends, teachers, attractive and abhorrent examples, crucial experiences, etc.) and, yes, your (more or less) intelligent reflection upon the facts as they impress *you-as-you-are*.

B.: If you are right, there are limits beyond which rational (i.e., logical and/or factual) argument cannot be extended. Intelligent reflection concerning means and ends, conditions and consequences operates within the frame of basic evaluations. Beyond those limits there could be only conversion by persuasion (rhetoric, propaganda, suggestion, promises, threats, re-education, psycho-therapy, etc.). There are also techniques of settlement of disagreements by way of compromise, segregation (separation, divorce) or higher synthesis. By 'higher synthesis' I mean, for example, the abandonment or severe restriction of the sovereignty of individual nations and a transfer of all sentiments of loyalty to a world government. Only if none of these techniques succeeds, then indeed coercion by violence, alas, seems inevitable. – (Universal pacificism is the only solution! But that's not my point at the moment.)

A.: You have expressed my point of view very well. But you are obviously unwilling to agree to it.

B.: Indeed not. Everything in me cries out for a belief in objectively and universally valid standards of moral evaluation.

A.: You will not get very far if you assume some theological or metaphysical absolutes. Any reference to the revealed commands of a divine authority is futile. For you would have to tell how you can know those imperatives as divine; and even if you were to know them as such you would have to state a reason as to why anybody should obey them. The same criticisms apply to any alleged metaphysical insight into what man ought to be. And if you dismiss theological and metaphysical foundations for morality you will find it difficult to argue for standards that are independent of human needs and interests.

B.: It's precisely human needs and interests that provide a solid foundation for moral standards. In all cultures that we call 'civilized' there are essentially the same ideals of cooperation (as opposed to conflict), of helpfulness (as opposed to harmfulness), of love (as opposed to hatred), of justice (as opposed to inequity), and of perfection and growth (as opposed to stagnation and decay). Cultural relativity and the variability of human nature have been exaggerated. There is a significant core of essential features shared by all human beings. Human nature as it is constituted biologically and psychologically, and as it finds its existence in a context of interdependence with other human beings, could scarcely fail to develop just those ideals of morality. I admit that these ideals are only rarely fulfilled or even approximated in actual conduct. But they are *the* standards of ethical evaluation. It is with reference to this frame that we make our judgments of 'good' and 'bad', 'right' and 'wrong.'

A.: Much as I share your ideals, I can't refrain from calling your attention to the fact that there are notable exceptions that restrict severely not only the universality of certain types of conduct (this is what you admitted), but also the universality of the very standards or ideals of morality. To many an ancient or Oriental culture the idea of perfection or progress remained completely strange. The prevailing ideologies of capitalism and nationalism basically extol the ideals of competition over those of co-

Operation. Only superficially and often hypocritically do they pay lip service to humanitarian or Christian ideals. And the very principle of justice (in the sense of equal rights for all) has been flouted not only by tyrants, aristocrats and fascists but also by such eminent philosophers as Plato and Nietzsche. Our own divergence on the issue of radical pacificism is equally a case in point. There are countless further, possibly secondary and yet radical divergencies as regards attitudes toward civil liberties, sex and marriage, birth control, euthanasia, the role of religion (church and state), animals (vegetarianism, vivisection), etc., etc.

B.: Disregarding the secondary divergencies, I must say that the deviations from the more fundamental and true moral ideals are simply perversions and corruptions. Whoever denies the principles of justice and neighborliness is immoral. Kant was essentially right and convincingly logical in defining moral conduct by his categorical imperative. Only a principle that is binding for all and excludes any sort of arbitrary privilege and partiality can justifiably be called ethical. The ideals that I enumerated are the very essence of what is meant by 'morality.' To be moral consists precisely in placing oneself in the service of interests and ideals that transcend purely selfish purposes.

A.: This is what *you* mean by 'morality.' (And, of course, it is in keeping with traditional morality.) But Nietzsche, for example, explicitly proposed a revolution in all traditional morality. Clearly, he considered his own value-system as the 'true ethics.' Are you not aware that you are begging the very question at issue? You speak of 'true moral ideals'; you call certain views 'immoral', 'perverse', 'corrupt'; you say that only certain types of principles can "justifiably be called ethical." You are using persuasive definitions[1] here. You call 'moral' or 'ethical' only such doctrines or principles as agree with your own convictions about what is *right*. The fascination with the '*logicality*' of Kant's categorical imperaive may in part lie in its implicit appeal to some version of the principle of sufficient reason: If there is no reason to discriminate (as regards rights and obligations) between two persons then such discrimination is willful, arbitrary, unjust. But far from involving strictly logical contradictions, such 'unjustifiable' discriminations would merely violate *one* (not as you would say '*the*') definition of justice. A reason for discrimination could always be found. That it may not be accepted as a 'good,' 'relevant' or 'sufficient' reason is but a consequence of the ethical principles or fundamental evaluations of some alternative system. Let me assure you again that I share your moral attitudes. But strongly as I feel about them,

I see no need for, and no profit in defending them with bad logic. You cannot by some verbal magic establish justifications for ideals which obviously are neither logically nor empirically unique. These ideals compete with genuine alternatives.

B.: I can't believe this. The ideals that I have listed are the ones that will benefit humanity in the long run. Not just a particular group, but all of mankind.

Moreover these ideals are comprised by the essence of *rationality*. Man, the rational animal, is by his very nature not only characterized by his capacity for adequate deductive and inductive thinking, but also by his sense of justice and his abhorrence of violence as a method for the settlement of disputes.

A.: You are still begging the question. Those who do not accept the principle of equality are not interested in *all* of mankind. Furthermore, your time-honored conception of human nature is clearly not an account of actual fact, but of an ideal (by no means universally shared) which you utilize for a persuasive definition of MAN. You won't convince any serious opponents by mere *definitions*. But you might try to entice, persuade, educate or reform them in other ways. You may also hope that the increasing interdependence of all mankind on this planet will eventually generate a fundamental uniformity in the principles of moral evaluation.

B.: You underestimate the role of experience in the settlement of moral conflicts and disputes. Those who have had an opportunity to experience different ways of life soon learn to discriminate between the better and the worse. Experience in the context of needs and interests, of claims and counter-claims, of existing and emerging rights and obligations in the social milieu soon enough mould the moral conscience of man. We do not live in a vacuum. The constant encouragements and discouragements of our actions and their underlying attitudes form the very atmosphere of the life in the family, the workshop, the market place, the tribunal, etc. Add to that the basic sympathy human beings feel for each other and you will have to admit that there is a large mass of empirical factors that operate in the direction of a common standard of social morality.

A.: If I may use a parallel drawn from the field of aesthetics, there are a great many people who prefer pulp-magazine stories to 'good' literature; or swing (jazz, jive or whatever is the fashion) to 'great' music. Similarly, there are plenty of people who have had an opportunity to experience both the ruthless and the kindly way of life and yet subscribe to the principles of the former. Kropotkin rightly, though somewhat sentimentally, pointed out that despite the cruel struggle for existence in the animal kingdom there is also a good deal of mutual help and self-sacrifice. If human sympathy were as fundamental as (he and) you claim it is, there could hardly be such views as those of Nietzsche, Hitler, and Mussolini on the 'greatness' of war. Only by endorsing one norm against other possible alternatives can you avail yourself of the premises by which to validate the special moral precepts which are dear to your heart.

B.: You still have failed to give me a single good reason why I or you or anyone should adhere to even those moral principles which we happen to share. Your position is a skepticism that could easily lead to moral indifference and cynicism.

A.: And what sort of reason do you expect me to give you? If I provided you with premises from which you could *deduce* our moral standards, you would ask me for a justification of those premises. And you surely don't want a reason in the sense of a motive. You are motivated already. You do not seriously entertain doubt as long as this motivation prevails. And nothing that I've said was intended to undermine it. The aim of my remarks was clarification; not education, fortification or edification. Too many philosophers have sold their birthright for a pot of message.

II. ANALYSIS AND CONCLUSIONS

The foregoing argument illustrates among other things the ever-present pitfalls of the *petitio principii* in the procedures of justification. If the radical pacifist is accused of an exaggerated value-fixation upon 'reverence for life' he is free to retort that his opponent has a hypertrophied value fixation upon liberty or upon the survival of the greater number of persons. In order to condemn some value-fixations as inhumane, immoral or perverse, it is necessary to invoke some ideals or standards of humaneness, morality or normality. It is only with reference to such ideals or standards that we can justify the *approval* of thrift, honesty, friendship, the devotion to science or art, etc. and the *disapproval* of avarice, hypocrisy, belligerence, sexual aberrations, etc. From a purely factual psychological or socio-psychological point of view *all* value-fixations may be explained in terms of some causal-genetical principles, such as Wundt's 'Heterogony of Purposes', Allport's 'Functional Autonomy', or some other laws of motivation as formulated in psychoanalytic or behavioristic theories.

Let us suppose that socio-historical and anthropological research could show that there are basic invariant moral ideals embodied in otherwise diverse cultures. Even then it cannot be denied that the rank-order of the normative force of these ideals has varied with time, clime, and cultural conditions. Wise moral philosophers along with the great dramatists and novelists of the ages have always known that moral problems in their most poignant and irresoluble form consist in the conflict of good with good or right with right. The understandable hope for the demonstration of one unique set of standards in terms of which an objective and universally binding adjudication of all moral issues could be achieved, may well turn out to be chimerical. Only if certain baisc – and to many people all-too-obvious-valuational premises are taken for granted, can we obtain the semblance of objective deducibility of more special moral rules. If, for example, we take for granted that the life of the species *homo sapiens* is to be preserved, that conflict and violence is to be minimized, then a great number of special precepts are derivable from these premises taken together with special facts and laws concerning human conditions and behavior. The truth implied in the critique of the 'naturalistic fallacy' reduces to the truism that factual statements alone cannot possibly entail normative conclusions. Some normative premises are indispensable.

If rational argument, criticism or justification is to be distinguished from persuasion by means of the emotional and motivational expressions and appeals of language, what are the forms of such reasoning and what are its criteria of validity?

The classical doctrine of self-evidence as a criterion of validity or truth still exerts its powerful influence. Brentano, Husserl and the phenomenologists; G. E. Moore, C. D. Broad, W. D. Ross, A. C. Ewing and other recent English intuitionists have revitalized this ancient (and Cartesian) tradition. There is scarcely any space here even to remind the reader that this philosophical point of view is open to the most serious objections. Its relevance to the truth of the axioms of geometry has become suspect since the developments of the non-Euclidean geometries and their application in modern physics and astronomy. More fundamentally, the recent developments in the philosophical foundations of logic and mathematics have shown that self-evidence is neither a necessary nor a sufficient condition for truth or validity. The better intentions of the intuitionists to the contrary notwithstanding, the doctrine of self-evidence is at fault precisely because it is psychologistic. The accent of self-evidence is a result of habituation. Basic principles or presuppositions which delimit a certain universe of discourse or specify a certain field of validating procedures acquire the appearance of absolute cogency and uniqueness, because they form the indispensable (and hence within this context unquestioned) *conditiones sine qua non* of the very enterprise which they make possible and for which they legislate. Finally, intuitive self-evidence cannot possibly be claimed to yield absolutely unique or indubitable knowledge. Notoriously and especially in regard to moral judgments (not to mention aesthetic evaluations) there is no unanimity on just *which* principles are self-evident. It requires some arrogance to claim one's own intuitions infallible, and the disagreeing intuitions of others as in need of revision (by 'deeper reflection', 're-education', 'enlightenment', etc.).

At this point one of the most crucial questions in all philosophy arises: Are the justifying principles of knowledge, i.e., the principles of deductive and inductive logic, as undemonstrable and as much lacking uniqueness as are the norms of moral judgments? If intuitive cogency is to be abandoned as a criterion of truth, are we not faced with an analogous plurality or relativity in regard to basic presuppositions in the field of cognition?

Only a few suggestions can here be made as regards these burdensome scruples.[2] Firstly, the validity of deductive or inductive inference is presupposed in ethical argument. But no distinctly ethical norms are required for the validation of knowledge-claims. Reasoning in matters of morality utilizes, as any reasoning must, principles of deductive inference when special cases are subsumed under general (in this case, moral) rules. And in any practical issue of moral choice, inductive inference is indispensable for the determination of the most likely consequences of actions. There is no question then, that in the context of validation, the principles of cognition are more fundamental than the norms of morality. In this sense we may safely claim the 'primacy of *pure* reason.' Secondly, despite the fashionable notions about 'alternative logics' it can be shown that at least the rules of deductive inference possess a uniqueness which, even if not present in the same degree, is also characteristic of the rules of inductive inference.

In order to grasp this situation clearly, a fundamental distinction, often badly neglected or blurred beyond recognition, must now be drawn: When we speak of 'justification' we may have reference to the legitimizing of a knowledge-claim; or else we may have in mind the justification of an action. The first case may be called '*justificatio cognitionis*' (validation) the second, '*justificatio actionis*' (vindication). The rules of deductive and inductive inference serve as the justifying principles in validation; purposes together with (inductively confirmed or at least confirmable) empirical knowledge concerning means-ends relations, or in the extreme, degenerate case with purely logical truths, serve as the basis of vindication (pragmatic justification). Only ends can justify means, even if in accordance with the well-known slogan it will be admitted that a given end may not justify the utilization of every means for its attainment.

The word 'reason' displays ambiguities similar to the word 'justification.' Besides naming a capacity of the human mind (part of which is the ability to state reasons) it is used in referring to causes and purposes, as well as to grounds of validation. Aristotle, Schopenhauer, and many thinkers between and after, have struggled to disentangle these and other meanings of 'reason.' Kant's distinction between the questions '*quid facti*' and '*quid juris*' has shed a flood of light on the basic issues of philosophy and has since become indispensable for the analysis of the problems of epistemology and ethical theory.

The justifying principles (*justificantia*) for the establishment of knowledge-claims have been retraced to their ultimate foundations in the rules of inference and substitution in deductive logic. We cannot without vicious circularity disclose any more ultimate grounds of validation here. Similarly the rules of maximal probability in inductive inference form the ultimate validating basis of all empirical reasoning. Correspondingly the supreme norms of a given ethical system provide the ultimate ground for the validation of moral judgments. No matter how long or short the chain of validating inferences, the final court of appeal will consist in one or the other type of justifying principles. Rational argument presupposes reference to a set of such principles at least implicitly agreed upon. Disagreement with respect to basic principles can thus only be removed if the very frame of validation is changed.[3] This can occur either through the disclosure and explication of a hitherto unrecognized common set of standards, i.e., still more fundamental validating principles to which implicit appeal is made in argument, or it can be achieved through the pragmatic justification of the adoption of an alternative frame, or finally, through sheer persuasion by means of emotive appeals.

Validation terminates with the exhibition of the norms that govern the realm of argument concerned. If any further question can be raised at all, it must be the question concerning the pragmatic justification (vindication) of the (act of) adoption of the validating principles. But this is a question of an entirely different kind. The answers we can give to this sort of question are apt to appear trivial, but for the sake of philosophical clarification they are nevertheless indispensable and illuminating. If the logical reconstruction of justification is pursued as here suggested, then even an obvious, not to say utterly trivial, vindication will at least make fully clear which aims are attained by means of the adoption of some specified validating principles. Thus it is quite plain that the adoption of the rules of deductive inference is pragmatically justifiable in that only reasoning which accords with them can insure the transition from true propositions to other true propositions. No vicious circularity is involved here. We are not attempting the (impossible) validation of ultimate validating principles. We can afford, and could not possibly refrain from, using logic in a vindicating argument, precisely because we are here concerned with arguments about means-ends relations. There is a similar vindication, formulated by H. Reichenbach [1928, Sections 38, 39; 1940; and 1949b] for the adoption of the principle of induction. The reasoning in both cases is purely deductive because of the extreme (degenerate) nature of the question at issue. In regard to induction the following holds: If there is an order of nature at all (and we don't know that there is and we don't know that there isn't – beyond the scope of actual observations), then the method of simplest generalization is the only method of which it can be demonstrated (deductively) that (1) it *can* (but of course need not) succeed in disclosing that order and (2) that is is self-corrective. This obvious, simple tautology provides a pragmatic justification of the adoption of the rule of induction for anyone who wishes to attain the two mentioned aims, namely to make true inductive inferences (e.g. predictions) and to be able to keep such inferences adaptable to the accumulating evidence.

It may be charged that our analysis is outrageously artificial; that we never have occasion to 'choose' a basis of validation; that in real-life-situations we find validating and vindicating arguments so intimately fused, that their separation distorts severely the dialectics of both cognitive and valuative arguments. My reply is, firstly, that all logical analysis from Aristotle through Descartes down to our time necessarily consists of an artificial and schematic reconstruction (see [Feigl, 1950c]) and its illuminating character depends precisely upon the disentanglement of factors or aspects which, though admittedly *fused* in ordinary argument, are in danger of being *con*fused in philosophical reflection. Secondly, I would say that those who make the charge under discussion, characteristically resolve the problems of justification simply by a *fiat* of definition. Induction, for example, is said to need no justification because the rule of induction defines (at least in part) *what we mean* by '*justifiable inference*.' Similarly, as we have remarked already, such moral principles as those of justice or benevolence may be claimed to constitute (at least in part) *what we mean* by a '*(rational) morality*.' If this sort of analysis results in a clear explication of the legislative principles of a given domain of validation, I should gladly admit that it is a helpful step in the clarification of philosophical perplexities. But it should be equally clear that this procedure is apt to rest its case simply with a persuasive definition of certain key-terms such as 'rational', 'valid', 'probable', 'morally right', etc. Once aware of the persuasive character of these definitions one should wish, in all candor, to state why one finds them persuasive. And the answer to this question must clearly refer to one's interests, purposes or ideals. Thus, while vindication can never prove (validate) any principles of validation, it can clarify their role in the context of human thought and action.

The validating principles of deductive and inductive logic do not seem at all to have any plausible alternatives or competitors. This is so, very likely because in this age of science our conception of the criteria of valid and reliable knowledge have already been so sharply focussed and so severely purged of pre-scientific (non-scientific and unscientific) elements. The purposes of the cognitive enterprise are today so clearly delimited that its basic criteria (but of course not its special methods and techniques) have attained practically universal consent.

It is only too tempting to hope for a similarly universal code of morals. But in view not only of the stark realities of group and culture-centered ethical standards, but also because of the ever present quandaries regarding the priority between the several supreme standards ('*prima facie* obligation', i.e., the validating principles of moral judgments) within a given group or culture we can scarcely expect a universal unanimity of purposes which would vindicate a set of unique standards and a rigid order of priority among them for any and all questions of moral decision. At this point, I must concede, that the relativism implicit in the emotivist analyses (of Stevenson, for example) may prove insuperable. But beyond this important concession I would stress that the emotivist assimilation of moral issues to questions of personal taste and preference does not even begin to do justice to the nature of argument and justification in the moral realm of discourse. There is a great deal of validation in ethical arguments which is only too easily lost sight of, if attention is primarily fixed upon persuasion or vindication.

In analogy to the analysis of justification in the cognitive domain I suggest that moral judgments are to be reconstructed as knowledge-claims and as subject to validation (or invalidation) by virtue of their accordance (or non-accordance) with the supreme norms of a given ethical system. In order to carry out this reconstruction, judgments of right and wrong, and likewise statements of obligation and of rights, must be construed as empirical propositions. This is possible only after these typically normative terms (and other relatives and derivatives 'good', 'evil', 'desirable', 'condemnable', etc.) have been given a factual reference in addition to their positive for negative emotive appeals. This means that we make, in this context deliberately a legitimate device of what in other contexts must indeed be repudiated as the 'naturalistic fallacy.'[4] This amounts to construing moral norms in the logical form of general laws. But in contradistinction to the general laws of the empirical sciences the moral laws are not subject to confirmation or disconfirmation by empirical evidence – at least and certainly not in the same sense. Their logical character is rather that of basic definitions or conventions for the use of normative terms with reference to empirical aspects of conduct, intentions, attitudes, personality traits and social objectives. In regard to the factual content as well as in their cultural function, normative moral terms are quite similar to such terms of medicine as 'healthy', 'diseased', 'normal', 'abnormal', 'well-functioning', 'mal-functioning', etc. Just as in questions regarding normality or abnormality in medicine, we require a factual content (in addition to emotive appeals) of these terms in ethics. We need likewise factual reference in order to break through the circle of formal tautologies (such as 'the good is that which it is right to accomplish') and to attach these formal-and-emotive terms to empirical aspects of the facts of individual and social life.

Only with a reconstruction of this sort can we escape the sterility of formalism in ethics. If we wish to know for example whether killing in self-defense is morally right, we cannot get an answer unless definite and empirically specified moral rules (including priority-rules as between standards) are provided as *justificantia cognitionis* of the *correctness* of the moral judgment at issue. Obviously the same considerations apply to questions of distributive and retributive justice, to the evaluation of the various virtues, of measures of social, legal, political reform, etc.

It is a simple consequence of the proposed analysis and reconstruction that it is futile to criticize one system of norms in terms of another which is logically incompatible with it. Validation of moral judgments always requires a set of given norms to which we must hold fast, at least temporarily, in order to examine the validity of more special moral judgments. As in the case of the justification of cognition, so here in the domain of ethics, the only further step concerns vindication. The purposes which may be adduced in vindicating arguments for a whole system of moral norms are embodied in the individual interests and social ideals which we have come to form in response to life experience. The principle of justice (the golden rule) or other implicit definitions of 'right actions' may, for example, be vindicated by reference to the ideal of a peaceful, harmonious and cooperative society. Or the principle of benevolence may be vindicated by reference to the ideal of the greatest happiness of the greatest number. We see then that the perennial dispute between deontological and teleological theories in ethics may perhaps be settled by the recognition that the former are concerned with validation, the latter with vindication.

The present approach differs from both the intuitionistic and the emotivist point of view (and is in more than one way closer to the Kantian) in that the great variety of self-evident *prima facie* obligations countenanced by the intuitionists and the corresponding equally great variety of interest-fixations allowed for by the emotivists are supplanted by a relatively small number of basic norms and priority rules. Naturally, the task of demonstrating that this is an adequate and feasible reconstruction is enormous and has here been barely suggested. In contradistinction to the Kantian metaphysics of morals a plurality of alternative ethical systems is here envisaged as a matter of historical and contemporary fact. As long as there are changing and divergent terminal purposes and ideals there will be different systems of moral validation. The moral approval of a given ideal is of course trivially validated by the system which that ideal vindicates; and, contrariwise, trivially invalidated by an alternative incompatible system. – But enough has been said about the dangers of the *petitio principii.*

One final question: Does the pluralism and relativism implied in the preceding remarks rule out *objectivity* in ethics? As may be expected by now, the answer depends upon the precise meaning which one is going to connect with the term 'objectivity.'

The objectivity of the truths of arithmetic lies in their logical necessity. Anyone who understands the postulates and definitions of arithmetic and complies with the rules of deductive logic will concede the universal validity of arithmetical truth. The objectivity of propositions of factual knowledge means something different: the intersensual and intersubjective confirmation of knowledge-claims – and everything that these phrases imply, especially the principles of confirmation. 'Objectivity' in the moral domain may mean a variety of aspects: (1) The logical necessity inherent in validation. (2) The logical consistency of the norms of one system. (3) The factual objectivity of the characterization of the empirical features of attitudes, conduct, etc. which are the subject of moral appraisal. (4) The factual objectivity of statements regarding conditions-consequences and means-ends relations. (5) The factual objectivity of statements concerning human needs, interests and ideals as they arise in the social context. (6) The conformity of the norms with the basic bio-psycho-social nature of man, especially as regards the preservation of existence, the satisfaction of needs, and the facts of growth, development and evolution.[5] (7) The degree of universality with which certain moral norms are actually or potentially embodied in the conscience of man within given cultural groups or perhaps even in cultural groups of all times and climes. (8) The equality of all individual persons before the moral laws – as conceived in the universal applicability of these laws.

Crucial questions arise only in regard to the factual truth of the seventh point and the significance of the eighth. The actuality of universally common standards, as we have pointed out, is problematic. Only at the price of a precarious attenuation of meaning, if not of the risk of tautological vacuity, could one defend this claim. As to the potential convergence towards common standards we may allow for a cautious optimism. Finally, and perhaps most critically, there remains the question whether we shall mean by an 'ethical norm' one which embodies a thorough-going impartiality. If so, then we have by definitional fiat implied the essence of the principle of justice, the very conception of 'moral law.' But when we speak of the 'ethics' of feudalism, or even of the 'ethics' of Nazism, along with the 'ethics' of Christianity, or the 'ethics' of democracy, we obviously utilize a different definition covering much more ground. Of course, we are free to declare that 'fascist ethics' is a contradiction in terms; i.e., we may decide that a code of norms for the appraisal of conduct is to be called 'ethics' only if it embodies at least the principles of benevolence and impartiality. But must we then not conclude that the word 'ethics' itself is subject to persuasive definition?

NOTES

* This essay is a revision of an earlier (hitherto unpublished and altogether different) version of my essay 'De Principiis Non Disputandum . . .?' included in [Feigl, 1950a]. In 'De Principiis . . . .' the problem of justification is discussed not only with reference to ethical principles but also in regard to the more fundamental principles of deduction, induction and the criterion of factual meaningfulness. For an important analysis of closely related issues, see also [Wilfrid Sellars, 1950].

1 This useful phrase was coined by C. L. Stevenson. In his book *Ethics and Language* [1944, p. 210], he explains it as follows: "In any 'persuasive definition' the term defined is a familiar one, whose meaning is both descriptive and strongly emotive. The purport of the definition is to alter the descriptive meaning of the term, usually by giving it greater precision within the boundaries of its customary vagueness; but the definition does *not* make any substantial change in the term's emotive meaning. And the definition is used, consciously or unconsciously, in an effort to secure, by this interplay between emotive and descriptive meaning, a redirection of people's attitudes."

2 For a fuller discussion, cf. [Feigl, 1950a].

3 For an extremely important and clarifying discussion of the distinction between questions within a presupposed frame and questions concerning the frame itself (in connection with closely related issues), cf. R. Carnap [1950a].

4 This fallacy was most infelicitously labeled by G. E. Moore. As I view the matter, Moore's criticism should have been directed against the confusion of motivative appeals

with factual meaning; or more closely in keeping with Moore's intentions, against the confusion of the phenomenological 'oughtness' (its relatives and opposites) with the empirical characteristics of conduct with which these intuitively given (and indeed phenomenologically unanalyzable) qualities of moral awareness are associated.

[5] This elementary but important point stressed in naturalistic ethics from Aristotle down to the philosophizing biologists of our time, is apt to be neglected by purely analytic philosophers.

22. EVERYBODY TALKS ABOUT THE TEMPERATURE

(*On the Nature of Value Judgments*)

THEME AND VARIATIONS

[1964c]

Difficile est satiram non scribere.
Juvenal

Common Man: It's too cold up here.*

Inquirer: I don't know about that. Are you trying to say that the temperature isn't what it 'ought' to be?

Plain Factualist: No, he means just what he said – it's too cold up here. It's a simple fact: take a look at the thermometer outside!

Inquirer: 'Too cold!' 'Too hot!' I suppose you think you'd know if it was 'just right'?

Plain Factualist: Sure, I would.

Inquirer: Well, *how*, for heaven's sake?

Plain Factualist: Just as I said – look at the thermometer. If it's between 68° and 74° Fahrenheit, then it's just right.

Inquirer: *Sancta Simplicitas*! The thermometer *by itself* can tell you that from 68° to 74° is 'just right'? I doubt if that's what you mean.

Common Man (*basso ostinato*): We must consult the people! Majority rule will decide!

Socialist: Yes, *if* the people are not already victims of the propaganda of the privileged few.

Aristocrat: Hmph! The large masses will never develop subtlety enough to discern the *true* standard of temperature.

Socialist (*con fuoco*): Intolerable conceit!

Inquirer: Hold on, now! I'm really interested in finding out how all of you go about deciding when a temperature is 'just right.' Tell me a little more. If you have a norm of some kind – and you apparently do – how are you going to justify it?

Plain Factualist: Oh, now I get you. You want to know what a *normal* temperature is. Obviously, it's the *average* temperature. You figure it out statistically.

Inquirer: All right. But the average temperature in Minnesota – I think you'll agree – isn't the same as the average temperature in California. How, may I ask, can statistics tell you which of the two is 'just right'?

Hedonist: Well, you've got to take account of human feelings. A temperature that's pleasant and comfortable is called normal.

Metaphysician: One man's heat is another man's cold. I mean to say that if you allow yourself to go by the subjective reactions of mere human beings you will inevitably get into a quagmire of relativities and never attain an absolute criterion valid for all.

Zen Buddhist: (Silently puts Metaphysician first on ice, then into a sauna; finally puts on his coat and goes about his business.)

Existentialist: (ruminatively) What is needed here is a clarification of categories. The coldness is *vorhanden*, the registered degree on the thermometer is *zuhanden*, but the real problem concerns *Dasein*: it is not what the temperature ought to be but what I, confronted by this particular *Vorhandenheit*, ought to be – how I ought to become *existenziell ursprünglich*.

Physiologist: I think there are objective criteria – maybe not absolute, but scientifically valid. For example, the best functioning of the human organism depends upon a certain range of temperature conditions. Outside that range, these functions are impaired, if not imperilled.

Inquirer: Yes, but just how does that help? All you've done is to define the 'best temperature' in terms of 'best functioning.' Now that you've done that, can you think of a good objective definition of 'best functioning'?

Evolutionist: Well, I can. An organism functions optimally if it is most favorably equipped for the struggle for existence. Consequently, a 'normal' temperature is one that is most conducive to the survival of the individual, and to the propagation of the species.

Psychoanalyst (Rank, perhaps, but not file): The craving for warmth, you see, is doubtless due to the unconscious wish to return to the maternal womb. It must be this prenatal temperature to which we refer as 'normal.'

Technocrat: Rubbish and nonsense! Every one of you fails to see that this is not a problem for theoretical speculation at all, but for practical action. Properly utilizing our technological know-how, we can change the climates of the earth! Temperatures everywhere can be made just right!

Biological Engineer: *Or*, by appropriate *biological* techniques, we could adjust everyone to the particular climate in which he has to live!

Common Man (*basso ostinato*): We must consult the people! Majority rule will decide!

Inquirer: For goodness' sake! Can't you wild men get it through your heads that this whole problem has to have some sort of logical clarification, before we tackle anything else?

Metaphysician: The trouble, as I see it, is that the question cannot be settled by appeal to experience. The *norm* we are looking for must be anchored to something basic, absolute, inherent in the true nature, in the transcendent essence, of the universe as it is in and by itself.

Empiricist: Oh, fine! I gather from what you say that nobody can *ever* know when the temperature is 'just right.' That kind of talk won't get us anywhere. Metaphysical verbiage acts like a sedative on the inquiring mind. For my part, I am still convinced that experience, and the experimental method, can determine . . .

Pragmatist: Precisely! An all-around investigation of the conditions and the consequences of various temperatures will illuminate the specific structure of the means-ends-continuum, and thus enable us to provide for an approximate and warranted tentative criterion.

Inquirer: Well, I suppose you know what you're trying to say. But if you're going to decide anything about the relation of means to ends, it seems to me you have to have a pretty clear idea of both. And don't the *ends* determine the standards for the means?

Pragmatist: Oh, I hope you aren't looking for anything final and unquestionable. Really, the quest for certainty and the search for some kind of absolute suggest an immature mind. And, in the long run, normative judgments, like any other hypotheses, must be corroborated by the operations of intelligent theory.

Inquirer: I'll go along with that, if you'll explain how 'intelligent inquiry' determines a norm. It looks to me as if you really can't 'inquire' into a norm, the way you can inquire whether some particular temperature is or isn't normal – you have to *presuppose* a norm.

Kantian: Quite so, of course. All of which goes to show that standards, themselves, cannot be established by any empirical investigation. Norms are presupposed: they are *a priori*; they are inherent in pure practical reason, and therefore, universally valid.

Conventionalist: Well, of course, I would agree that your norm has to be presupposed. But you must realize that it's going to be *a priori* in just the same way an arbitrary definition is *a priori*.

Hedonist, Physiologist, Evolutionist, Empiricist, Pragmatist (*unisono fortissimo*): Arbitrary!? Arbitrary!? Far from it!! The standard is imposed by certain *facts* – facts about man and about his environment!

Interest Theorist: *Now* we're getting somewhere! I've been waiting for a chance to point out that the standard of the 'just right temperature' is determined by human needs, human wants, and human interests.

Inquirer: Maybe so, but aren't there all sorts of human needs and human interests? Don't we think of some of them as justified, or 'normal', and some of them as unjustified, or 'abnormal'? All of you scientists insist on begging the question!

Gestaltist: Ah! My friends, you will never arrive at any resolution of our problem, it is clear. The reason is that you are bent on analyzing our needs and interests in atomistic, splendid *isolation*. But the truth, of course, is that only the *total configuration* of the *interdependent* needs and interests, the field of vectors and valencies, with its varying degrees of harmony, can determine the standard of normality – in the matter of temperature, or of anything else. I think, perhaps, this is what our friend Pragmatist dimly had in mind.

Inquirer: Maybe it is, but you seem to think there are universally valid, objective criteria of 'harmony.' What makes you so sure?

Relativistic Autobiographist: Let's be frank about this. We can't justify absolutism – now, can we? Empirical definitions of normality are *always* relative. Let's admit, here and now, that all you and I can do in these questions of value is describe our own private personal individual attitudes and preferences. When I say that such and such a temperature is what it 'ought' to be, all I mean is that I *like* that particular temperature.

Inquirer: Maybe we're not talking about the same thing. You're talking about a statement of fact; I'm talking about a value judgment. How do you come to use the word 'ought' in the first place?

Common Man (*basso ostinato*): We must consult the people! Majority rule will decide!

Intuitionist: Look here, all of you. Just suppose, for the sake of argument, we say that the temperature is 'just right' when it is between 68° and 74° Fahrenheit. Now, this is obviously a synthetic proposition. But our naive friend, Plain Factualist, misinterprets it: he thinks it is a tautology. And I suspect that Conventionalist, here, is making the same mistake. Yet there are two concepts involved in this statement of ours: one of them, the concept of temperature, is empirical – it's *factual*; you might call it 'natural.' The other concept, 'just right', is valuational-normative; you might call that one 'non-natural.' Now, our original statement connects the two concepts, but doesn't say they're the same. In fact, it leaves the question completely open whether the temperature range, 68° to 74°, is 'really' a normal temperature.

Inquirer: All right, but how would you go about answering that question? Or *would* you?

Intuitionist: You understand, we have a direct intuition of the quality that makes a temperature 'just right.' That quality, gentlemen – that quality is indefinable, unanalyzable, irreducible. It is a simple quality, like *yellow* – except that it is not a sensory quality.

Empiricist: But even if you choose to believe in this kind of intuition, it still comes under the heading of 'direct experience'; and as such, it must have a perfectly natural explanation.

Intuitionist: So what? What if you could explain the experience, and track down its psychological origin? You would still have to *have* that kind of direct apprehension, in order to make any final decision about the truth of a value judgment.

Inquirer: But what makes you think that a subjective intuition like that could possibly have anything approaching universal validity? What do you say to all these relativists around you? It is pretty clear that two different people might have hopelessly conflicting intuitions.

Intuitionist: Ah, but we know that in logic, and in mathematics, there is agreement – potentially, anyhow. Wouldn't we be fools to deny that such agreement is at least *possible* in the matter of what temperatures are 'just right'?

Common Man (*basso ostinato*): We must consult the people! Majority rule will decide!

Radical Emotivist: It's plain that if value judgments like this one aren't *empirical* statements – if they aren't physical statements, or biological, or psychological, or even autobiographical – then they aren't propositions at all. They may look like propositions, but they obviously can't be either true or false. They only express likes and dislikes. You wouldn't call them knowledge claims; they're only symptoms of attitudes. When you say it's 'too cold' or 'too warm', your statement has about as much propositional meaning as chilblains, or a heat rash. The best you can hope for is that it might make a listener *do* something to change the temperature.

Critical Emotivist: But that's a rather crude way of looking at it. There's something else. Besides functioning as expressions or appeals, judgments of values can also communicate a factual content, *if* we are careful to specify some factual application for our critical statements. And we must make this factual application in terms of certain empirical features that we can readily identify. If our standards change, as our experiences change – well, then, we have to give the persuasive terms, which express our interests, some new factual content. You probably see that it is precisely this double nature of value judgments that has caused most of our confusion, and has sent us through the merry chase of our discussion.

Common Man: Well, maybe so; but I stick to what I said in the first place: it really *is* too cold up here. Let's move to some place where it's warmer.

Sapienti sat

NOTE

* The following imaginary symposium and its heated arguments were inspired by the real frigidity of a Minnesota winter night.

23. IS SCIENCE RELEVANT TO THEOLOGY?

[1966a]

The following remarks intend to outline what I consider the major points at issue. It seems to me that most of the previous contributions to this topic published in *Zygon* have dealt with the implications of current science for theology and religion mainly by way of rather gingerly, halfhearted allusions. The straightforward spirit of the eighteenth-century Enlightenment (e.g., Hume, Kant) needs reviving and 'updating.' Surely, no one can claim to 'know all the *answers*', and – in all humility (I trust this is still regarded as a virtue!) – I wish to set out what strike me as, at least, some of the pertinent questions. And I shall also attempt to give some tentative *answers*. Since I have been asked to do this in very brief compass, the harsh tone and terse style of my presentation will make my contentions appear more dogmatic and intransigent than I should wish them to be.

1. The term 'religion' is used with such a great breadth of meaning as to include even atheistic or non-theistic 'philosophies of life'; it covers anything from a *Gefühlsreligion* (i.e., a sentiment of awe; and/or a deep commitment to certain ideals) to the theistic religions that are usually conceived within a theological frame. Hence it is imperative to be quite clear as to whether the questions concern a religion *with* or *without* a theology.

2. There can be little doubt that the experiencing of religious sentiments, or a deep 'engagement' either on aesthetic or ethical grounds, can in no way conflict with either the methods or the results of science. Speaking for myself, and quite personally, I have occasionally been emotionally *overwhelmed* by the beautiful in nature (e.g., the scenery of mountains, or of the ocean); by the love, friendship, and kindness of fellow human beings; by the greatness of the heroes of true humanity (humane-ness!); by artistic or scientific creativity; by the 'message' of great poetry; and in my case, most especially, by the 'message' of great music. Indeed to my way of feeling and thinking, some of the works of such deeply 'religious' (in the wider sense) masters as Bach, Mozart, Schubert, Beethoven, Franck, Bruckner, Mahler, Poulenc (and a few others) are among the highest achievements of the human spirit. Several of my well-meaning friends keep telling me that this shows that 'deep down' I am a truly religious person. But since I see not the slightest reason for accepting *any* sort of theology (and plenty of reasons for rejecting *all* of them), it can

at most be granted that I have 'religious' *experiences*. In my younger days I felt just a little guilty for 'sponging' on the emotions of devoutly religious (i.e., theistic) composers. But I now think that the expression and evocation of feelings, sentiments, and moods is tied to a deeper level of experience than is formulated ideationally in a theologically based religion. The occurrence of such experiences (from the aesthetic through the moral to the deeply 'mystical') *can* of course (but need not always) be made the subject of scientific (e.g., psychological or sociopsychological) inquiry. But the mere enjoyment (or suffering) of these experiences can in no way be incompatible with the knowledge-claims of science (or, for that matter, of theology). It is the *interpretation* put on those experiences that makes a difference in this regard.

3. Nietzsche's phrase "God is dead" is now bandied about even in the daily papers and in the periodicals. My first reaction was that neither the truly theistic theologians nor the devoutly religious believers need pay any attention to this recent insurgence of disbelief. At least in the more or less orthodox Judeo-Christian tradition (should one not include that of Islam here too?) the ground of religious (theistic!) *faith* is *Revelation* (as transmitted in the Scriptures); and it has been fairly generally agreed that revelation is totally, fundamentally, different in kind from any sort of empirical evidence that supports the knowledge claims of science. The first move, then, should plausibly be: Let there be 'peaceful coexistence' between not only (non-theistic) religious experience, but also, and especially, theistic (theologically interpreted) religion and the enterprise of science.

4. But the situation is not quite so simple. Under the influence of the Age of Enlightenment and of the current Age of Analysis we have come to ask more incisive questions; and we have arrived at a higher level of aspiration as regards clarity, candor, and honesty of thought. This is, indeed, at least partly a result of the propagation of the spirit of the rational, critical approach that is the outstanding, paramount feature of the modern scientific method. The knowledge-claims of science are such that they are in principle susceptible to testing. The best justification for accepting scientific theories (as in physics, biology, psychology, or the social sciences) is – as Sir Karl Popper so persuasively puts it – that even the most strenuous and severe attempts at refuting them (by experiments, statistical designs, i.e., ultimately by observation) have failed! There is of course no such thing as a *definitive* verification of any theory in the empirical sciences. Such theories must forever be kept, in principle, open to criticism, modification, revision – to the point of total refutation and replacement by an alternative theory. Now, as Bernard Shaw (with some pointed exaggeration) once said: "Religion is always right. Religion

solves every problem and thereby abolishes problems from the universe. Religion gives us certainty, stability, peace and the absolute. It protects us against progress which we all dread. Science is the very opposite. Science is always wrong. It never solves a problem without raising ten more problems." I take this to mean that for orthodox, theologically based religion, there are not – and cannot be – any other standards of critical appraisal than those stemming from an understanding of Revealed Truth; and, in some instances of religious (including mystical) experience. This, however, is no longer the case with the *modernists* in theology – from Tillich, Niebuhr, Bultmann, Bonhoeffer, the Bishop of Woolwich, *et al.*, through the entire spectrum of the demythologizers to the 'God is dead' theologians (?!) Altizer, van Buren, *et al.*

5. It seems obvious that the spirit of the Enlightenment, of analysis, and of the critical approach has been, and continues to be, a strong stimulus in the modernistic movements. The crucial questions that any honest scientist must constantly ask himself are: '*What do I mean*?' (by the words or symbols I use); and '*How do I know*?' (what I claim to be true – or well confirmed). Now, of course these questions, asked and answered in the empirical sciences in one characteristic way, may or may not be understood in a similar way in theology. The orthodox theologian may well answer that his concept of the deity can be grasped only by analogy, metaphor, or allegory. The *via negativa* of St. Thomas Aquinas and his latter-day disciples clearly imposes severe limitations on the analogical conceptions of a personal God and his attributes of omniscience, omnipotence, and benevolence. Notoriously, the perennial enigma of evil required logical devices by which the theological dogma was made proof against disproof, immune to empirical test, and hence absolutely irrefutable. Even the 'man in the street' seems satisfied with the contention that 'God's existence can neither be proved nor disproved'; hence that it is a matter of faith. But do not the words 'faith' or 'belief' connote that some proposition is held as *true*? And must one not, in all candor, ask as to whether even the meaning of such words as 'belief' or 'truth' is the same in theology and in science?

6. Indeed, so great is the positive emotional (expressive and/or evocative) appeal of the words 'belief' and 'truth' that equivocations of their various (and quite different) respective meanings are almost the order of the day. 'Belief' may mean the ordinary, empirically grounded attitude we take toward well-confirmed propositions – be it in everyday life, or in science. 'Belief' may mean a radically transempirical faith or creed as in the orthodox theistic religions. And 'belief' may also mean a wholehearted commitment to

an ideal, a 'cause', or – as in the 'I-Thou' relationship – an attitude of trust, love, and understanding.

I am not saying that everybody is obligated to ask himself what he means by 'belief', 'faith', 'truth.' But for those of us who aspire to some measure of intellectual honesty and conceptual clarity, I must say that we have had the tools of analytic philosophy, of the clarification of ideas with us – with ever increasing incisiveness and effectiveness – ever since the times of Socrates, Plato, and Aristotle. Current philosophical analysis – despite many unresolved controversies – at least has awakened us most forcefully to the need of making distinctions wherever there is a danger of confusions, i.e., a danger of giving credit where it is not due (or contrariwise, not giving it where it is due).

7. The crucial issue is, of course, what to do about *transempirical belief*. By this I mean *radically* transempirical belief such as is involved in the doctrines of *Deus revelatus* and *Deus absconditus*. May I reassure my readers that I am not invoking here the notorious positivist (or logical empiricist) criterion of factual meaningfulness. Although I am still of the opinion that this criterion can be formulated in a way that is logically defensible and that enables us to distinguish between (good!) science and (pernicious!) metaphysics, I shall not use it in my critique of theology, because I think I have more effective critical weapons at my disposal. I shall not either regress to the 'Warfare between Science and Theology' – as it was understood in the nineteenth century. This concerned mainly the clash of the *literal* interpretation of the Scripture with the 'facts' of science. This is old hat, passé – and largely uninteresting today – for two reasons: (1) The modernist theologians have largely abandoned all fundamentalism and literalism. They have demythologized. (2) Truly open-minded scientists, without in the least giving quarter to obscurantism, readily admit that science (e.g., the theory of evolution) is far from finished; it explains scarcely any of the fascinating and marvelous *specifics* of the pageant of the species throughout the succession of its stages.

Moreover, the majority of the modernists (certainly the demythologizers) among the theologians no longer look for the 'chinks in the armor of science.' This sport is quite out of fashion these days, though purely logically I would be the first to admit that a well-reasoned teleological argument should not be dismissed on *a priori* grounds. I would even go so far as to grant that a sufficiently modernized argument *à la* Paley (or Lecomte du Noüy) would be at least worthy of consideration by scientists, philosophers, and theologians. I say all this, I hope it is clear, not for some obscurantist reasons, but rather because I sense an all-too-sanguine attitude in the neo-Darwinists. As long as the explanations offered by the theories of genetics, population pressure, mutations, and survival of the fittest are as sketchy as they are still today, I would plead that the biologists supplement (or supplant) their current theories by better (scientific!) theories. And even if this should become hopeless, recourse to supernaturalistic 'explanations' is futile, precisely because 'the inscrutable will of god' (just like the Absolutes of Metaphysics, or the Entelechies of the Vitalists) explains *too much*. The facts and regularities of the world as we come to know them in empirical science are equally compatible with monotheistic, polytheistic, Manichaeistic, or Zoroastrian (God and Satan interfering with each other) 'explanations.' These are mere verbal sedatives, pseudo-explanations – radically different from responsible scientific explanations.

Some parts of Freudian metapsychology are methodologically objectionable for the same reason: No matter what the behavior, the dreams, the conscious experience of a given person, it can all be 'explained' in terms of the life force (eros) and the death instinct (thanatos) by suitable (mythological) interpretation of their combined (or separate) effects. This is not science but untestable metaphysics. It amounts to no more than a pictorially and emotionally appealing verbal gloss on the description of the ascertainable facts. The same holds for Teilhard de Chardin's rather poetic rendering of biological, psychological, and social evolution. Père Teilhard, a respected paleontologist, in his *The Phenomenon of Man* has not contributed anything to either the theories of evolution or to the philosophy of biology.

8. Nevertheless, I am far from saying that the concept of a *Deus revelatus* or a *Deus absconditus* is meaningless. I would grant it at least a vague analogical significance (cognitive, indeed – not just emotive!), but I would insist on asking as to *whether there are any good reasons* for believing *that such a deity exists*. Mere tradition (no matter how glorious); mere emotional need; the message of ancient documents; the persuasive oratory of preachers; the wishful thinking about a hereafter; about a heavenly arbiter of right and wrong; etc., etc., are obviously *no good reasons whatever* for accepting the belief in the existence of a personal god. In fact, to the scientifically sophisticated mind, the more anthropomorphic the conception of the deity, the less credible it becomes. Why should wishful, all-too-human thinking be more trustworthy here than in the modes of primitive magical and animistic thought? Of course, purely personally speaking, I am not unacquainted with the emotional hankering for a heavenly father (why not mother? Roman Catholicism at least provides for that too!); surely, I can feel the tug of the wish for survival after bodily death, especially when thinking of the possibility of a reunion

with beloved parents, relatives, friends. I have even a modicum of understanding (empathy and sympathy) of the wish for reincarnation, and even for the ultimate dissolution and universal union in the Nirvana, as they are so beautifully presented in some of the Oriental religions. But just because we are now able to 'see through' (psychologically, psychoanalytically, etc.) the motivations of these forms of wishful thinking, they become (I am not saying absolutely disproved) extremely suspect and implausible.

9. A few words now about a demythologized theology. Does it still contain any sort of truth-claim? If we go to the extreme limits of demythologization (and must we not, in all consistency, do just that?), what else is left but the *moral message* of religion? Certainly the stern, or alternatively, forgiving, fatherly judge is retained only as a parable or allegory. Jesus, along with Moses, the Prophets, and Mohammed, is then to be viewed as an – indeed exceptional – but still entirely human and highly progressive teacher of morality.

10. Let us finally look at the rather obscure and devious ways in which some present-day scientists and philosophers of science attempt to achieve a rapprochement of science and theology. We are told that the old 'objectivist' view of science is obsolete; that the alleged interaction between the human observer and the observed physical situation resembles the I-Thou relationship (Buber) between man and fellowman, and – by timid suggestion – perhaps also the I-Thou relation between man and God!

The observer-observed relation in quantum mechanics has been a highly controversial matter for forty years. Many critical voices (of physicists, as well as of logicians of science) have been raised against the 'Copenhagen' (complementarity and duality) interpretation of quantum physics. The interpretation actually used in the work of the physicists is guided by the rule due to Max Born, according to which the (square of the modulus of the) ψ in the Schrödinger equation represents a probability for the occurrence of certain micro-events. These probabilities depend quite objectively upon the physical conditions of the experimental arrangements. The outstanding quantum physicist, Alfred Landé, has in many publications attempted (I think with some success) to 'demythologize' quantum mechanics – to eliminate the (unnecessary) mysteries of complementarity and of the duality of waves and particles. More significantly, it has long been recognized that the alleged 'intervention by the human observer' (let alone his –immaterial?? – mind!) is a gross misinterpretation. (The interaction between measuring instrument and the measured magnitudes of the physical situation is, of course, important, but the reading of the measuring instrument and its result – as on tape or film – can be performed by the human observer at any time, even long after the 'act of measurement.) A similar misinterpretation of the role of the observer in relativistic physics was exposed and removed many years ago.

11. What could possibly be gained for theology even if these were *not* misinterpretations? Are these good scientists clinging to straws while they are sinking in the ocean of a demythologized theology? Are they trying to tell us that the much vaunted 'objectivity' of science does not exist; that even scientific truth rests on subjective passion (Polanyi?), subjective estimates (subjectivist probability theory?), in other words on 'belief' or 'faith'? Are they trying to tell us that science and theology are, 'in the last analysis', in the same boat? If so, I recommend that they perform the 'last analysis' a little more consistently and conscientiously. I can hardly suppress the thought that these recent forms of obscurantism had better be understood on the basis of the social psychology of current science. Just as the entirely unilluminating dialectical materialism of Russian philosophers is to be understood on historical and political grounds, so perhaps the subjectivistic obscurantism, with its vague flirtations with a theology (of which it is even left unclear to what extent it is demythologized), will finally be explained as one of the minor symptoms of the exasperating tension between East and West.

I shall refrain from repeating what many scientifically oriented humanists have been saying for quite some time. There is no question in my mind that the intellectual vanguard of mankind is already able to adopt an ethics without supernatural foundation or supernatural sanctions. Such a scientific humanism must, however, emancipate itself completely from the dogmas of nineteenth-century materialism. A new view of 'place of mind in nature' is in the making. Through the collaboration of scientists and (scientifically trained) philosophers, this perennial perplexity and central issue of modern thought may finally be overcome. (But that is another – and by far too long – story for this occasion.)

Surely, we have to live – and get used to so living – with an unfinished view of the world. We shall never cease to ask new questions, we shall – in all probability – continue to be confronted with all sorts of new problems, theoretical as well as practical. There is no philosopher's stone, there are no ultimate answers. But the endless quest of science; the candid search for knowledge and clarity; and the morally humane application of scientific knowledge are surely something that mankind can and, in all humility, should pursue. If we are not to exterminate ourselves on this planet by the immoral and stupid use of our scientific and technological power; if we are to use this power for the betterment of the human condition – surely, a new age of enlightenment, scientific as well as ethical, is our most imperative need.

CONCLUSIONS

Theology, inasmuch as it is literalistic (fundamentalistic), is incompatible both with the most assured results and the most reliable methods of science. As far as I know, modern Protestant theologies have not produced a coherent account of revelation. The Scripture is not even internally consistent, let alone compatible with the best-confirmed propositions of astronomy, geology, and biology. Hence deviations from the literal interpretation are nowadays clearly unavoidable. What then are the criteria that theologians are willing to use in their reinterpretations of the Scripture? I cannot see even a sketch of a blueprint of a promissory note in this regard. (In science we have at least that much at the beginning of the development of a new theory.) Partly demythologized theology is a questionable halfway house, unclear in content, intent, or truth-claim. Theology completely demythologized is no longer a theology at all; it reduces to a moral message, formulated by the use of allegorical, but essentially exhorting, consoling, edifying, or fortifying language. The current fashion of 'God is dead' demythologization is just one of the effects of our age of science and philosophical analysis. Half-hearted attempts to show that the 'subjective' element in science opens the door to theology rest on grave misunderstandings of the scientific method or of the conceptual frame of modern scientific theories.

The existentialist phrases of 'ultimate concern', 'ground of being', 'dread of nothingness', have not been used in any clear and unambiguous way. To my (no doubt, rather simple) mind, 'dread of nothingness' means fear of death (some have it, others don't). 'Ultimate concern' – even for many church-or synagogue-attending members of our industrial society– is basically the chasing of the dollars! 'Ground of being' is a metaphorical phrase that has (to some) a vivid pictorial and/or emotional appeal; but what sort of cognitive meaning is to be attached to it? (Analogical conception in science is legitimate, fruitful, and crystal clear in comparison.)

The much-referred-to religion of great scientists often consists – as in the case of Einstein – in the belief in the order of nature. Einstein's God ("subtle but not malicious") never had anything to do with the moral commandments. Sir James Jeans conceived of God as a super-mathematician. His reasoning was: Since the laws of nature are mathematical in form; and since laws presuppose a lawgiver, therefore . . . (!). This is not only a glaring fallacy of four terms ('law' – prescriptive versus descriptive) but highly anthropomorphic to boot. Jeans surely created God in his (the mathematician's) image! An assertion frequently made by quite a few scientists who wish to avoid anthropomorphism runs something like this: 'There must, after all, be a *Power* behind (beyond, back of) the Universe that is responsible for all its marvelous features that we observe and study in the sciences.' But this notion of a 'Power' is then left entirely unclear; nothing is said as to how the 'Power' produces the phenomena of nature. Other current watered-down conceptions of God, like those of some modernists, are merely a metaphorical symbol for something greatly cherished. (For Wieman and Dewey, God is the symbol for the highest ethical and social values.) Full intellectual honesty demands a wholehearted acceptance of a scientifically oriented and philosophically clarified humanism.

24. ETHICS, RELIGION, AND SCIENTIFIC HUMANISM

[1969e]

The kind of world we are facing nowadays is one of deep concern. There are so many horrible things going on that it is psychologically understandable that a kind of moral disillusionment has set in, particularly among the younger generation. In my own feeble way I hope to retain some moral backbone in spite of all the developments that might make one skeptical, if not completely disheartened. The ethical outlook that I represent is, if you want a label for it, *scientific humanism*. What I wish to present is more along the lines of a sober philosophical analysis rather than mere preachment. The present generation doesn't like to be preached at any more than I do. I think that there is a fundamental difference between the analytic clarification of ideas and the communication of scientific knowledge claims on the one hand, and the edification, exhortation, and consolation that belongs to religious language on the other.

Traditional religions in their orthodox forms breed in their own way more problems than they solve, and people who might originally have had open minds, close them and never open them again. There is a difference between the scientific attitude and the dogmatic theological attitude. We have two extremes as far as intellectual attitudes are concerned. On the one hand there is dogmatism, and on the other hand extreme skepticism. The dogmatic attitude is the attitude of the closed mind; the extreme skeptic has his mind open on both ends; everything flows through and nothing sticks. Clearly the critical approach that is so typical of the scientific attitude is the attitude of the open mind, but not open on both ends. In other words, no scientist in his right mind will doubt everything equally strongly. He will have doubts on some things; and while he might admit that in principle all scientific knowledge is open to revision, that does not mean that he is questioning everything all of the time or some of the things all of the time. So I do think that the critical attitude exemplified by some of our greatest scientists proves that they observe the golden mean between the dogmatic attitude that often goes with theological fixations and the extreme skepticism that philosophers sometimes cook up for their own amusement. Actually you cannot live as an extreme skeptic because you then would have to doubt every step you take in any and all of your activities.

We all live by 'faith' or 'belief' of some kind, but it is vitally important to distinguish different kinds of belief. Before I go into the various meanings of the word 'belief', let me illustrate by means of an allegory: Religious faith is like a cane, a walking stick. Some carry the cane because they could not walk without it; it serves them as a crutch; that is, religion in their case upholds them. They could not live without an orthodox kind of religious faith, and such people should not be deprived of it if that is what they need. Other people carry the cane because father did and father must have known what was right. Then there are those who carry the cane because it is a sort of personal ornament. It gives them admission to the kind of club and social organization that they want to be in. There are still others who use the cane as a magic wand – in faith-healing, for example. And some carry the cane only to use it in steep places of the road. (I'm not sure whether God will pay attention to people who turn to Him only in their distress.) Finally there are some people who carry the cane to whack other people over the head with it. Now this allegory is cheap and crude. I realize this, and I want to say immediately that if we want to be broad-minded and open-minded philosophers we have to understand the psychology of religion much better. In our age this is facilitated by new developments in psychology and also by such developments in our philosophical outlook as have resulted from the clarification of ideas carried on by empiricists, logicians, and linguistic analysts.

As I see it, we are living in a new age of enlightenment in which we ask persistently, and we hope with good results, two major questions: 'What do you mean?' and 'How do you know?' What do you mean by words that you see, and how do you know your assertions to be true or probable, i.e., on the basis of what evidence, grounds, or reasons? In asking these questions – especially the first question – we go beyond the so-called warfare between science and theology that had been written about many years ago – for example, by Andrew White. The major concern then was the incompatibilities of the best-established results of science with the dogmas contained in the Scriptures. On the whole we feel today that this is not the most prominent issue. We are no longer primarily concerned with the discrepancies between the modernists and theologians, nor to revise theological doctrines in such a fashion as to make them compatible with science and to reconstrue them in the light of the best knowledge available in astronomy, geology, biology, theory of evolution, etc. Rather, the question, 'What do you mean?' really is the central issue; namely, the meaning of the word 'belief' or of any cognate terms. I'll put it in quotes because I'm now talking about the word 'belief', and it seems to me that it has at least three different meanings. Unless you recognize this you will never get clear in your discussion of science versus theology or theologically framed religious faith.

Now taking our start from everyday language we find, first, that each of us is perfectly clear that we use the word 'belief' in a sense that we could call empirical, on the lowly level of common life, as when I say I believe that there might be a rainstorm tonight (I may be right, I may be wrong, but this is my belief); or when I say I believe that I have two dollars in my pocket. It has been pointed out that even the best-established scientific theories are objects of belief in the sense that whatever evidence speaks for them does not do so conclusively or with finality. This is very simple logic, in that any scientific theory that has been formulated is in the form of a universal proposition: under these circumstances, no matter where and when, such and such conditions will follow. In the more developed sciences this is expressed in mathematical language such as in differential equations. In science we tentatively put forth knowledge claims which have a universal range, and which are in principle refutable by bona fide instances to the contrary. So whether it is on the elementary level of common life or whether it is on the level of scientific theorizing, as in recent theoretical physics, we are dealing with beliefs that are capable of tests; and we know what we are talking about because we can tell what kind of difference it would make. This principle is well known in American philosophy due to the work of Charles S. Peirce and William James. I think that it was William James who said that a difference must make a difference if there is to be a difference – meaning that if there is no difference between the affirmation and the denial of a certain proposition, that proposition did not really have scientific significance because it is completely removed from any test.

Now some forms of theology have precisely built into their conceptual structure the idea of absolute untestability. Since I am referring to *some* forms, but by no means to all, I will call this second kind of belief simply 'trans-empirical.' Usually people in this connection speak of the 'supernatural', but that is a cloudy notion because you never know how to draw the line when something has not as yet been scientifically explained. The boundary between the supernatural and the natural and between what can be explained scientifically and what cannot is often unclear. However, if you take some typical examples from traditional theology, then very often, after a lot of discussion and many dialectical moves, the final answer is that 'this is a matter of faith.' 'There is nothing that can be proved or disproved'; at least, that is what the man in the street will tell you. I'll wager that if you took a Gallup poll of the opinion of the man in the street on theological matters, he would say, 'Don't confuse this with science.' In science you can prove or disprove, in the sense of confirming or disconfirming, at least indirectly and incompletely, no matter how strongly or weakly. But it requires an act of faith to uphold the theological dogma, and this is entirely different from belief in the empirical sense. Thus testability is ruled out because in the end we are told that there are mysteries that are unfathomable to the human mind.

William James tried to help us in this connection with his famous story about the cat in the library. A cat in the library that knows all the cozy nooks and corners and knows how to get in and out might 'think', 'I know all about the library!' But the cat wouldn't have the slightest idea what people are doing there; taking down books, making notes, scanning the indexes, etc. Those things would escape the cat completely. Now the conceited scientist, says William James, is like that conceited cat in the library. He might say, 'Here is the universe with its regularities described in our scientific laws, there are no mysteries any more. Well, there may be a few open questions and a few obscure corners in the universe, but by and large we know the hang of it all because of the basic laws we have found in physics, biology, etc.; by and large we have learned what most of our natural universe is like.' There actually was an article a few years ago by an outstanding Harvard scientist who said that we had the main problems solved; the rest required a little mathematical refinement and experimental clarification, but by and large science is near its end. Contrast this with what Newton said when he compared himself with a boy who was picking up shells at the seashore and thinking of the tremendous ocean of the unknown and unclarified thinking before him. True, Newton lived three centuries ago, when science was just beginning to formulate its first great synthesis. But surely we cannot decide by *a priori* reasoning between those two views, because all we have to go by is inductive reasoning. New facts crop up in observations and experiments over and over again as the scientists test their theories. A humble attitude like Newton's is more commendable than the attitude of that scientist who was smug enough to think that science was about to come to an end. If you want my personal opinion, I think that science is an endless quest: the universe has many more aspects than we are aware of or that we have formulated in our science thus far.

Now a few more words about the trans-empirical aspects of belief. Take the famous problem of evil. As I see it in the Judeo-Christian tradition we have trans-empirical faith in a God who is all-knowing, all-powerful, who is all-good or benevolent. If there is such a God, then there should be no evil. There should be no evil in the world because, if there is, He would certainly know about it. If He is benevolent, He would want to eliminate evil; and if

He is all-powerful, He has the ability to do so. However, unless you are a Christian Scientist, you will admit that there is evil in the world, for it stares you in the eye. After all, there is much that is unfortunate. There are wars, racial tensions, poverty, deprivation, injustice, suffering, misery, and disease in the world. All this we call evil; and it takes a lot of verbal juggling to talk yourself out of it. Elementary logical reasoning directly reveals that, if there is a God, then He could not be all three: benevolent, omniscient, and omnipotent. This, of course, does not disprove the existence of God: it just disproves that at least one of the three attributes of God would not be fulfilled because the conjunction holds only if all members of the conjunction are true. It was William James who attempted to resolve the paradox of evil by arguing that God is surely omniscient, and He is surely benevolent, but not all-powerful. And James thought that was an inspiring message for all human beings, because then we have to help the good Lord with our own efforts.

But James was by no means attached to any kind of orthodox religion. And the *usual* answer to the problem of evil that we get from theologians is that there are things (as there are for the cat in the library) that are absolutely 'beyond our understanding'; that we simply must have 'faith' in the existence of God and His goodness and greatness and power, and that ultimately the problem is an 'unsolvable mystery.'

Now here you have a sharp difference between the scientific attitude and the theological one. Modest scientists will admit that of course there are unsolved problems, but if they are faced with a problem that they can recognize and analyze as unsolvable, then they get suspicious that they may not even have a good question. Let me briefly illustrate the point by an example that is partly off the track, but it will help us to see that borderline questions of this sort have occasionally come up, even in science. Take, for instance, Newton's doctrine of absolute space. For those of you who know the details, I should add that I am talking merely about the kinematics of absolute space and not about the dynamical arguments of Newton. Newton was influenced by Platonic doctrines, according to which space is reality in and by itself, existing independently, and whatever exists in the world has its place in absolute space. In fact there was a basis for Newton's thinking this because he considered space a sense organ of God. God, as it were, has a finger in every pie and is omnipresent because space is his sense organ. During Newton's lifetime, this doctrine was attacked by Leibniz, Newton's great contemporary in Germany, who challenged not Newton himself but Samuel Clarke, Newton's faithful disciple. Leibniz asked questions that are very close to the kind of questions that later pragmatists and positivists were to ask: 'What is the difference that makes the difference?' For instance, Leibniz asked in his correspondence: Suppose we take all of the particles in the universe, stars, atoms, etc., in their momentary orientation and distances, etc., and leave everything intact while moving all the contents of the universe two miles to the east. Would that make a difference that makes a difference? Obviously not. There would be nothing testable about that difference. So, as long as space is considered as an absolutely independent reality, it could not possibly be testable. The critique by Leibniz of absolute space and time is similar to the one that I think has influenced some of the 'God is dead' theologians in recent time. They are bothered not so much by the conflict between the best scientific evidence and the Scriptural dogmas and assertions about Genesis, and so on, as by the question, 'Do we still mean anything if we remove our assertions entirely from empirical testability?'

We will return to this in a moment. The third meaning of the word 'belief' is perhaps best called 'commitment.' If I got up before you on a soap box and said, "I believe in human equality," you would immediately realize that I am not trying to tell you that we are all equal, because we are all more or less unequal, physically, intellectually, etc. Clearly what I was expressing was my commitment to a doctrine of human rights, human equality before the law, equality of opportunity, and so on. In other words, the statement involves taking a firm attitude and is itself not a knowledge claim at all; it is a matter of putting yourself on record and vowing a certain attitude.

Now it seems to me that some of the humanly more valuable elements in religion come precisely from the connection of transempirical faith with a commitment. Do we have to have transempirical faith in order to have the kind of moral and social commitment that normal, sane, and humane human beings generally have, at least, as a standard of criticism? Quite true, we do not always behave according to the ideals to which we have committed ourselves, and we may sometimes feel guilty about it; but when it comes to criticism of others, or even to criticism of ourselves, we do use moral standards which we all have to some extent in common, at least at a given stage in human development and civilization.

Implict in what I am saying here is, of course, the controversial philosophical conviction that moral principles are essentially in the imperative form, and that we are deluding ourselves if we give them a purely descriptive, declarative form. To be sure this is an often-used device in the scientific age; people are often so fact-minded that they insist on statements that are couched in the form of declarative statements capable of empirical evidence. But I ask you, could you, for instance, provide any kind of empirical evidence for a doctrine of human rights? You can talk about human nature. As long as you merely state the facts about human nature, individually, psychologically, anthropologically, or socio-psychologically, you will not get an imperative out of it that would be the expression of a commitment of how we ought to think or act. In other words I do not see how you could arrive from premises, that contain only statements concerning what *is* the case, or what happens under what circumstances, to moral imperatives that tell you what *ought* to be done or what we ought to strive for.

Hence, we have commitments of this sort. I wish to stress immediately how tremendously important they are in the whole business and art of living together and what terrible blunderers we are still in this art of living together, as witness all the upheavals, riots, and wars that are going on. I will later try to show you in what way we can come to a philosophical understanding of commitments.

At this point I wish to insert a brief discussion of some theological arguments that try to show that just as there is a kind of experience that supports the statements of everyday life, or the factual knowledge claims of the scientist, so there is a kind of experience that similarly supports the religious beliefs. There are some theologians who do not pay much attention to this kind of argument, but you often hear it from philosophers that empiricists are too narrow-minded. They pay attention only to sensory experience, we are told. It is fairly obvious that scientific knowledge, as well as the knowledge of everyday life, is based on evidence that comes to us through the senses – seeing, hearing, touching, smelling, etc. But there are other forms of experience; and many of us know from first hand that there is such a thing as religious experience, as in the actual act of worship or prayer. There is the experience which the mystics refer to as the 'mystical experience', and which seems exceptional. But still in extreme forms of religious experience the mystics usually tell us that they cannot really describe these experiences or give us a clear-cut interpretation of them. They can circumscribe them, they can hint at them, they can allude to them. But we are told by some philosophers that you cannot be so narrow-minded as to say that trans-empirical faith has no factual meaning, nor can it have any factual truth because it cannot be based on sensory experience. This is the complaint.

Now here I have to be very brief. It seems to me that you don't have to subscribe to any dogmatic Freudian psychoanalytic theories in order to see that the human situation is such that we do not have to explain mystical experience in the way which the mystic himself or the religious person takes it; namely, as given by the grace of a deity. We can have a perfectly natural explanation, just as we have a perfectly natural explanation about the origins of our sensory experiences that serve as evidence for our knowledge claims in everyday life and the factual knowledge claims of the advanced scientist. Now it seems to me that the human situation being what it is, and human childhood and infancy being what they are, it should not be surprising that the notion of a superior power would arise; and if it is available in the culture, it would be readily adopted by the person growing up through adolescence. In other words, the greater power of parents, educators, and elders in comparison with the helpless infant brings about an image of a superior being to which he can appeal, which may forgive, or which may punish. Accordingly, in some of the monotheistic religions the idea of a personal God is usually couched in terms of the father image. We then are told we must not take this too literally. But then the question is what does demythologizing leave over if it is carried to the bitter end? The situation of the infant is such that his very physical existence depends upon the sustenance and assistance rendered by the father and mother. Very soon the 'do's' and 'don't's' of education take over – in other words, imperatives reinforced by rewards or punishments. What could be more natural than that the infant would develop the idea that perhaps there is a superior power of the paternal or maternal type, somewhere but invisible, who watches over us, to whom we are morally responsible for our actions, and with regard to whom we may feel proud or guilty, as we do toward our natural parents. Now this is a piece of psychology that I don't wish to present dogmatically. It is just to show that the usual argument from mystical religious experience claims to find an explanation analogous to but different from scientific explanation, that it is the alleged result of grace coming down upon us from the deity, and that it is not accessible to sensory observation.

I'm usually told at this juncture of the argument that one had better not explain the issues psychologically. One could say in reply that, according to the Freudians, a person who rejects God is a typical father-killer with a strong Oedipus complex and that accounts for his attitude. So they say that what is sauce for the goose is sauce for the gander. I don't see that this reply is to the point of the original argument because the question was as to whether we need a trans-empirical explanation of mystical experience; and I have suggested what I consider to be at least a beginning or blueprint or promissory note for a negative answer. Of course, I would also want an explanation of the behavior of the raging atheist. All I wanted to show is that we do not need the theistic kind of reasoning in order to account for the phenomenon of religious experience. The influence of the cultural and moral traditions as well as psychology can provide a very plausible naturalistic explanation of religious experience. As Voltaire once put it, if God didn't exist, he would have to be invented.

In our age of scientific enlightenment we often ask ourselves what kind of foundation we can give to morality, to our basic commitments, to human rights and human equality? This is, of course, not an easy question to answer. It seems to some people much more convenient to appeal to trans-empirical authority. Often the question is raised, 'What lends authority to the ethical principles without which we could not survive?' I believe that the cynicism and anarchism we see in so many young people today represents a danger not only for the survival of civilization and for human well-being, but even for the psychological well-being of the individual concerned. Now what can we say in this connection? Philosophers have been in a quandary for centuries as to how to justify fundamental principles of morality.

This invites a little logical analysis to find out what it is that we mean by justification. If we justify knowledge claims in mathematics it is done by proof; that is, by deductive derivation. If we justify knowledge claims in empirical science, it is done by observation, experimentation, statistical design, call it inductive, or what you will. There are also principles that at least pragmatically have proven their worth and that is the very reason why we keep them around and live in accordance with them. In the ethical domain it is more difficult to derive what you want to derive, and I find that it is easier to beg the question by putting an ethical or moral accent in your principles. Perhaps we are all suffering from the philosophical demand for demonstration. There are limits to demonstration. From this point of view I am quite inclined to say that you won't get any place with ethical justification unless you start with certain commitments. The adoption of those commitments can be made palatable, but there is nothing that we can prove or disprove about them. To be sure, utilitarians and hedonists try to show that if we are to have the greatest happiness for the greatest number, then these moral precepts commend themselves for adoption; but of course this idea presupposes already that the major aim, the greatest good of the greatest number, is itself morally desirable.

Some clarification would help here. Kant, the German philosopher of the eighteenth century, introduced the distinction between hypothetical and categorical imperatives, in plainer language, conditional and unconditional imperatives. If your doctor tells you that *if* you want to improve your health, you'd better take the following medicines or change your diet or stop smoking, then this is a conditional imperative. It assumes that you accept the end and that he with his professional knowledge can give you the kind of advice that you can adopt in order to attain this end. This is the purpose of the applied sciences, such as technology, engineering, medicine, and so on. It is the technological sciences that give us information about the best means that will most economically conclude the attainment of those ends. These are hypothetical or conditional imperatives. *If* you want this end, you'd better avail yourself of the following means. You'd better do such and such, in order to gain what you most deeply desire. By contrast, ethical imperatives must be based upon a fundamental or supreme ethical imperative that is unconditional; that is, if your intuitions are correct, because somewhere in the analysis of moral behavior and moral experience we do come upon certain terminal values, but some that furnish the basis to the rest of our values. And these are the ones to which we commit ourselves, at least as principles of criticism, even if we do not always obey them. Just remember a situation when you are dealt with unjustly or unfairly. You cry, 'Unfair! Unjust!' Whenever we are in danger, we appeal to such principles. We find them even on the highest level: for example, in the Supreme Court and in the Congress of the United States and the United Nations, where one appeals to fairness and justice.

Now it is true that practically all of these ethical terms are emotionally laden and are open to persuasive definitions. Charles L. Stevenson's remarkable book *Ethics and Language* gives an example of such a definition which illustrates the point better than an abstract explication. A liquor advertisement, for instance, says 'that true temperance is a cocktail before dinner, a glass of wine with dinner and a cordial or brandy after dinner!' Thus, you can see what goes on in persuasive definitions. Words have a certain emotive appeal. 'Freedom' and 'slavery', for instance, have certain opposite emotive appeals. It is quite characteristic that in international discussions we deal out to each other labels that have emotive appeals. The totalitarian countries say, 'Capitalism in America is pure slavery; look at the poor, exploited worker.' We look at the fellow in the totalitarian countries and we say, 'You think you have freedom? My goodness, with all those restrictions!' and so on and so forth. It is necessary that we look below the surface of the verbal games that people play and penetrate to the definitions that tamper with the factual meaning of words but retain their emotive appeal.

Accordingly when I claim that there are ethical ideals that seem to be fairly basic in human concerns, I allow for a little latitude that of course can be exploited by 'persuasive' definitions. Nevertheless, I think that there is a tenable middle ground between the relativists and the absolutists in morals. The relativists, for instance the anthropologists of the last century, have left their mark by maintaining a pluralism of moral values. On the other side, the absolutists, for instance theologically or metaphysically inclined philosophers, have given us a dogmatic monism in claiming that there is only one set of moral standards common to all mankind. The relativists hold an empirically discernible idea, namely that human values are relative to human needs, interests, and desires. Whereas some of the absolutists, including Kant, say that basic human values are independent of human interests and needs in that we may have to act even against our interests to do the morally right thing. This latter point has been exaggerated I think.

In my view we may well reject extreme relativism. The relativist, especially the anthropologist, has confused mores and folkways with morality. If we dig deeper into human nature we find that in some social contexts certain moral ideals inevitably work themselves out. So I think that a unified set of supreme moral values can be empirically discerned as inherent in the conscience of man, even if it is not always displayed in his behavior. We cannot, however, get away from the fact that human needs and interests and human nature in general are highly relevant for human values or moral ideals. I assume a sort of synthesis between a 'nothing but' and a 'something more' view of morality; namely, morality on the one hand is relative to human interests, and moral values neither come down from on high nor are dictated by the deity. There is a golden mean that combines the valid element of monism – i.e., that ethical principles are universally applicable – with the empiricism of relativism which teaches that human values are related to human nature. If you want a label for this call it 'scientific humanism.'

Now what are these moral values that we share in common? Very likely they have had a development. It is perfectly clear, for instance, that the ancient enlightened philosophers were perfectly satisfied to exploit men as slaves and had no particular compunction about it since they distinguished between 'superior' and 'inferior' human beings. In this regard a tremendous transformation has occurred. And we can certainly be thankful to the great religions for their ethical contributions. We do not have to accept their trans-empirical method in order to support their commitment to the ideals of the brotherhood of man or equality before the law. I have already said we cannot demonstrate anything in this field because all such demonstrations would ultimately be a matter of a vicious circle. But what we can show is that in the development of human civilization certain transformations have taken place. There are certain activities and abilities of the human animal that are essential for his survival. But when civilization takes over, something else supervenes in addition to what was a purely biological function in the first place. For example, our eyes are clearly outposts in the preservation of existence. We recognize dangers, we recognize our food by eyesight; but seeing can become 'beholding' the beautiful in nature and art. So here a certain sublimation has taken place. Similarly, hearing, a similar biological function in the first place, may become a means to listening to great music; walking may become hiking or dancing. The original functions, you understand, do not fade out but are supervened by the further functions that represent our higher cultural activities. Speech as a means of communication for very practical purposes is indispensable, but it also may become poetry and song. Sexuality may become love. I don't recommend that sexuality fade out, but love is something more than mere sexuality. Hard toil and burdensome work may become creative work, as we see it in the works of artists and scientists, and may be intrinsically enjoyable. Schopenhauer spoke of human intelligence as a lantern in the light of which we look for food and avoid dangers. Surely it has its biological and practical functions, but it may be sublimated into scientific activities. The great scientists pursue science for its own sake, not merely for its practical applications.

If human nature had been as bad as Hobbes believed, and everyone had been the enemy of everyone else, we would soon, in the natural necessities and exigencies of living together, arrive at least at what you might call 'business reciprocity.' 'I'll do something for you because I expect you can do something for me.' That's the lowest level. Of course, even beneath that level is the commitment not to do harm to anyone else, because if you do there are bound to be repercussions. So from these lowly origins we may learn and arrive at higher levels of morality. Take the Hippocratic oath, for example, which young medics have to swear. The oath is essentially the commitment not to do harm, but further it is the commitment to help patients. This is quite elementary. The knowledge that the doctor acquires in his study of anatomy, physiology and medicine would enable him to make his patient gravely ill and kill him. But he does not use his knowledge this way if he is an ethical medical man and an ethical human being. The Hippocratic oath is clearly a matter of commitment. One can, however, arrive at higher levels with the principles of justice and fairness. In spite of the fact that they are all subject to persuasive definitions, there is a core there that we understand and that we all appeal to when we are pressed.

I don't think that we have to go to the New Testament to find this. One can find it in the Old Testament, especially among the Jewish prophets: the principles of kindness, fairness, equality. And I think among all people, but in very diverse forms, you have ideals closely approximating these. Now these moral commitments, or principles, or whatever you wish to call them, may well have come out of the natural development of human beings in the social context. What I am trying to maintain is that in this age of scientific enlightenment, I think we can avoid regressing to a completely cynical, anarchistic view of the world. I think that there are scientifically evident features of the development of the human race and of individuals. Psychology points in the direction of human needs and interests that must be satisfied. And in the social context, certain traffic rules simply have to be obeyed if we are to survive as a society. Thus we have ideals of justice and equality.

What if you don't have such an ideal, as for instance in the case of what a clinical psychologist or psychiatrist calls a 'psychopathic personality'? Here, unfortunately, there is nothing that you can do with such a person; not even therapeutic means have been found, and reeducation seems difficult or impossible. Psychopathic people, if they have any regrets (not repentance), regret only that they were not clever enough to avoid being found out. In regard to the person who is absolutely without conscience, I don't think that any logical or empirical demonstration would help. I think that all we can do in human society is to avoid the preaching of morality. Instead we should educate by example, and especially in regard to our children. Parents should show the children how things can be done, and in a very gentle way correct them.

If I had any reason to believe that orthodox religions would promote peace and justice in this world I wouldn't criticize them at all. The empirical evidence seems to speak against them. Much as I appreciate the deep moral concern of truly religious persons, I think that institutional religions have often encouraged wars and cruelty of one kind or another. The flame and sword of Islam is one example, the Crusades another. Preachers who bless the arms of their country have their counterparts in the preachers who bless the arms of the other country. All are examples of hypocrisy and injustice. If I had any evidence that traditional religion, implemented by a conceptual theological framework, was effective along the lines that every humanist would like to have effected, I would cease and desist in my criticism spoken from a logical point of view. My analysis has been addressed to those of our generation who are looking for the kind of clarity and intellectual responsibility in morality that we have achieved and continue to achieve in the sciences. From the point of view of the scientific humanist I think that there is still some hope that mankind will grow up. But whether or not mankind will actually learn to use scientific and untried intelligence in the treatment of moral problems depends upon the cooperative efforts of everyone. The opportunity for the development of a scientific humanism is very great indeed. But the challenges of our day are so tremendous that unless mankind develops soon the ability to resolve his moral problems with clarity and intelligence, the consequences of his folly will be too terrible to contemplate.

BIBLIOGRAPHY OF WORKS CITED BY HERBERT FEIGL IN *INQUIRIES AND PROVOCATIONS*

Abraham, Leo [1933], 'The Logic of Ethical Intuitionism', *Intl. J. of Ethics* **44**, pp. 37–55.

Achinstein, P. and S. F. Barker (eds.) [1969], *The Legacy of Logical Positivism* (The Johns Hopkins Press, Baltimore).

Aldrich, V. C. [1938], 'Messrs. Schlick and Ayer on Immortality', *Philosophical Rev.* **47**; reprinted in [Feigl and Sellars 1949b], pp. 171–175.

Armstrong, D. M. [1968], *A Materialist Theory of the Mind* (Humanities Press, New York).

Aune, Bruce [1967], *Knowledge, Mind and Nature* (Random House, New York).

Ayer, A. J. [1936], *Language, Truth and Logic* (Oxford University Press, London); second revised edition (Gollancz, London, 1946).

Ayer, A. J. [1940], *The Foundations of Empirical Knowledge* (Macmillan, New York).

Ayer, A. J. [1946], 'Other Minds', *Proc. Aristot. Soc. Suppl. Vol.* **20**, pp. 188–197.

Ayer, A. J. [1947], 'Phenomenalism', *Proc. Aristot. Soc.* **47**, pp. 163–196.

Ayer, A. J. [1953], 'One's Knowledge of Other Minds', *Theoria* **19**, pp. 1–20.

Ayer, A. J. [1954], 'Can There Be a Private Language?', *Proc. Aristot. Soc., Suppl. Vol.* **28**, pp. 63–76.

Ayer, A. J. [1956], *The Problem of Knowledge* (Macmillan, London).

Ayer, A. J. (ed.) [1959], *Logical Positivism* (The Free Press, Glencoe, Illinois).

Bar-Hillel, Y. [1954], 'Indexical Expressions', *Mind* **63**, pp. 359–379.

Barrett, William [1939], 'On the Existence of an External World', *J. Phil.* **36**, pp. 346–354.

Bavink, B. [1932], *The Natural Sciences* (Century, New York).

Beck, Lewis W. [1950], 'Constructions and Inferred Entities', *Phil. of Sci.* **17**, pp. 74–86.

Benjamin, A. C. [1937], *An Introduction to the Philosophy of Science*, (Macmillan, New York).

Bergmann, Gustav [1940a], 'On Physicalistic Models of Non-Physical Terms', *Phil. of Sci.* **7**, pp. 151–158.

Bergmann, Gustav [1940b], 'On Some Methodological Problems of Psychology', *Phil. of Sci.* **7**, pp. 205–219.

Bergmann, Gustav [1940c], 'The Subject Matter of Psychology', *Phil. of Sci.* **7**, pp. 415–433.

Bergmann, Gustav [1943a], 'Outline of an Empiricist Philosophy of Physics', *Am. J. of Physics* **2**, pp. 248–258 and 335–342.

Bergmann, Gustav [1943b], 'Psychoanalysis and Experimental Psychology', *Mind* **52**, pp. 122–140.

Bergmann, Gustav [1944], 'Holism, Historicism and Emergence', *Phil. of Sci.* **11**, pp. 209–221.

Bergmann, Gustav [1946], 'Remarks on Realism', *Phil. of Sci.* **13**, pp. 261–273.
Bergmann, Gustav [1947], 'The Logic of Quanta', *Am. J. of Physics* **15**, pp. 397–408 and 497–508.
Bergmann, Gustav and K. W. Spence [1941], 'Operationism and Theory in Psychology', *Psych. Rev.* **48**, pp. 1–13.
Bergmann, Gustav and K. S. Spence [1944], 'The Logic of Psychophysical Measurement', *Psych. Rev.* **51**, pp. 1–24.
Black, Max [1933], *The Nature of Mathematics* (Kegan Paul, London).
Black, Max [1939], 'Relations between Logical Positivism and the Cambridge School of Analysis', *J. of Unified Sci.* **8**, pp. 24–35.
Black, Max [1948], 'Linguistic Method in Philosophy', *Phil. and Phenom. Res.* **8**, pp. 635–649.
Black, Max [1949], *Language and Philosophy* (Cornell University Press, Ithaca, New York).
Black, Max (ed.) [1950], *Philosophical Analysis* (Cornell University Press, Ithaca, New York).
Black, Max [1962], *Models and Metaphors* (Cornell University Press, Ithaca, New York).
Bloomfield, Leonard [1939], 'Linguistic Aspects of Science', *Intl. Ency. of Unif. Sci.* **1**, No. 4 (University of Chicago Press, Chicago).
Boas, George [1939], 'Habit, Fact and Value', *J. Phil.* **36**, pp. 526–530.
Boring, Edwin G. [1933], *The Physical Dimensions of Consciousness* (Century Co., New York).
Boring, Edwin G. [1936], 'Psychophysiological Systems and Isomorphic Relations', *Psych. Rev.* **43**, pp. 565–587.
Boring, Edwin G. [1937], 'A Psychological Function is the Relation of Successive Differentiations of Events in the Organism', *Psych. Rev.* **44**, pp. 445–461.
Borst, C. V. [1970], *The Mind Brain Identity Theory* (Macmillan, London).
Braithwaite, R. B. [1937–38], 'Propositions about Material Objects', *Proc. Aristot. Soc.* **38**, pp. 269–290.
Braithwaite, R. B. [1953], *Scientific Explanation* (Cambridge University Press, Cambridge).
Bridgman, P. W. [1928], *The Logic of Modern Physics* (Macmillan, New York).
Bridgman, P. W. [1936], *The Nature of Physical Theory* (Princeton University Press, Princeton).
Broad, C. D. [1914], *Perception, Physics and Reality* (Cambridge University Press, Cambridge).
Broad, C. D. [1923], *Scientific Thought* (Kegan Paul, London and Harcourt Brace, New York).
Broad, C. D. [1925], *The Mind and Its Place in Nature* (Kegan Paul, London and Harcourt Brace, New York, 1929).
Broad, C. D. [1927–28], 'The Principles of Problematic Induction', *Proc. Aristot. Soc.* **28**, pp. 1–46.
Broad, C. D. [1933], 'The "Nature" of a Continuant' in *Examination of McTaggart's Philosophy* **1** (Cambridge University Press, Cambridge), pp. 264–278; reprinted in [Feigl and Sellars, 1949b], pp. 472–481.
Broad, C. D. [1949], 'Review of Julian S. Huxley's *Evolutionary Ethics*', *Mind* **53**, pp. 344–267; reprinted in [Feigl and Sellars, 1949b], pp. 564–586.

Brody, Nathan and Paul Oppenheim [1969], 'Application of Bohr's Principle of Complementarity to the Mind-Body Problem', *J. Phil.* **66**, pp. 97–113.

Brunswik, Egon [1939], 'The Conceptual Focus of Some Psychological Systems', *J. of Unif. Sci.* **8**, pp. 36–49.

Brunswik, Egon [1943], 'Organismic Achievement and Environmental Probability', *Psych. Rev.* **50**, pp. 255–272.

Brunswik, Egon [1946], 'Points of View', in *Encyclopedia of Psychology*, P. L. Harriman (ed.) (Philosophical Library, New York).

Brunswik, Egon [1947], *Systematic and Representative Design of Psychological Experiments* (University of California Press, Berkeley and Los Angeles).

Brunswik, Egon [1952], 'The Conceptual Framework of Psychology', *Intl. Ency. of Unif. Sci.* **1**, No. 10 (University of Chicago Press, Chicago).

Bures, C. E. [1938], 'The Concept of Probability', *Phil. of Sci.* **5**, pp. 1–20.

Bures, C. E. [1940], 'Operationism, Construction and Inference', *J. Phil.* **37**, pp. 393–401.

Burks, A. W. [1951], 'Reichenbach's Theory of Probability and Induction', *Rev. of Metaphysics* **4**, pp. 377–393.

Burks, A. W. [1953], 'The Presupposition Theory of Induction', *Phil. of Sci.* **20**, pp. 177–197.

Campbell, N. R. [1921], *Physics: The Elements* (Cambridge University Press).

Carnap, R. [1923], 'Über die Aufgabe der Physik', *Kant-Studien* **28**, pp. 90–117.

Carnap, R. [1926], *Physikalische Begriffsbildung* (Verlag G. Braun, Karlsruhe).

Carnap, R. [1927], 'Eigentliche und uneigentliche Begriffe', *Symposium: Philosophische Zeitschrift für Forschung und Aussprache* (Berlin-Schlachtensee), pp. 355–374.

Carnap, R. [1928a], *Scheinprobleme in der Philosophie: Das Fremdpsychische und der Realismusstreit* (Leipzig F. Meiner, Berlin). For Eng. tr., see [1969].

Carnap, R. [1928b], *'Der logische Aufbau der Welt'* (Weltkreis-Verlag, Berlin-Schlachtensee). For Eng. tr., see [1969].

Carnap, R. [1929], *Abriss der Logistik* (J. Springer, Vienna).

Carnap, R. [1931], 'Die physikalische Sprache als Universalsprache der Wissenschaft', *Erkenntnis* **2**, pp. 432–465. For Eng. tr. see [1934].

Carnap, R. [1932a], 'Psychologie in physikalischer Sprache', *Erkenntnis* **3**, pp. 107–142. For Eng. tr. see [1959b].

Carnap, R. [1932b], 'Über Protokollsätze', *Erkenntnis* **3**, pp. 215–228.

Carnap, R. [1933], 'Erwiderung auf die Aufsätze von E. Zilsel und K. Duncker', *Erkenntnis* **3**, pp. 177–188.

Carnap, R. [1934], *The Unity of Science* (Kegan Paul, London). Tr. of [1931].

Carnap, R. [1935a], *Philosophy and Logical Syntax* (Kegan Paul, London).

Carnap, R. [1935b], 'Les concepts psychologiques et les concepts physiques sont-ils foncièrement différents?', *Revue de synthèse* **10**, pp. 43–53.

Carnap, R. [1936a and 1937], 'Testability and Meaning', *Phil. of Sci.* **3** and **4** pp. 419–471 and 1–40.

Carnap, R. [1936b], 'Existe-t-il des prémisses de la science qui soient incontrôlables?', *Scientia* **60**, pp. 129–135.

Carnap, R. [1937], *The Logical Syntax of Language* (Kegan Paul, London).

Carnap, R. [1938], 'Logical Foundations of the Unity of Science', *Intl. Ency. of Unif.*

Sci. **1**, No. 1 (University of Chicago Press, Chicago), pp. 42–62; reprinted in [Feigl and Sellars, 1949b], pp. 408–423.

Carnap, R. [1939], 'The Foundations of Logic and Mathematics', *Intl. Ency. of Unif. Sci.* **1**, No. 3 (University of Chicago Press, Chicago).

Carnap, R. [1945a], 'On Inductive Logic', *Phil. of Sci.* **12**, pp. 72–97.

Carnap, R. [1945b], 'The Two Concepts of Probability', *Phil. and Phenom. Res.* **5**, pp. 513–532. Reprinted in [Feigl and Sellars, 1949b], pp. 330–348.

Carnap, R. [1949], 'Truth and Confirmation', in [Feigl and Sellars, 1949b], pp. 119–127.

Carnap, R. [1950a], 'Empiricism, Semantics and Ontology', *Revue Internationale de Philosophie* **11**, pp. 20–40.

Carnap, R. [1950b], *Logical Foundations of Probability* (University of Chicago Press, Chicago).

Carnap, R. [1951a], 'The Nature and Application of Inductive Logic', in Carnap [1950b].

Carnap, R. [1951b], *The Continuum of Inductive Methods* (University of Chicago Press, Chicago).

Carnap, R. [1956], 'The Methodological Character of Theoretical Concepts', in [Feigl and Scriven, 1956], pp. 38–76.

Carnap, R. [1959a], *Introduction to Semantics*, and *Formalization of Logic* [His *Studies in Semantics*, Volumes 1 and 2] (Harvard University Press, Cambridge, Mass.); (originally published in 2 volumes, 1942 and 1943, respectively).

Carnap, R. [1959b], 'Psychology in Physical Language' in [A. J. Ayer, 1959]. Tr. of [Carnap, 1932].

Carnap, R. [1963], 'The Philosopher Replies', in *The Philosophy of Rudolf Carnap*, P. A. Schilpp (ed.), (Open Court, New York), pp. 859ff.

Carnap, R. [1966], *Philosophical Foundations of Physics*, Martin Gardner (ed.), (Basic Books, New York).

Carnap, R. [1969], *The Logical Structure of the World* and *Pseudoproblems in Philosophy: The Heteropsychological and the Realism Controversy*, trans. by R. A. George (University of California Press, Berkeley and Los Angeles). Tr. of [1928a] and [1928b].

Cattell, R. B. [1946], *Description and Measurement of Personality* (World Book Co., Yonkers-on-Hudson).

Chisholm, R. M. [1946], 'The Contrary-to-Fact-Conditional', *Mind* **55**, pp. 289–307.

Church, Alonzo [1936], 'A Bibliography of Symbolic Logic', *J. of Symbolic Logic* **1**, No. 4; also **3**, No. 4 (1938).

Churchman, C. W. [1948], *Theory of Experimental Inference* (Macmillan, New York).

Cohen, Felix [1933], *Ethical Systems and Legal Ideals* (Harcourt Brace, New York).

Cohen, M. R. [1931], *Reason and Nature* (Harcourt Brace, New York).

Collingwood, R. G. [1940], *An Essay on Metaphysics* (Oxford University Press).

Colodny, Robert G. (ed.) [1962], *Frontiers of Science and Philosophy. University of Pittsburgh Series in the Philosophy of Science*, Volume 1 (University of Pittsburgh Press, Pittsburgh).

Colodny, Robert G. (ed.) [1965], *Beyond the Edge of Certainty: Essays in Contemporary Science and Philosophy. University of Pittsburgh Series in the Philosophy of Science*, Vol. 2 (Prentice-Hall, Englewood Cliffs, New Jersey).

Colodny, Robert G. (ed.) [1966], *Mind and Cosmos: Essays in Contemporary Science and Philosophy. University of Pittsburgh Series in the Philosophy of Science*, Vol. 3 (University of Pittsburgh Press, Pittsburgh).

Colodny, Robert G. (ed.) [1970], *The Nature and Function of Scientific Theories: Essays in Contemporary Science and Philosophy. University of Pittsburgh Series in the Philosophy of Science*, Vol. 4 (University of Pittsburgh Press, Pittsburgh).

Comte, Auguste [1877], *Cours de philosophie positive*. 4th edition 6 vols., (Paris).

Cornman, James [1970], 'The Identity of Mind and Body' in [Borst, 1970].

Dennes, W. R. [1939], 'The Appeal to Reason', *University of California Publications in Philosophy* **21**.

Dennes, W. R. [1946], 'Conflict', *Phil. Rev.* **55**, pp. 343–376.

Dewey, John [1938], *Logic, the Theory of Inquiry* (Holt, New York).

Dewey, John [1939], 'Theory of Valuation', *Intl. Ency. of Unif. Sci.* **2** No. 4 (University of Chicago Press, Chicago).

Dingle, H. [1932], *Science and Human Experience* (Macmillan, New York).

Drake, Durant [1925], *Mind and Its Place in Nature* (Macmillan, New York).

Drake, Durant [1933], *Invitation to Philosophy* (Houghton Mifflin, New York).

Dubislav, Walter [1931], *Die Definition* (F. Meiner, Leipzig).

Dubs, H. H. [1930], *Rational Induction* (University of Chicago Press, Chicago).

Ducasse, C. J. [1945], 'On the Method of Knowledge in Philosophy', *University of California Publications in Philosophy* **16**.

Ducasse, C. J. [1951], *Nature, Mind and Death* (Open Court, LaSalle, Illinois).

Eaton, R. M. [1931], *General Logic* (Charles Scribner's Sons, New York).

Eddington, A. S. [1928], *The Nature of the Physical World* (Cambridge University Press, Cambridge).

Eddington, A. S. [1929], *Science and the Unseen World* (Macmillan, New York).

Eddington, A. S. [1958], *The Philosophy of Physical Science* (Ann Arbor Paperbacks, University of Michigan Press, Ann Arbor) [originally published 1939, Cambridge University Press, Cambridge – Ed.].

Eddington, A. S. [1959], *New Pathways of Science* (Ann Arbor Paperbacks, University of Michigan Press, Ann Arbor) [originally published 1935, Cambridge University Press, Cambridge – Ed.].

Edwards, Paul [1949], 'Russell's Doubts about Induction', *Mind* **58**, pp. 141–163.

Einstein, Albert [1922], *Sidelights on Relativity* (Methuen, London).

Einstein, Albert [1933], *On the Method of Theoretical Physics* (The Clarendon Press, Oxford).

Einstein, Albert [1936], 'Physics and Reality', *J. of the Franklin Instit.* **221**, pp. 313–347.

Einstein, Albert and L. Infeld [1938], *The Evolution of Physics* (Simon & Schuster, New York).

Farrell, B. A. [1950], 'Experience', *Mind* **59**, pp. 170–198.

Feyerabend, Paul [1965], 'Problems of Empiricism', in [Colodny, 1965], pp. 145–260.

Feyerabend, Paul [1970a], 'Problems of Empiricism, Part II', in [Colodny, 1970], pp. 275–353.

Feyerabend, Paul [1970b], 'Philosophy of Science: A Subject with a Great Past', in [Stuewer, 1970], pp. 172–183.

Feyerabend, Paul [1970c], 'Against Method: Outline of an Anarchistic Theory of Knowledge', in [Radner & Winokur, 1970], pp. 17–130.

Feyerabend, Paul and Grover Maxwell (eds.) [1966], *Mind, Matter and Method* (University of Minnesota Press, Minneapolis).

Frank, Philipp [1932], *Das Kausalgesetz und seine Grenzen* (J. Springer, Vienna); Eng. tr. forthcoming, *Causality and Its Limits*, tr. Marie Neurath, in *Vienna Circle Collection* (D. Reidel Publishing Company, Dordrecht, Holland and Boston, Mass.).

Frank, Philipp [1941], *Between Philosophy and Physics* (Harvard University Press, Cambridge, Mass.).

Frank, Philipp [1949], *Modern Science and Its Philosophy* (Harvard University Press, Cambridge, Mass.) (revised and enlarged edition of [Frank, 1941]).

Frank, Philipp [1950], 'Comments on Realistic versus Phenomenalistic Interpretations', *Phil. of Sci.* 17, pp. 166–168.

Frank, Philipp [1957], *Philosophy of Science: The Link between Science and Philosophy* (Prentice-Hall, Englewood Cliffs, New Jersey).

Frechet, M. [1939], 'The Diverse Definitions of Probability', *Erkenntnis* 8, pp. 7–23.

Frege, Gottlob [1884], *Die Grundlagen der Arithmetik* (Koebner, Breslau); reprinted in 1934. (Eng. tr., *The Foundations of Arithmetic: A Logico-Mathematical Inquiry into the Concept of Number*, trans. by J. L. Austin. Blackwell, Oxford; Philosophical Library, New York, 1950.)

Frege, Gottlob [1893, 1903], *Die Grundgesetze der Arithmetik*, Vols. I and II (H. Pohle, Jena). (Eng. tr. of Introduction, opening sections and appendices as *The Basic Laws of Arithmetic: Exposition of the System*, tr. by Montgomery Furth, University of California Press, Berkeley, 1964.)

Galilei, Galileo [1914], *Dialogues concerning Two New Sciences*, tr. by Henry Crew and Alfonso de Salvio (Macmillan, New York). [See also Stillman Drake tr. (Univ. of Wisconsin Press, 1974) – Ed.]

Gätschenberger, F. [1920], *Symbole* (G. Braun, Karlsruhe).

Gödel, Kurt [1931], 'Über formal unentscheidbare Sätze der Principia Mathematica und verwandter Systeme I', *Monatshefte für Mathematik und Physik* 38, pp. 173–198. For Eng. tr., see *On Formally Undecidable Propositions of Principia Mathematica and Related Systems*, tr. by B. Meltzer (Oliver and Boyd, Edinburgh and London, 1962). (A new translation by J. van Heijenoort was published in van Heijenoort's *From Frege to Gödel: A Sourcebook in Mathematical Logic, 1879–1931*, Harvard University Press, Cambridge, Mass., 1967.)

Gomperz, Heinrich [1937], 'Some Simple Thoughts on Freedom and Responsibility', *Philosophy* 12, pp. 61–76.

Gomperz, Heinrich [1941], 'The Meanings of "Meaning" ', *Phil. of Sci.* 8, pp. 157–183.

Goodman, Nelson [1940], 'A Study of Qualities: An Essay in Elementary Constitutional Theory' (Doctoral Dissertation); "Entirely rewritten for publication" as [Goodman, 1966].

Goodman, Nelson [1947], 'The Problem of Counterfactual Conditionals', *J. Phil.* 44, pp. 113–128.

Goodman, Nelson [1966], *The Structure of Appearance* (Bobbs-Merrill, Indianapolis).

Revised and reprinted in *Boston Studies in the Philosophy of Science*, Vol. 53, Robert S. Cohen and Marx W. Wartofsky (eds.) (D. Reidel Publ. Co., Dordrecht, Holland and Boston, Mass.).

Grelling, K. [1928], 'The Philosophy of the Exact Sciences: Its Present Status in Germany', *Monist* **38**, pp. 97–119.

Grünbaum, Adolf [1963], *Philosophical Problems of Space and Time* (Alfred A. Knopf, New York); 2nd enlarged edition published in *Boston Studies in the Philosophy of Science*, Vol. 12, Robert S. Cohen and Marx W. Wartofsky (eds.) (D. Reidel Publ. Co., Dordrecht, Holland and Boston, Mass., 1973).

Grünbaum, Adolf [1971], 'Falsification and Rationality', 1971 International Colloquium on Issues in Contemporary Physics and Philosophy of Science (Penn. State University), revised and enlarged as 'Popper vs Inductivism', in *Progress and Rationality in Science*, Gerard Radnitzky and Gunnar Anderson (eds.), *Boston Studies in the Philosophy of Science*, Vol. 58 (D. Reidel Publ. Co., Dordrecht, Holland and Boston, Mass., 1978).

Gunderson, Keith [1970], 'Asymmetries and Mind-Body Perplexities', in [Radner and Winokur, 1970], pp. 273–309.

Haeckel, Ernst [1900], *The Riddle of the Universe at the Close of the 19th Century* (Harper, New York).

Hampshire, Stuart [1952], 'The Analogy of Feeling', *Mind* **61**, pp. 1–12.

Hampshire, Stuart [1966], *Philosophy of Mind* (Harper & Row, New York).

Hanson, N. R. [1969], 'Logical Positivism and the Interpretation of Scientific Theories', in [Achinstein and Barker, 1969], pp. 57–84.

Hardie, C. D. [1939], 'Our Knowledge of Other Minds', *Phil. of Sci.* **6**, pp. 309–317.

Heider, R. [1939], 'Environmental Determinants in Psychological Theories', *Psych. Rev.* **46**, pp. 383–410.

Helmholtz, H. von [1873], *Popular Lectures on Scientific Subjects* (D. Appleton and Co., New York).

Helmholtz, H. von [1921], *Schriften zur Erkenntnistheorie* (J. Springer, Berlin); Eng. tr., *Epistemological Writings*, in Robert S. Cohen and Y. Elkana (eds.), *Boston Studies in the Philosophy of Science*, Vol. 37 (D. Reidel Publ. Co., Dordrecht, Holland and Boston, Mass. 1977).

Hempel, C. G. [1935], 'On the Logical Positivists' Theory of Truth', *Analysis* **2**, pp. 49–59.

Hempel, C. G. [1939], 'Vagueness and Logic', *Phil. of Sci.* **6**, pp. 163–180.

Hempel, C. G. [1942], 'The Function of General Laws in History', *J. Phil.* **39**, pp. 35–48.

Hempel, C. G. [1945], 'Studies in the Logic of Confirmation', *Mind* **54**, Part I, pp. 1–26; Part II, pp. 97–121.

Hempel, C. G. [1949], 'The Logical Analysis of Psychology', in [Feigl and Sellars, 1949b], pp. 373–384. [Originally published as 'Analyse logique de la psychologie', *Revue de Synthèse* **10** (1935), pp. 27–42. – Ed.].

Hempel, C. G. [1950a], 'Problems and Changes in the Empiricist Criterion of Meaning', *Revue Internationale de Philosophie* **11**, pp. 41–63.

Hempel, C. G. [1950b], 'A Note on Semantic Realism', *Phil. of Sci.* **17**, pp. 169–173.

Hempel, C. G. [1952], 'Fundamentals of Concept Formation in the Empirical Sciences', *Intl. Ency. of Unif. Sci.* **2**, No. 7 (University of Chicago Press, Chicago).
Hempel, C. G. [1965], *Aspects of Scientific Explanation* (The Free Press, New York).
Hempel, C. G. [1966], *Philosophy of Natural Science* (Prentice-Hall, Englewood Cliffs, New Jersey).
Hempel, C. G. [1970], 'On the "Standard Conception" of Scientific Theories', in [Radner and Winokur, 1970], pp. 142–163.
Hempel, C. G. and P. Oppenheim [1936], *Der Typusbegriff im Lichte der neuen Logik* (Sijthoff, Leiden).
Hempel, C. G. and P. Oppenheim [1948], 'Studies in the Logic of Explanation', *Phil. of Sci.* **15**, pp. 135–175.
Hilbert, David [1930], *Die Grundlagen der Geometrie* (enlarged edition) (Teubner, Leipzig). For Eng. tr., see *Foundations of Geometry*, 2nd ed., tr. by Leo Unger (Open Court, LaSalle, Illinois, 1971).
Hilbert, David and W. Ackermann [1938], *Grundzüge der theoretischen Logik*, 2nd ed. (J. Springer, Berlin). For Eng. tr., see *Principles of Mathematical Logic*, tr. by L. M. Hammond, G. G. Leckie and F. Steinhardt (Chelsea, New York, 1950).
Hobart, R. E. [1930], 'Hume without Scepticism', *Mind* **39**, Part I, pp. 273–301; Part II, pp. 409–425.
Hobart, R. E. [1934], 'Free Will as Involving Determination and Inconceivable without It', *Mind* **43**, pp. 1–27.
Hofstadter, Albert [1938], 'On Semantic Problems', *J. Phil.* **35**, pp. 225–232.
Hofstadter, Albert and J. C. C. McKinsey [1939], 'On the Logic of Imperatives', *Phil. of Sci.* **6**, pp. 446–457.
Hook, Sidney [1940], *Reason, Social Myths and Democracy* (John Day Co., New York).
Hook, Sidney (ed.) [1950], *John Dewey: Philosopher of Science and Freedom* (The Dial Press, New York).
Hull, C. L. [1937], 'Mind, Mechanism, and Adaptive Behavior', *Psych. Rev.* **44**, pp. 1–32.
Hull, C. L. [1940], *Mathematico-Deductive Theory of Rote Learning* (Yale University Press, New Haven).
Hull, C. L. [1943a], 'The Problem of Intervening Variables in Molar Behavior Theory', *Psych. Rev.* **50**, pp. 273–291.
Hull, C. L. [1943b], *Principles of Behavior* (Appleton-Century-Crofts, Inc., New York).
Hume, David [1928], *A Treatise of Human Nature* (The Clarendon Press, London).
Hutchison, T. W. [1938], *The Significance and Basic Postulates of Economic Theory* (Macmillan, London).
Huxley, Julian [1947], *Touchstone for Ethics* (Harper, New York).

Jacobs, N. [1937], 'Physicalism and Sensation Sentences', *J. Phil.* **34**, pp. 602–611.
James, William [1907], *Pragmatism* (Longmans Green & Co., New York).
Jørgenson, Jørgen [1931], *A Treatise of Formal Logic* (Levin and Munksgaard, Copenhagen).

Kaila, Eino [1936], 'Über das System der Wirklichkeitsbegriffe', *Acta Philos. Fennica* **2** (Helsinki); Eng. tr. as 'On the System of the Concepts of Reality', in *Reality and Experience*, Robert S. Cohen (ed.), *Vienna Circle Collection*, Vol. 12 (D. Reidel Publ. Co., Dordrecht, Holland and Boston, Mass., 1979).

Kaplan, Abraham [1946], 'Definition and Specification of Meaning', *J. Phil.* **43**, pp. 281–288.

Katz, Jerrold J. [1962], *The Problem of Induction and Its Solution* (University of Chicago Press, Chicago).

Kaufmann, Felix [1936], *Methodenlehre der Sozialwissenschaften* (Vienna). Eng. tr. forthcoming in *Vienna Circle Collection* (D. Reidel Publ. Co. Dordrecht, Holland and Boston, Mass.).

Kaufmann, Felix [1944], *Methodology of the Social Sciences* (Oxford University Press, New York).

Kekes, John [1966], 'Physicalism, the Identity Theory and the Doctrine of Emergence', *Phil. of Sci.* **33**, pp. 360–375.

Kekes, John [1970], 'Theoretical Identity', *Southern J. of Phil.* **8**, pp. 25–36.

Keynes, J. M. [1904], *Studies and Exercises in Formal Logic*, 4th ed. (Macmillan, London).

Keynes, J. M. [1921], *A Treatise on Probability* (Macmillan, London).

Kneale, William C. [1949], *Probability and Induction* (The Clarendon Press, London).

Kneale, William C. [1953], 'Induction, Explanation and Transcendent Hypotheses', in [Feigl and Brodbeck, 1953a].

Koch, Sigmund [1941], 'The Logical Character of the Motivation Concept', *Psych. Rev.* **48**, pp. 15–38 and 127–154.

Koch, Sigmund [1944], 'Hull's Principles of Behavior', *Psych. Bull.* **41**, pp. 269–286.

Köhler, Wolfgang [1938], *The Place of Values in a World of Facts* (Liveright, New York).

Köhler, Wolfgang [1940], *Dynamics in Psychology* (Liveright, New York).

Kordig, Carl R. [1971], 'The Theory-Ladenness of Observation', *Rev. of Metaphysics* **24** (March), pp. 448–484.

Kraft, Viktor [1937], *Die Grundlagen einer wissenschaftlichen Wertlehre* (J. Springer, Vienna). Eng. tr. forthcoming as *Foundations for a Scientific Analysis of Value*, Henk L. Mulder ed., *Vienna Circle Collection*, (D. Reidel Publishing Company, Dordrecht, Holland, and Boston, Mass.).

Lakatos, Imre (ed.) [1968], *The Problems of Inductive Logic* (North-Holland Publ. Co., Amsterdam).

Lakatos, Imre and Alan Musgrave (eds.) [1960], *Problems in the Philosophy of Science* (North-Holland Publ. Co., Amsterdam).

Lakatos, Imre and Alan Musgrave (eds.) [1970], *Criticism and the Growth of Knowledge* (Cambridge University Press, Cambridge).

Langer, Suzanne K. [1942], *Philosophy in a New Key* (Harvard University Press, Cambridge, Mass.).

Lashley, K. S. [1923], 'The Behavioristic Interpretation of Consciousness', *Psych. Rev.* **30**, Part I, pp. 237–272; Part II, pp. 329–353.

Lenzen, Victor F. [1931], *The Nature of Physical Theory* (J. Wiley, New York).

Lenzen, Victor F. [1939a], 'Procedures of Empirical Science', *Intl. Ency. of Unif. Sci.* **1**, No. 5 (University of Chicago Press, Chicago).

Lenzen, Victor F. [1939b], 'The Hypothesis of Dualism', *Phil. of Sci.* **6**, pp. 254–256.

Lenzen, Victor F. [1945], 'The Concept of Reality in Physical Theory', *Phil. Rev.* **18**, pp. 321–344.

Lewin, Kurt [1935], *A Dynamic Theory of Personality* (McGraw-Hill, New York).

Lewin, Kurt [1938], 'The Conceptual Representation and the Measurement of Psychological Forces', *Contr. to Psych. Theory* **1**, No. 4.

Lewin, Kurt [1943], 'Defining the "Field at a Given Time" ', *Psych. Rev.* **50**, pp. 292–310.

Lewin, Kurt [1949], 'Cassirer's Philosophy of Science and the Social Sciences', in [Schilpp, 1949], pp. 269–288.

Lewis, C. I. [1918], *Survey of Symbolic Logic* (University of California Press, Berkeley).

Lewis, C. I. [1929], *Mind and the World Order* (Charles Scribner's Sons, New York).

Lewis, C. I. [1934], 'Experience and Meaning', *Phil. Rev.* **43**; reprinted in [Feigl and Sellars, 1949b], pp. 128–145.

Lewis, C. I. [1941], 'Some Logical Considerations Concerning the Mental', *J. Phil.* **38**, pp. 225–233; reprinted in [Feigl and Sellars, 1949b], pp. 385–392.

Lewis, C. I. [1946], *An Analysis of Knowledge and Valuation* (Open Court, LaSalle, Illinois).

Lewis, C. I. and C. H. Langford [1932], *Symbolic Logic* (Century Co., New York).

Lewy, C. [1969], 'Is the Notion of Disembodied Existence Self-Contradictory?', *Proc. Aristot. Soc., N.S.* **43**.

Lindsay, R. B. and H. Margenau [1936], *Foundations of Physics* (J. Wiley, New York).

Linsky, L. (ed.) [1952], *Semantics and Philosophy of Language* (University of Illinois Press, Urbana, Illinois).

Littman, R. A. and A. Rosen [1950], 'Molar and Molecular', *Psych. Rev.* **57**, pp. 58–65.

London, I. D. [1944], 'Psychologists' Misues of the Auxiliary Concepts of Physics and Mathematics', *Psych. Rev.* **51**, pp. 266–291.

London, I. D. [1946], 'Some Consequences for History and Psychology of Langmuir's Concept of Convergence and Divergence of Phenomena', *Psych. Rev.* **53**, pp. 170–188.

Lovejoy, A. O. [1930], *The Revolt against Dualism* (W. W. Norton & Co., New York).

Lundberg, G. A. [1939], *Foundations of Sociology* (Macmillan, New York).

MacCorquodale, K. and P. E. Meehl [1948], 'On a Distinction between Hypothetical Constructs and Intervening Variables', *Psych. Rev.* **55**, pp. 95–107.

Mach, Ernst [1896], *Die Principien der Wärmelehre. Historisch-kritisch entwickelt* (J. A. Barth, Leipzig); Eng. tr. forthcoming, Brian McGuinness (ed.), *Vienna Circle Collection* (D. Reidel Publ. Co., Dordrecht, Holland, and Boston, Mass.).

Mach, Ernst [1897], *Contributions to the Analysis of Sensations* (Open Court, Chicago).

Mach, Ernst [1898], *Popular Scientific Lectures* (Open Court, Chicago).

Mach, Ernst [1905], *Erkenntnis und Irrtum* (J. A. Barth, Leipzig). For Eng. tr., see *Knowledge and Error*, Brian McGuinness (ed.), *Vienna Circle Collection*, Vol. 3 (D. Reidel Publ. Co., Dordrecht, Holland, and Boston, Mass., 1976).

Mach, Ernst [1914], *Analysis of Sensations* (Open Court, Chicago); 2nd edition of [1897].

Mach, Ernst [1919], *The Science of Mechanics*, 4th edition (Open Court, Chicago).

Mach, Ernst [1921], *Die Prinzipien der physikalischen Optik* (J. A. Barth, Leipzig). For Eng. tr., see *The Principles of Physical Optics* (Methuen, London, 1926).

Malcolm, Norman [1970], 'Scientific Materialism and the Identity Theory', in [C. V. Borst, 1970].

Margenau, Henry [1950], *The Nature of Physical Reality* (McGraw-Hill, New York).

Maxwell, Grover [1960a], 'Scientific Methodology and the Causal Theory of Perception', in [Lakatos and Musgrave, 1960], pp. 148–160.
Maxwell, Grover [1960b], 'Reply to Professors Quine, Ayer, Popper and Kneale', in [Lakatos and Musgrave, 1960], pp. 167–177.
Maxwell, Grover [1969], Review of Bruce Aune's *Knowledge, Mind and Nature. Phil. Rev.* 78, pp. 392–397.
Maxwell, Grover [1970], 'Theories, Perception, and Structural Realism', in [Colodny, 1970], pp. 3–34.
Maxwell, Grover [1974], 'Corroboration without Demarcation', in [P. A. Schilpp, 1974], pp. 293–321.
Medlin, Brian [1967], 'Ryle and the Mechanical Hypothesis', in [Presley, 1967].
Meehl, Paul E. [1950a], 'On the Circularity of the Law of Effect', *Psych. Bull.* 47, pp. 52–74.
Meehl, Paul E. [1950b], 'A Most Peculiar Paradox', *Phil. Studies* 1, pp. 47–48.
Meehl, Paul E. [1958], *what, Then, Is Man?*, Graduate Study Number III (Concordia Publishing House, St. Louis).
Meehl, Paul E. [1966], 'The Compleat Autocerebroscopist', in [Feyerabend and Maxwell, 1966].
Meehl, Paul E. [1970], 'Psychological Determinism and Human Rationality', in [Radner, 1970], pp. 310–372.
Meehl, Paul E. and W. Sellars [1956], 'The Concept of Emergence', in [Feigl and Scriven, 1956d], pp. 239–252.
Mehlberg, Henry [1948], 'Positivisme et science', *Studia Philosophica* 3.
Mehlberg, Henry [1958], *The Reach of Science* (University of Toronto Press, Toronto).
Menger, Karl [1934], *Moral, Wille, Weltgestaltung* (Springer, Vienna). For Eng. tr., see *Morality, Decision, and Social Organization: Toward a Logic of Ethics. Vienna Circle Collection*, Vol. 6 (D. Reidel Publ. Co., Dordrecht, Holland, and Boston, Mass., 1974).
Menger, Karl [1937], 'The New Logic', *Phil. Sci.* 4, pp. 299–336.
Mill, John Stuart [1930], *A System of Logic*, 8th edition (Harper, New York).
Mises, Richard von [1939a], *Kleines Lehrbuch des Positivismus* (W. P. Van Stockum an Zoon, The Hague). For Eng. tr., see *Positivism: A Study in Human Understanding*, Jerry Bernstein and Roger Newton, trs. (Harvard University Press, Cambridge, Mass., 1951).
Mises, Richard von [1939b], *Probability, Statistics, and Truth* (W. Hodge and Co., London); [2nd revised Eng. ed., tr. of 3rd definitive German ed. (Allen and Unwin, London and Macmillan, New York, 1957) – Ed.].
Morris, C. W. [1932], *Six Theories of Mind* (University of Chicago Press, Chicago).
Morris, C. W. [1937], *Logical Positivism, Pragmatism and Scientific Empiricism* (Hermann, Paris).
Morris, C. W. [1939], 'Foundations of the Theory of Signs', *Intl. Ency. of Unif. Sci.* 1, No. 2 (University of Chicago Press, Chicago).

Nagel, Ernest [1939], 'The Formation of Modern Conceptions of Formal Logic in the Development of Geometry', *Osiris* 7.
Nagel, Ernest [1950], 'Science and Semantic Realism', *Phil. of Sci.* 17, pp. 174–181.
Nagel, Ernest [1956], *Logic without Metaphysics* (The Free Press, Glencoe, Ill.).
Nagel, Ernest [1971], 'Theory and Observation', in *Observation and Theory in Science,*

E. Negal, Sylvain Bromberger and Adolf Grünbaum (eds.) (The John Hopkins Press, Baltimore), pp. 15–43.

Nelson, E. J. [1936], 'The Inductive Argument for an External World', *Phil. of Sci.* 3, pp. 237–249.

Neurath, Otto [1931], *Empirische Soziologie* (J. Springer, Vienna). For Eng. tr., see [1973b].

Neurath, Otto [1935], *Le dèveloppement du cercle de Vienne et l'avenir de l'empirisme logique* (Hermann, Paris).

Neurath, Otto [1973a], *Empiricism and Sociology*, Marie Neurath and Robert S. Cohen (eds.), *Vienna Circle Collection*, Vol. 1 (D. Reidel Publ. Co., Dordrecht, Holland and Boston, Mass.).

Neurath, Otto [1973b], 'Empirical Sociology', in [Neurath, 1973a], pp. 319–421.

Neurath, Otto, Hans Hahn, and Rudolf Carnap [1929], *Wissenschaftliche Weltauffassung: Der Wiener Kreis*, Veröffentlichungen des Vereines Ernst Mach (Artur Wolf Verlag). For Eng. tr. see ['The Scientific Conception of the World: The Vienna Circle'], in [Neurath, 1973a], pp. 299–318.

Nicod, Jean [1930], *Foundations of Geometry and Induction* (Kegan Paul, London).

Northrop, F. S. C. [1947], *The Logic of the Sciences and the Humanities* (Macmillan, New York).

Pap, Arthur [1946], *The* A Priori *in Physical Theory* (King's Crown Press, New York).

Pap, Arthur [1949a], *Elements of Analytic Philosophy* (Macmillan, New York).

Pap, Arthur [1949b], 'Does Science Have Metaphysical Presuppositions?', in [1949a], Chapter 16. Reprinted in [Feigl and Brodbeck, 1953a].

Pap, Arthur [1951], 'Other Minds and the Principle of Verifiability', *Revue Internationale de Philosophie* 5, pp. 280–306.

Pap, Arthur [1952], 'Semantic Analysis and Psycho-Physical Dualism', *Mind* 61, pp. 209–221.

Pap, Arthur [1962], *Introduction to the Philosophy of Science* (Macmillan, The Free Press, New York).

Pearson, Karl [1911], *The Grammar of Science*, 3rd ed. (A. and C. Black, London).

Peirce, Charles S. [1878], 'How to Make Our Ideas Clear', *Popular Science Monthly* 12.

Peirce, Charles S. [1931], *Collected Papers*, Vol. 5, Charles Hartshorne and Paul Weiss (eds.) (Harvard University Press, Cambridge).

Pepper, S. C. [1967], *Concept and Quality* (Open Court, LaSalle, Illinois).

Perry, Ralph B. [1910], 'The Egocentric Predicament', *J. Phil., Psych. and Sci. Methods* 7, No. 1, pp. 5–14.

Perry, Ralph B. [1916], *Present Philosophical Tendencies* (Longmans Green & Co., New York).

Perry, Ralph B. [1926], *General Theory of Value* (Longmans Green & Co., New York).

Poincaré, Henri [1929], *The Foundations of Science* (The Science Press, New York).

Popper, Karl R. [1935], *Logik der Forschung* (J. Springer, Vienna). For Eng. tr., see [1959].

Popper, Karl R. [1940], 'What is Dialectic?', *Mind* 49, pp. 403–426.

Popper, Karl R. [1949], 'A Note on Natural Laws and So-Called "Contrary-to-Fact Conditionals"', *Mind* 58, pp. 62 66.

Popper, Karl R. [1959], *The Logic of Scientific Discovery* (Basic Books, New York).

Popper, Karl R. [1962], *Conjectures and Refutations* (Basic Books, New York).

Popper, Karl R. [1971], 'Conjectural Knowledge: My Solution of the Problem of Induction', *Revue Internationale de Philosophie* **95–96**, pp. 167–197.

Presley, C. F. (ed.) [1967], *The Identity Theory of Mind* (University of Queensland Press, St. Lucia).

Pratt, C. C. [1939], *The Logic of Modern Psychology* (Macmillan, New York).

Pratt, C. C. [1945], 'Operationism in Psychology', *Psych. Rev.* **52**, pp. 262–269.

Quine, W. V. O. [1940], *Mathematical Logic* (W. W. Norton, New York).

Quine, W. V. O. [1951], 'Semantics and Abstract Objects', *Proceedings of the American Academy of Arts and Sciences* **80**, p. 91.

Radner, Michael and Stephen Winokur (eds.) [1970], *Analyses of Theories and Methods of Physics and Psychology. Minnesota Studies in the Philosophy of Science*, Volume 4 (University of Minnesota Press, Minneapolis).

Ramsey, F. P. [1931], *The Foundations of Mathematics and Other Logical Essays* (Harcourt Brace, New York and Kegan Paul, London).

Ramsperger, A. G. [1942], *Philosophies of Science* (Croft & Co., New York).

Reichenbach, Hans [1924], *Axiomatik der relativistischen Raum-Zeit-Lehre*, (F. Vieweg, Braunschweig). For Eng. tr., see *Axiomatization of the Theory of Relativity*, tr. and ed. by Maria Reichenbach (University of California Press, Berkeley, 1969).

Reichenbach, Hans [1928], *Philosophie der Raum-Zeit-Lehre* (W. de Gruyter, Berlin). For Eng. tr., see *The Philosophy of Space and Time*, tr. by Maria Reichenbach (Dover, New York, 1958).

Reichenbach, Hans [1932], 'Axiomatik der Wahrscheinlichkeitsrechnung', *Mathematische Zeitschrift* **34**, pp. 568–619.

Reichenbach, Hans [1935], *Wahrscheinlichkeitslehre* (Sijthoff, Leyden). For Eng. tr., see [1949b].

Reichenbach, Hans [1936], 'Logistic Empiricism in Germany and the Present State of Its Problems', *J. Phil.* **33**, pp. 141–160.

Reichenbach, Hans [1938], *Experience and Prediction* (University of Chicago Press, Chicago).

Reichenbach, Hans [1940], 'On the Justification of Induction', *J. Phil.* 37, pp. 97–103. Reprinted in [Feigl and Sellars, 1949b], pp. 324–329.

Reichenbach, Hans [1944], *Philosophical Foundations of Quantum Mechanics* (University of California Press, Berkeley).

Reichenbach, Hans [1947], *Elements of Symbolic Logic* (Macmillan, New York).

Reichenbach, Hans [1948], 'Rationalism and Empiricism', *Phil. Rev.* 57, pp. 330–346.

Reichenbach, Hans [1949a], 'The Logical Foundations of the Concept of Probability', in [Feigl and Sellars, 1949b], pp. 305–323; also in [Feigl and Brodbeck, 1953a].

Reichenbach, Hans [1949b], *Theory of Probability* (University of California Press, Berkeley). Rev. tr. of [Reichenbach, 1935].

Riehl, Alois [1894], *Science and Metaphysics* (Kegan Paul, London).

Ritchie, B. F. and A. Kaplan [1940], 'A Framework for an Empirical Ethics', *Phil. of Sci.* 7, pp. 476–491.

Rosenblueth, A. and N. Wiener [1943], 'Behavior, Purpose and Teleology', *Phil. of Sci.* **10**, pp. 18–24.

Rosenblueth, A. and N. Wiener [1950], 'Purposeful and Non-Purposeful Behavior', *Phil. of Sci.* 17, pp. 318–326.

Russell, Bertrand [1921], *The Analysis of Mind* (Macmillan, London).

Russell, Bertrand [1929a], *Our Knowledge of the External World*, 2nd edition (W. W. Norton, New York).

Russell, Bertrand [1929b], *Mysticism and Logic* (W. W. Norton, New York).

Russell, Bertrand [1938], *The Principles of Mathematics*, 2nd edition (W. W. Norton, New York).

Russell, Bertrand [1940], *An Inquiry into Meaning and Truth* (W. W. Norton, New York).

Russell, Bertrand [1948], *Human Knowledge* (Simon & Schuster, New York).

Ryle, Gilbert [1949], *The Concept of Mind* (Hutchinson's University Library, London).

Rynin, D. [1947], 'Probability and Meaning', *J. Phil.* 44, pp. 589–597.

Salmon, Wesley C. [1957], 'Should We Attempt to Justify Induction?', *Phil. Studies* 8, pp. 33–48.

Salmon, Wesley C. [1966], 'The Foundations of Scientific Inference', in [Colodny, 1966], pp. 135–275.

Schächter, Josef [1935], *Prolegomena zu einer kritischen Grammatik* (J. Springer, Vienna). For Eng. tr. see *Prolegomena to a Critical Grammar*, trans. by Paul Foulkes, *Vienna Circle Collection*, Vol. 2 (D. Reidel Publ. Co., Dordrecht, Holland and Boston, Mass., 1973).

Schaffner, Kenneth [1967], 'Approaches to Reduction', *Phil of Sci.* 34, pp. 137–147.

Scheffler, Israel [1963], *The Anatomy of Inquiry* (Alfred A. Knopf, New York).

Schilpp, Paul A. (ed.) [1949], *The Philosophy of Ernst Cassirer. The Library of Living Philosophers* (Evanston, Illinois).

Schilpp, Paul A. (ed.) [1963], *The Philosophy of Rudolf Carnap. The Library of Living Philosophers* (Open Court, LaSalle, Illinois).

Schilpp, Paul A. (ed.) [1974], *The Philosophy of Karl Popper.* 2 vols. *The Library of Living Philosophers* (Open Court, LaSalle, Illinois).

Schlick, Moritz [1918], *Allgemeine Erkenntnislehre*, 1st edition (J. Springer, Berlin); 2nd edition (1925). For Eng. tr., see [1974].

Schlick, Moritz [1920], *Space and Time in Contemporary Physics* (Oxford University Press, New York).

Schlick, Moritz [1926], 'Erleben, Erkennen, und Metaphysik', *Kant-Studien* 31, pp. 146–158. For Eng. tr., see [Schlick (1980)].

Schlick, Moritz [1932a], 'Causality in Everyday Life and in Science', *University of California Publications in Philosophy* 15 (Berkeley, California), pp. 99–125.

Schlick, Moritz [1932b], 'The Future of Philosophy', in *College of the Pacific Publications in Philosophy*, Volume 1 (Stockton, California), pp. 45–62. Reprinted in [Schlick (1980)].

Schlick, Moritz [1932c], 'A New Philosophy of Experience', in *College of the Pacific Publications in Philosophy*, Volume 1 (Stockton, California), pp. 63–78. Reprinted in [Schlick (1980)).

Schlick, Moritz [1936], 'Meaning and Verification', *Phil. Rev.* 45, pp. 339–369.

Schlick, Moritz [1938a], *Gesammelte Aufsätze, 1926–1936* (Gerold, Vienna). (For Eng. tr. see Schlick [1979] and [1980].)

Schlick, Moritz [1938b], 'Positivism and Realism', in [1938a].

Schlick, Moritz [1939], *Problems of Ethics* (Prentice-Hall, New York).

Schlick, Moritz [1974], *General Theory of Knowledge*, tr. by A. E. Blumberg, with an Introduction by A. E. Blumberg and Herbert Feigl (Springer, New York); (Eng. tr. of [1918]).
Schlick, Moritz [1979], *Moritz Schlick: Philosophical Papers, 1909–1922*, Volume 1, Henk L. Mulder and Barbara F. B. van de Velde-Schlick (eds.), *Vienna Circle Collection*, Vol. 11 (D. Reidel Publ. Co., Dordrecht, Holland and Boston, Mass.).
Schlick, Moritz [1980], *Moritz Schlick: Philosophical Papers, 1923–1936*, Volume 2, Henk L. Mulder and Barbara F. B. van de Velde-Schlick (eds.), *Vienna Circle Collection*, Vol. 11 (D. Reidel Publ. Co., Dordrecht, Holland and Boston, Mass.).
Schrödinger, Erwin [1944], *What is Life?* (Cambridge University Press, Cambridge).
Scriven, Michael [1953], 'The Mechanical Concept of Mind', *Mind* **62**, pp. 230–240.
Sellars, R. W. [1932], *The Philosophy of Physical Realism* (Macmillan, New York).
Sellars, W. S. [1947a], 'Pure Pragmatics and Epistemology', *Phil. of Sci.* **14**, pp. 181–202.
Sellars, W. S. [1947b], 'Epistemology and the New Way of Words', *J. Phil.* **44**, pp. 645–660.
Sellars, W. S. [1948a], 'Concepts as Involving Laws and Inconceivable without Them', *Phil. of Sci.* **15**, pp. 287–315.
Sellars, W. S. [1948b], 'Realism and the New Way of Words', *Phil. and Phenom. Res.* **8**, pp. 601–634. Reprinted in [Feigl and Sellars, 1949b], pp. 424–456.
Sellars, W. S. [1949], 'Acquaintance and Description Again', *J. Phil.* **46**, pp. 496–504.
Sellars, W. S. [1950], 'Language, Rules and Behavior', in [Hook, 1950], pp. 289–315.
Sellars, W. S. [1952], 'Mind, Meaning and Behavior', *Phil. Studies* **3**, pp. 83–95.
Sellars, W. S. [1953a], 'Inference and Meaning', *Mind* **62**, pp. 313–338.
Sellars, W. S. [1953b], 'Is There a Synthetic *A Priori*?', *Phil. of Sci.* **20**, pp. 121–138.
Sellars, W. S. [1953c], 'A Semantical Solution of the Mind-Body Problem', *Methodos* **5**.
Sellars, W. S. [1954], 'Some Reflections on Language Games', *Phil. of Sci.* **21** (1954).
Sellars, W. S. and J. Hospers (eds.) [1952], *Readings in Ethical Theory* (Appleton-Century-Crofts, New York).
Shapere, Dudley [1966], 'Meaning and Scientific Change', in [Colodny, 1966], pp. 41–85.
Singer, E. A. [1924], *Mind as Behavior* (Adams & Co., Columbus, Ohio).
Skinner, B. F. [1938], *The Behavior of Organisms* (Appleton-Century Crofts, New York).
Skinner, B. F. [1945], 'The Operational Analysis of Psychological Terms', *Psych. Rev.* **52**, pp. 270–277; reprinted in [Feigl and Brodbeck, 1953a].
Skinner, B. F. [1953], *Science and Human Behavior* (Macmillan, New York).
Smart, J. J. C. [1959], 'Sensations and Brain Processes', *Phil. Rev.* **68**, No. 2, pp. 141–156.
Smart, J. J. C. [1963], *Philosophy and Scientific Realism* (Routledge & Kegan Paul, London).
Smart, J. J. C. [1968], *Between Science and Philosophy* (Random House, New York).
Smith, Kendon [1958], 'The Naturalistic Conception of Life', *Am. Scientist* **46**, No. 4, pp. 413–423.
Sommerhoff, G. [1950], *Analytical Biology* (Oxford University Press, London).
Spence, K. W. [1942], 'Theoretical Interpretations of Learning', in *Comparative Psychology* (Chapter 11), (Prentice-Hall, New York).

Spence, K. W. [1944], 'The Nature of Theory Construction in Contemporary Psychology', *Psych. Rev.* **51**, pp. 47–68.

Spence, K. W. [1948], 'The Postulates and Methods of Behaviorism', *Psych. Rev.* **55**, pp. 67–77.

Stace, W. T. [1932], *Theory of Knowledge and Existence* (Oxford University Press, London).

Stace, W. T. [1934], 'Refutation of Realism', *Mind* **43**, pp. 145–155. Reprinted in [Feigl and Sellars, 1949b], pp. 364–372.

Stace, W. T. [1944], 'Positivism', *Mind* **53**, pp. 215–237.

Stebbing, L. S. [1937], *Philosophy and the Physicists* (Methuen, London).

Stevens, S. S. [1935], 'The Operational Definition of Psychological Concepts', *Psych. Rev.* **42**, pp. 517–527.

Stevens, S. S. [1939], 'Psychology and the Science of Science', *Psych. Bull.* **36**, pp. 221–263.

Stevens, S. S. [1945], 'Symposium on Operationism', *Psych. Rev.* **52**, pp. 241–294.

Stevenson, C. L. [1937], 'The Emotive Meaning of Ethical Terms', *Mind* **46**, pp. 14–31.

Stevenson, C. L. [1938a], 'Ethical Judgments and Avoidability', *Mind* **47**, pp. 45–57.

Stevenson, C. L. [1938b], 'Persuasive Definitions', *Mind* **47**, pp. 331–350.

Stevenson, C. L. [1944], *Ethics and Language* (Yale University Press, New Haven).

Storer, T. [1953], 'A Note on Empiricism', *Phil. Studies* **4**, No. 5, p. 78.

Strawson, P. F. [1952], *Introduction to Logical Theory* (Methuen, London).

Strawson, P. F. [1958], 'Persons', in [Feigl, Maxwell and Scriven, 1958a], pp. 330–353.

Stuewer, Roger (ed.) [1970], *Historical and Philosophical Perspectives of Science. Minnesota Studies in the Philosophy of Science*, Volume 5 (University of Minnesota Press, Minneapolis).

Tarski, Alfred [1936], 'Der Wahrheitsbegriff in den formalisierten Sprachen', *Studia Philos.* **1** [Eng. tr., 'The Concept of Truth in Formalized Languages', Chapter 8 of Tarski, *Logic, Semantics, Metamathematics* (Clarendon Press, Oxford, 1956) – Ed.].

Tarski, Alfred [1941], *Introduction to Logic and to the Methodology of the Deductive Sciences* (Oxford University Press, New York).

Teilhard de Chardin, Pierre [1959], *The Phenomenon of Man* (Harper & Row, New York).

Tolman, E. C. [1932], *Purposive Behavior in Animals and Men* (The Century Co., New York).

Tolman, E. C. [1935], 'Psychology versus Immediate Experience', *Phil. of Sci.* **2**, pp. 356–380.

Tolman, E. C. [1938], 'Physiology, Psychology and Sociology', *Psych. Rev.* **45**, pp. 228–241.

Tolman, E. C. and E. Brunswik [1935], 'The Organism and the Causal Texture of the Environment', *Psych. Rev.* **42**, pp. 43–77.

Tolman, E. C., B. F. Ritchie, and D. Kalish [1946], 'Studies in Spatial Learning I. Orientation and the Short-Cut', *J. Exper. Psych.* **36**, pp. 13–24.

Toulmin, Stephen E. [1958], *The Uses of Argument* (Cambridge University Press, Cambridge).

Turing, A. M. [1950], 'Computing Machinery and Intelligence', *Mind* **59**, pp. 433–460.

Waismann, F. [1936], *Einführung in das mathematische Denken* (Gerold, Vienna). For Eng. tr., see *Introduction to Mathematical Thinking*, tr. by T. J. Benac (Hafner, New York, 1951).

Wann, T. W. (ed.) [1964], *Behaviorism and Phenomenology* (University of Chicago Press, Chicago).

Weinberg, J. R. [1936], *An Examination of Logical Positivism* (Harcourt Brace, New York).

Weinschenk, Curt [1936], *Das Wirklichkeitsproblem der Erkenntnistheorie und das Verhältnis des Psychischen zum Physischen* (Reisland, Leipzig).

Weiss, A. P. [1925], *The Theoretical Basis of Human Behavior*, 1st edition (R. G. Adams & Co., Columbus, Ohio); 2nd edition (1929).

Weyl, Hermann [1949], *Philosophy of Mathematics and Natural Science* (Princeton University Press, Princeton, New Jersey).

White, Morton G. [1950], 'The Analytic and the Synthetic: an Untenable Dualism', in [Hook, 1950], pp. 316–330.

Whitehead, A. N. and B. Russell [1925, 1927], *Principia Mathematica*. 2nd edition. 3 vols. (Cambridge University Press, Cambridge).

Wiener, Norbert [1948], *Cybernetics* (J. Wiley & Sons, New York).

Wiener, P. P. (ed.) [1953], *Readings in Philosophy of Science* (Charles Scribner's Sons, New York).

Will, Frederick [1950], 'Generalization and Evidence', in [Black, 1950], pp. 384–413.

Williams, D. C. [1934], 'The Argument for Realism', *Monist* **44**, pp. 186–209.

Williams, D. C. [1947], *The Ground of Induction* (Harvard University Press, Cambridge, Mass.).

Wisdom, J. O. [1947], *The Metamorphosis of Philosophy* (Al-Maaref Press, Cairo).

Wisdom, J. O. [1952a], *Foundations of Inference in Natural Science* (Methuen, London).

Wisdom, John [1952b], *Other Minds* (Blackwell, Oxford).

Wittgenstein, Ludwig [1922], *Tractatus Logico-Philosophicus* (Kegan Paul, London).

Wittgenstein, Ludwig [1953], *Philosophical Investigations* (Macmillan, New York and Blackwell, Oxford).

Witt-Hansen, J. [1958], *Exposition and Critique of the Conceptions of Eddington Concerning the Philosophy of Physical Science* (G. E. C. Gads Verlag, Copenhagen).

Woodger, J. H. [1929], *Biological Principles* (Harcourt Brace, New York).

Woodger, J. H. [1937], *The Axiomatic Method in Biology* (Cambridge University Press, Cambridge).

Woodger, J. H. [1939], 'Technique of Theory Construction', *Intl. Ency. of Unif. Sci.* **2**, No. 5 (University of Chicago Press, Chicago).

Zilsel, Edgar [1916], *Das Anwendungsproblem* (J. A. Barth, Leipzig).

BIBLIOGRAPHY OF THE WRITINGS OF HERBERT FEIGL

1927 and 1928: 'Zufall and Gesetz' in *Wissenschaftlicher Jahresbericht der Philosophischen Gesellschaft an der Universität zu Wien*. (A summary of H. Feigl's doctoral dissertation. Complete manuscript available at University of Vienna.)

1929: *Theorie und Erfahrung in der Physik* (G. Braun, Karlsruhe). Eng. tr. of Ch. III, 'Sinn und Gültigkeit der physikalischen Theorien', in [1980], pp. 116–144.

1930: 'Wahrscheinlichkeit und Erfahrung', *Erkenntnis* **1**, 249–259. Eng. tr. in [1980], pp. 107–115.

1931: (with A. E. Blumberg) 'Logical Positivism', *Journal of Philosophy* **28**, 281–296.

1934: (a) 'The Logical Character of the Principle of Induction', *Philosophy of Science* **1**, 20–29. Reprinted in H. Feigl and W. Sellars [1949b]. Reprinted in [1980], pp. 153–163.

(b) 'The Logical Analysis of the Psycho-Physical Problem', *Philosophy of Science* **1**, 420–445.

(c) Discussion 'The Principle of Induction' (letter replying to Professor Homer H. Dubs), *Philosophy of Science* **1**, 484–486.

1935: Discussion 'Spatial Location and the Psychophysical Problem' (a reply to V. C. Aldrich), *Philosophy of Science* **2**, 257–261.

1936: 'Sense and Nonsense in Scientific Realism', in *Actes du Congrès international de philosophie scientifique*. Vol. 3: *Langage et pseudo-Problèmes*, (Hermann, Paris), pp. 50–56.

1939: (a) 'Moritz Schlick', *Erkenntnis* **7**, 393–419. Reprinted in *Moritz Schlick, Philosophical Papers*, H. L. Mulder and B. F. B. van de Velde-Schlick (eds.), *Vienna Circle Collection*, Volume 11 (D. Reidel Publ. Co., Dordrecht, Holland, 1978).

(b) 'The Significance of Physics in Man's Philosophy', *American Physics Teacher* **7**, 324–327.

1942: 'The Meaning of Freedom', *The Interpreter* (University of Minnesota) **17**, 1–2 (December).

1943: 'Logical Empiricism', in *Twentieth Century Philosophy*, D. D. Runes (ed.) (Philosophical Library, New York), pp. 371–416. Reprinted with omissions in H. Feigl and W. Sellars (eds.) [1949b]; and in *Philosophic Problems*, M. Mandelbaum, F. W. Gramlich, and A. R. Anderson (eds.) (Macmillan, New York, 1957). Reprinted in part under the title 'Meanings in Ethical Discourse', in *Contemporary Philosophy*, J. L. Jarrett and S. M. McMurrin (eds.) (Henry Holt, New York, 1954); and also in part as 'The Meaning of Positivism', in *Perspectives in Philosophy*, R. N. Beck (ed.) (Holt, Rinehart, and Winston, New York, 1961).

1945: (a) 'Operationism and Scientific Method', *Psychological Review* **52**, 250–259.

Also in H. Feigl and W. Sellars (eds.) [1949b]. Reprinted in [1980], pp. 171–185.

(b) 'Rejoinder and Second Thoughts' (to 'Operationism and Scientific Method'), *Psychological Review* 52, 284–288. Reprinted with slight alterations as 'Some Remarks on the Meaning of Scientific Explanation', in H. Feigl and W. Sellars (eds.) [1949b]. Reprinted in [1980], pp. 185–191.

1949: (a) 'Naturalism and Humanism: An Essay on Some Issues of General Education and a Critique of Current Misconceptions Regarding Scientific Method and the Scientific Outlook in Philosophy', *American Quarterly* 1, 135–148. Reprinted as 'The Scientific Outlook: Naturalism and Humanism', in H. Feigl and M. Brodbeck [1953a]. Reprinted in [1980], pp. 366–377.

(b) (co-editor with W. Sellars) *Readings in Philosophical Analysis* (Appleton-Century-Crofts, New York).

1950: (a) 'De Principiis non Disputandum . . . ? On the Meaning and the Limits of Justification', in *Philosophical Analysis*, Max Black (ed.) (Cornell University Press, Ithaca, New York), pp. 119–156 (new ed., Prentice-Hall, Englewood Cliffs, N. J., 1963). Reprinted in part in Y. H. Krikorian and A. Edel (eds.), *Contemporary Philosophic Problems* (Macmillan, New York, 1959). Also in *Readings in Analytic Philosophy*, John Hospers (ed.) (The Free Press, Glencoe, Illinois, 1970) [in an early dialogue version – Ed.]. Also reprinted in [1980], pp. 237–268.

(b) 'Existential Hypotheses: Realistic versus Phenomenalistic Interpretations', *Philosophy of Science* 17, 35–62. Reprinted in [1980], pp. 192–223.

(c) 'Logical Reconstruction, Realism and Pure Semiotic', *Philosophy of Science* 17, 186–195. Reprinted in [1980], pp. 224–236.

(d) 'The Mind-Body Problem in the Development of Logical Empiricism', *Revue Internationale de Philosophie* 4, 64–83. Reprinted in H. Feigl and M. Brodbeck [1953a]; and in [1980], pp. 286–301.

(e) 'Bibliography of Logical Empiricism', *Revue Internationale de Philosophie* 4, 95–102.

(f) 'Felix Kaufmann's Conception of Philosophy as Clarification', *Twelfth Street* 3, 12–13.

(g) Review of K. R. Popper's 'A Note on Natural Laws and Contrary to Fact Conditionals', *Journal of Symbolic Logic* 15, 144–145.

(h) 'The Difference between Knowledge and Valuation', *Journal of Social Issues* 6, 39–44.

(i) 'Logical Positivism', in *Collier's Encyclopedia.*

1951: (a) 'Confirmability and Confirmation', *Revue Internationale de Philosophie* 5, 268–279. Reprinted in *Readings in Philosophy of Science*, P. P. Wiener (ed.) (Charles Scribner's Sons, New York); and also in [1980], pp. 145–152.

(b) 'Principles and Problems of Theory Construction in Psychology', in *Current Trends in Psychological Theory*, W. Dennis (ed.) (University of Pittsburgh Press, Pittsburgh), pp. 179–213.

1952: 'Validation and Vindication: An Analysis of the Nature and the Limits of Ethical Arguments', in *Readings in Ethical Theory*, W. Sellars and J. Hospers (eds.) (Appleton-Century-Crofts, New York), pp. 667–680. Reprinted in [1980], pp. 378–392.

1953: (a) (co-editor with May Brodbeck) *Readings in the Philosophy of Science* (Appleton-Century-Crofts, New York).

(b) 'Unity of Science and Unitary Science', in H. Feigl and M. Brodbeck [1953a].

(c) 'Notes on Causality', in H. Feigl and M. Brodbeck [1953a].

1954: 'Scientific Method without Metaphysical Presuppositions', *Philosophical Studies* **5**, 17–31. Reprinted in [1980], pp. 95–106.

1955: (a) 'Aims of Education for Our Age of Science: Reflections of a Logical Empiricist', in *The Fifty-Fourth Yearbook of the National Society for the Study of Education*, Part I (University of Chicago Press, Chicago), pp. 304–341. Reprinted in part in *American Physics Teacher* **2**, 295–296 (1964).

(b) 'Functionalism, Psychological Theory, and the Uniting Sciences', *Psychological Review* **62**, 232–235.

(c) 'Some Major Issues and Developments in the Philosophy of Science of Logical Empiricism', *Proceedings of the Second International Congress of the International Union for Philosophy of Science* (Neuchâtel). Reprinted in H. Feigl and M. Scriven, (eds.) [1956d], pp. 3–37.

(d) 'Physicalism and the Foundations of Psychology' (résumé), in *Proceedings of the Second International Congress of the International Union for the Philosophy of Science* (Neuchâtel).

1956: (a) Review of J. R. Newman (ed.), *What Is Science*? in *Contemporary Psychology* **1**, 275–276.

(b) 'Das hypothetisch-konstruktive Denken: Zur Methodologie der Naturwissenschaft', *Deutsche Universitätszeitung* **23/24**, 8–13. Reprinted in *Die Philosophie und die Wissenschaften: Simon Moser zum 65. Geburtstag* (Anton Hain, Meisenheim am Glan, 1967).

(c) 'Levels of Scientific Inquiry', *University of Minnesota Medical Bulletin* **28**, 90–97. Reprinted with minor additions in *Bulletin of the Bell Museum of Pathobiology*, No. 2 (University of Minnesota, 1972).

(d) (co-editor with M. Scriven) *Minnesota Studies in the Philosophy of Science.* Volume I: *The Foundations of Science and the Concepts of Psychology and Psychoanalysis* (University of Minnesota Press, Minneapolis).

1957: (a) 'Comments', in R. Lepley (ed.), *The Language of Value* (Columbia University Press, New York).

(b) 'Empiricism versus Theology', in *A Modern Introduction to Philosophy*, A. Pap and P. Edwards (eds.) (The Free Press, Glencoe, Illinois), pp. 533–538.

1958: (a) (co-editor with M. Scriven and G. Maxwell) *Minnesota Studies in the Philosophy of Science.* Volume II: *Concepts, Theories, and the Mind-Body Problem* (University of Minnesota Press, Minneapolis).

(b) 'The "Mental" and the "Physical" ', in [1958a], pp. 370–497. (For *Postscript after Ten Years*, see [1967a]).

(c) 'Other Minds and the Egocentric Predicament', *Journal of Philosophy* **55**, 978–987.

(d) 'A Note on Justification and Reconstruction', *Philosophical Studies* **9**, 70–72.

(e) 'Critique of Intuition According to Scientific Empiricism', *Philosophy East-West* **8**, 1–16.

1959: 'Philosophical Embarrassments of Psychology', *American Psychologist* **14**, 115–128. Reprinted in *Psychologische Beiträge* **6**, 340–364 (1962); and in *The Science of Psychology: Critical Reflections*, Duane P. Schultz (ed.) (Appleton-Century-Crofts, New York, 1970).

1960: 'Mind-Body, Not a Pseudoproblem', in *Dimensions of Mind*, S. Hook, (ed.) (New York University Press, New York), pp. 24–36. Reprinted in *Theories of the Mind*, Jordan Scher (ed.) (The Free Press of Glencoe, New York, 1962); and in [1980], pp. 342–350.

1961: (a) (co-editor with G. Maxwell) *Current Issues in the Philosophy of Science* (Holt, Rinehart and Winston, New York).

(b) 'Philosophical Tangents of Science', Vice Presidential Address, *Proceedings of Section L of the American Association for the Advancement of Science*, in H. Feigl and G. Maxwell (eds.) [1961a], pp. 1–17.

(c) 'On the Vindication of Induction', *Philosophy of Science* **28**, 212–216.

(d) (with G. Maxwell) 'Why Ordinary Language Needs Reforming', *Journal of Philosophy* **58**, 488–498.

(e) Review of *Logical Positivism*, A. J. Ayer (ed.), in *Contemporary Psychology* **6**, 88–89.

1962: (a) 'Matter Still Largely Material' (symposium paper written for the St. Louis meeting of the American Philosophical Association, May 5, 1961), in *Philosophy of Science* **29**, 39–46. Reprinted in *The Concept of Matter*, E. McMullin (ed.) (University of Notre Dame Press, Notre Dame, Indiana, 1963).

(b) (co-editor with G. Maxwell) *Minnesota Studies in the Philosophy of Science*. Volume III: *Scientific Explanation, Space, and Time* (University of Minnesota Press, Minneapolis).

1963: (a) 'Modernized Theology and the Scientific Outlook', *The Humanist* **23**, 74–80. (For German tr., see [1965a].)

(b) 'The Power of Positivistic Thinking: An Eassy on the Quandaries of Transcendence', in the *Proceedings and Addresses of the American Philosophical Association* **36**, 21–41. Reprinted in [1980], pp. 38–56.

(c) 'Towards a Philosophy for Our Age of Science', in *Proceedings of the XIIIth International Congress of Philosophy* (Mexico City) **4**, 101–110.

(d) 'Physicalism, Unity of Science and the Foundations of Psychology', in *The Philosophy of Rudolf Carnap*, P. A. Schilpp (ed.) (Open Court, LaSalle, Illinois), pp. 227–268. Reprinted in [1980], pp. 302–341.

1964: (a) 'What Hume Might Have Said to Kant', in *The Critical Approach to Science and Philosophy*, M. Bunge (ed.) (The Free Press, New York), pp. 45–51. Reprinted in [1980], pp. 164–170.

(b) 'From Logical Positivism to Hypercritical Realism', in *Proceedings of the XIIIth International Congress of Philosophy* (Mexico City) **5**, 427–436.

(c) 'Everybody Talks about the Temperature: Satirical Theme and Variations on the Nature of Value Judgments', in *Danish Yearbook of Philosophy*, Vol. 1 (volume dedicated to Jørgen Jørgensen on the occasion of his 70th birthday). (Munksgaard, Copenhagen). Reprinted in [1980], pp. 393–398.

(d) 'Some Remarks on the Logic of Scientific Explanation', in *A Broader View of Research in the University* (Mimeographed), (Department of Chemical Engineering, University of Minnesota, Minneapolis).

(e) 'Logical Positivism after Thirty-Five Years', in *Philosophy Today* 8, 228–245. (For Spanish tr., see [1965c].)

1965: (a) 'Modernisierte Theologie und wissenschaftliche Weltanschauung', in *Club Voltaire: Jahrbuch für kritische Aufklärung* (Szczesny Verlag, Munich). (Tr. of [1963a].)

(b) 'Models of Man', *The Humanist* **25**, 260–261.

(c) 'Positivismo logico despues 35', in *Revista Occidentale* (Madrid). (Tr. of [1964e].)

1966: (a) 'Is Science Relevant to Theology?', in *Zygon* **1**, No. 2, 191–199 (University of Chicago Press, Chicago). Reprinted in [1980], pp. 399–407.

(b) 'The Outlook of Scientific Humanism', in *Proceedings of the Fourth Congress of the International Humanist and Ethical Union* 1966 Secretariat (Utrecht, Netherlands), pp. 49–54; reprinted in *The Abdication of Philosophy: Philosophy and the Public Good*, Eugene Freeman (ed.) (Open Court, LaSalle, Illinois, 1976), pp. 73–79.

(c) 'The Distinction between Analytic and Synthetic Statements as a Cornerstone of Logical Empiricism', in *Deskription, Analytizität und Existenz*, Paul Weingartner (ed.) (Pustet, Salzburg, Munich), pp. 175–181 (reprinted as excerpt from [1955c]).

1967: (a) *The 'Mental' and the 'Physical'* with a *Postscript after Ten Years* (University of Minnesota Press, Minneapolis). [Reprint of [1958b] (pp. 3–116) with the original bibliography (pp. 117–132), the new *Postscript after Ten Years* (pp. 133–160) and a supplement to the bibliography (pp. 161–170].

(b) 'Prospects and Challenges: Contemporary Science and Philosophy', in *Science and Contemporary Society*, Frederick J. Crosson (ed.) (University of Notre Dame Press, Notre Dame, Ind.), pp. 129–153.

1969: (a) 'The Origin and Spirit of Logical Positivism', in *The Legacy of Logical Positivism: Studies in the Philosophy of Science*, P. Achinstein and S. F. Barker (eds.) (The Johns Hopkins Press, Baltimore), pp. 3–24. Reprinted in [1980], pp. 21–37.

(b) 'The Philosophy of Science: Causality, Chance, and Scientific Explanation', in *The Great Ideas Today, 1969*, R. M. Hutchins, M. Adler and O. Bird (eds.) (Encyclopedia Britannica, Inc., Chicago), pp. 146–189.

(c) 'Reduction of Psychology to Neurophysiology?', in *Philosophy of Science*, Philosophy of Science Society, Japan (ed.) (Risoska, Tokyo), pp. 163–184.

(d) 'The *Wiener Kreis* in America', in *The Intellectual Migration 1930–1960*, D. Fleming and B. Bailyn (eds.) (Harvard University Press, Cambridge, Mass.), pp. 630–673. Reprinted in [1980], pp. 57–94.

(e) 'Ethics, Religion and Scientific Humanism', in *Moral Problems in Contemporary Society*, Paul Kurtz (ed.) (Prentice-Hall, Englewood Cliffs, N. J.), pp. 48–64. Reprinted in [1980], pp. 408–421.

1970: (a) (with Charles Morris) *Bibliography and Index* for *International Encyclopedia of Unified Science: Foundations of the Unity of Science. Toward *an International Encyclopedia of Unified Science*, Vol. II, No. 10, O. Neurath, R. Carnap, C. Morris (eds.) (University of Chicago Press, Chicago), pp. 947–1023.

(b) 'The "Orthodox" View of Theories: Remarks in Defense as Well as Critique', in *Minnesota Studies in the Philosophy of Science*. Volume IV: *Analyses of Theories and Methods of Physics and Psychology*, M. Radner and S. Winokur (eds.) (University of Minnesota Press, Minneapolis), pp. 3–16.

(c) 'Beyond Peaceful Coexistence', in *Minnesota Studies in the Philosophy of Science*. Volume V: *Historical and Philosophical Perspectives of Science*, R. H. Stuewer (ed.) (University of Minnesota Press, Minneaplis), pp. 3–11.

(d) 'Critique of Dialectical Materialism', in *Dialogues on the Philosophy of Marxism* from the Proceedings of the Society for the Philosophical Study of Dialectical Materialism, John Somerville and Howard Parsons (eds.) (Greenwood Press, Westport, Ct., 1974).

(e) 'Memorial Minute: Rudolf Carnap', in *Proceedings and Addresses of the American Philosophical Association* **44** (1970–71), pp. 204–205.

1971: (a) 'Some Crucial Issues of Mind-Body Monism', in *Synthese* **22**, 295–312. (D. Reidel Publ. Co., Dordrecht, Holland and Boston). Reprinted in [1980], pp. 351–365.

(b) 'Mach's Views' (book review: *Ernst Mach. Physicist and Philosopher*, R. S. Cohen and R. J. Seeger, eds.), *Science* **171**, 471–472.

(c) 'Research Programmes and Induction', in *PSA 1970*, R. C. Buck and R. S. Cohen (eds.) (*Boston Studies in the Philosophy of Science*, Volume VIII), (D. Reidel Publ. Co., Dordrecht-Holland), pp. 147–150.

(d) 'Homage to Rudolf Carnap', *PSA 1970*, R. C. Buck and R. S. Cohen (eds.) (*Boston Studies in the Philosophy of Science*, Volume VIII), (D. Reidel Publ. Co., Dordrecht-Holland), pp. xi–xv.

(e) 'Empiricism at Bay?', in *Methodological and Historical Essays in the Natural and Social Sciences*, R. S. Cohen and M. W. Wartofsky (eds.) (*Boston Studies in the Philosophy of Science*, Volume XIV), (D. Reidel Publ. Co., Dordrecht-Holland and Boston, Mass.), pp. 1–20. Reprinted in [1980], pp. 269–285.

1972: (a) (with Millard S. Everett) *A Philosophy for Our Age of Science* (unpublished manuscript).

(b) (co-edited with W. Sellars and K. Lehrer) *New Readings in Philosophical Analysis* (Appleton-Century-Crofts, New York).

1973: (a) 'Positivism in the 20th Century (Logical Empiricism)', in *Dictionary of the History of Ideas*, Vol. 3, P. P. Wiener (ed.) (Charles Scribner's Sons, New York), pp. 545–551.

1974: (a) 'No Pot of Message', in *Mid-Twentieth Century Philosophy*: *Personal Statements*, P. Bertocci (ed.) (Humanities Press, New York), pp. 120–139. Reprinted in [1980], pp. 1–20.

(b) (with P. E. Meehl) 'The Determinism-Freedom and Body-Mind Problems', in *The Philosophy of Sir Karl Popper*. Vol. 1. P. A. Schilpp (ed.) (*Library of Living Philosophers*), (Open Court, LaSalle, Illinois), pp. 520–559.

(c) 'Critique of Dialectical Materialism', in *Dialogues on the Philosophy of Marxism*, from the Proceedings of the Society for the Philosophical Study of Dialectical Materialism, J. Somerville and H. Parsons (eds.) (Greenwood Press, Westport, Ct.).

1975: 'Russell and Schlick: A Remarkable Agreement on a Monistic Solution of the Mind-Body Problem', in *Erkenntnis* 9, No. 1, 11–34.

1978: 'Positivism and Logical Empiricism', in *Encyclopedia Britannica*, 15th ed.

1980: *Inquiries and Provocations: Selected Writings 1929–1974*, Robert S. Cohen (ed.), *Vienna Circle Collection*, Vol. XIV (D. Reidel Publ. Co., Dordrecht, Holland and Boston, Mass.)

(1) 'No Pot of Message', pp. 1–20; reprinted from [1974a].
(2) 'The Origin and Spirit of Logical Positivism', pp. 21–37; reprinted from [1969a].
(3) 'The Power of Positivistic Thinking', pp. 38–56; reprinted from [1963b].
(4) 'The *Wiener Kreis* in America', pp. 57–94; reprinted from [1969d].
(5) 'Scientific Method without Metaphysical Presuppositions', pp. 95–106; reprinted from [1954a].
(6) 'Probability and Experience', pp. 107–115; English trans. of [1930a].
(7) 'Meaning and Validity of Physical Theories', pp. 116–144; English trans. of [1929a].
(8) 'Confirmability and Confirmation', pp. 145–152; reprinted from [1951a].
(9) 'The Logical Character of the Principle of Induction', pp. 153–163; reprinted from [1934a].
(10) 'What Hume Might Have Said to Kant', pp. 164–170; reprinted from [1964a].
(11) 'Operationism and Scientific Method', pp. 171–185; reprinted from [1945a].
'Rejoinder and Second Thoughts', pp. 185–191; reprinted from [1945b].
(12) 'Existential Hypotheses', pp. 192–223; reprinted from [1950b].
(13) 'Logical Reconstruction, Realism and Pure Semiotic', pp. 224–236; reprinted from [1950c].
(14) 'De Principiis Non Disputandum ... ?', pp. 237–268; reprinted from [1950a].
(15) 'Empiricism at Bay?', pp. 269–285; reprinted from [1971e].
(16) 'The Mind-Body Problem in the Development of Logical Empiricism', pp. 286–301; reprinted from [1950d].
(17) 'Physicalism, Unity of Science and the Foundations of Psychology', pp. 302–341; reprinted from [1963d].
(18) 'Mind-Body, *Not* a Pseudoproblem', pp. 342–350; reprinted from [1960].
(19) 'Some Crucial Issues of Mind-Body Monism', pp. 351–365; reprinted from [1971a].
(20) 'Naturalism *and* Humanism', pp. 366–377; reprinted from [1949a].
(21) 'Validation and Vindication: An Analysis of the Nature and the Limits of Ethical Arguments', pp. 378–392; reprinted from [1952].

(22) 'Everybody Talks about the Temperature', pp. 393–398; reprinted from [1964c].
(23) 'Is Science Relevant to Theology?', pp. 399–407; reprinted from [1966a].
(24) 'Ethics, Religion, and Scientific Humanism', pp. 408–421; reprinted from [1969e].

INDEX OF NAMES

VIENNA CIRCLE COLLECTION

1. OTTO NEURATH, *Empiricism and Sociology*. Edited by Marie Neurath and Robert S. Cohen. With a Section of Biographical and Autobiographical Sketches. Translations by Paul Foulkes and Marie Neurath. 1973, xvi + 473 pp., with illustrations. ISBN 90-277-0258-6 (cloth), ISBN 90-277-0259-4 (paper).

2. JOSEF SCHÄCHTER, *Prolegomena to a Critical Grammar*. With a Foreword by J. F. Staal and the Introduction to the original German edition by M. Schlick. Translated by Paul Foulkes. 1973, xxi + 161 pp. ISBN 90-277-0296-9 (cloth), ISBN 90-277-0301-9 (paper).

3. ERNST MACH, *Knowledge and Error. Sketches on the Psychology of Enquiry*. Translated by Paul Foulkes. 1976, xxxviii + 393 pp. ISBN 90-277-0281-0 (cloth), ISBN 90-277-0282-9 (paper).

4. MARIA REICHENBACH and ROBERT S. COHEN, *Hans Reichenbach: Selected Writings, 1909-1953 (Volume One)*. 1978, in press. ISBN 90-277-0291-8 (cloth), ISBN 90-277-0292-6 (paper). *Hans Reichenbach: Selected Writings, 1909-1953 (Volume Two)*. 1978, in press. ISBN 90-277-0909-2 (cloth), ISBN 90-277-0910-6 (paper). Sets: ISBN 90-277-0892-4 (cloth), ISBN 90-277-0893-2 (paper).

5. LUDWIG BOLTZMANN, *Theoretical Physics and Philosophical Problems. Selected Writings*. With a Foreword by S. R. de Groot. Edited by Brian McGuinness. Translated by Paul Foulkes. 1974, xvi + 280 pp. ISBN 90-277-0249-7 (cloth), ISBN 90-277-0250-0 (paper).

6. KARL MENGER, *Morality, Decision, and Social Organization. Toward a Logic of Ethics*. With a Postscript to the English Edition by the Author. Based on a translation by E. van der Schalie. 1974, xvi + 115 pp. ISBN 90-277-0318-3 (cloth), ISBN 90-277-0319-1 (paper).

7. BÉLA JUHOS, *Selected Papers on Epistemology and Physics*. Edited and with an Introduction by Gerhard Frey. Translated by Paul Foulkes. 1976, xxi + 350 pp. ISBN 90-277-0686-7 (cloth), ISBN 90-277-0687-5 (paper).

8. FRIEDRICH WAISMANN, *Philosophical Papers*. Edited by Brian McGuinness with an Introduction by Anthony Quinton. Translated by Hans Kaal (Chapters I, II, III, V, VI and VIII and by Arnold Burms and Philippe van Parys. 1977, xxii + 190 pp. ISBN 90-277-0712-X (cloth), ISBN 90-277-0713-8 (paper).

9. FELIX KAUFMANN, *The Infinite in Mathematics. Logico-mathematical writings.* Edited by Brian McGuinness, with an Introduction by Ernest Nagel. Translated from the German by Paul Foulkes. 1978, xviii + 236 pp. ISBN 90-277-0847-9 (cloth), ISBN 90-277-0848-7 (paper).

10. KARL MENGER, *Selected Papers in Logic and Foundations, Didactics, Economics.* 1978, in press. ISBN 90-277-0320-5 (cloth), ISBN 90-277-0321-3 (paper).

11. HENK L. MULDER and BARBARA F. B. VAN DE VELDE-SCHLICK, *Moritz Schlick: Philosophical Papers Volume I (1909-1922).* Translated by Peter Heath. 1978, xxxviii + 376 pp. ISBN 90-277-0314-0 (cloth), ISBN 90-277-0315-9 (paper).

12. EINO SAKARI KAILA, *Reality and Experience. Four Philosophical Essays.* Edited by Robert S. Cohen. 1978, in press. ISBN 90-277-0915-7 (cloth), ISBN 90-277-0919-X (paper).

13. HANS HAHN, *Empiricism, Logic, and Mathematics, Philosophical Papers.* Edited by Brian McGuinness. 1980. ISBN 90-277-1056-1 (cloth), ISBN 90-277-1066-X (paper).

GPSR Compliance

The European Union's (EU) General Product Safety Regulation (GPSR) is a set of rules that requires consumer products to be safe and our obligations to ensure this.

If you have any concerns about our products, you can contact us on ProductSafety@springernature.com

In case Publisher is established outside the EU, the EU authorized representative is:

Springer Nature Customer Service Center GmbH
Europaplatz 3
69115 Heidelberg, Germany

Batch number: 10815964

Printed by Printforce, the Netherlands